Climate Change, Science, and the Politics of Shared Sacrifice

Climate Change, Science, and the Politics of Shared Sacrifice

FIRST EDITION

TODD A. EISENSTADT
American University

STEPHEN E. MACAVOY
American University

New York Oxford
OXFORD UNIVERSITY PRESS

Oxford University Press is a department of the University of Oxford.
It furthers the University's objective of excellence in research, scholarship,
and education by publishing worldwide. Oxford is a registered trademark of
Oxford University Press in the UK and certain other countries.

Published in the United States of America by Oxford University Press
198 Madison Avenue, New York, NY 10016, United States of America

For titles covered by Section 112 of the US Higher Education Opportunity
Act, please visit www.oup.com/us/he for the latest information
about pricing and alternate formats.

Library of Congress Cataloging-in-Publication Data

Names: Eisenstadt, Todd A., author. | MacAvoy, Stephen E., author.
Title: Making climate change cool : climate change, science, and the
politics of shared sacrifice / Todd A. Eisenstadt, American University,
Stephen E. MacAvoy, American University.
Description: First edition. | New York : Oxford University Press, [2022] |
Includes bibliographical references and index. |
Identifiers: LCCN 2020047688 (print) | LCCN 2020047689 (ebook) | ISBN
9780190063696 (paperback) | ISBN 9780190063856 (epub)
Subjects: LCSH: Climatic changes. | Climatic changes—Government policy. |
Climatic changes—International cooperation. | Climate change
mitigation.
Classification: LCC QC903 .E425 2022 (print) | LCC QC903 (ebook) | DDC
363.738/74—dc23
LC record available at https://lccn.loc.gov/2020047688
LC ebook record available at https://lccn.loc.gov/2020047689

Printing number: 9 8 7 6 5 4 3 2 1
Printed by LSC Communications, Inc., United States of America

For Natalia and Paola, with love and to contribute to the solution before passing the problem on to you.
—T.E.

For Emma and Evelyn, with love and hope.
—S.M.

For our students and all who seek sound policies based on empirical evidence and knowledge.
—T.E. and S.M.

Brief Contents

Contents

About the Authors

Todd A. Eisenstadt is a political science professor and the Director of Research at the Center for Environmental Policy (CEP) at American University's School of Public Affairs and the author/editor of nine previous books. Believing the social sciences needed to better engage with climate change and environmental politics more broadly, he shifted his research agenda midcareer, after several decades of studying democratization and political development. In this new line of research, he and Karleen Jones West published *Who Speaks for Nature? Indigenous Movements, Public Opinion, and the Petro-State in Ecuador* (Oxford University Press, 2019) and he has published a range of related articles in peer review journals. In 2018–2019, he received a Council on Foreign Relations International Affairs Fellowship (IAF) for Tenured International Relations Scholars sabbatical at the World Bank, where he took the advice of one of his own professors that the best way to learn about something is to research and write about it. Formerly an award-winning print journalist who has worked on six continents, Eisenstadt grew up in Corrales, New Mexico.

Stephen E. MacAvoy is a biogeochemist and has been particularly interested in how both nutrients and contaminants flow through, and are incorporated by, aquatic ecosystem components. Research and consulting activities have taken him to the Gulf of Mexico hydrocarbon seeps, Barbados, eleven US states, and other locations. MacAvoy obtained his PhD from the University of Virginia in 2000 and conducted postdoctoral work at the University of Georgia. He joined the faculty of American University in 2003 and has directed the Graduate Program in Environmental Science since then. In 2016, he became Chair of the Department of Environmental Science. Recent publications have appeared in *Applied Geochemistry*, *Environmental Science and Pollution Research*, *Ecological Engineering*, and *Marine Mammal Science*. He is currently conducting research on the inorganic and organic geochemistry of the Anacostia River in Washington, DC.

Preface

In our first meeting of the honors class "Climate Change Science and Politics" after the 2016 US presidential election, all our students attended, but a half dozen spread themselves out on the floor in the front of the classroom, completely demoralized. The two authors, co-teaching the class, tried to pep them up and tell them what they could do as citizens, consumers, and activists to keep a focus on climate. Several of them absorbed our pep talk for a few minutes and then turned it back on us. "Well, that's fine that we have work to do," they proclaimed. "But what are you going to do to step up your efforts?"

First, we shrugged our shoulders, looked at each other, and grinned sheepishly. Then, we got our footing and said that our college student protest days were largely over, but that we were researchers and writers and that a single text did not really exist for the course we were teaching. This book tries to solve that problem. In it, we combine basic discussions of some of the physical, geological, and chemical explanations (and technological solutions) of climate change, and we consider some of the political, economic, and social consequences, based on basic theories of international relations, natural resource economics, and interest group politics. It was a challenging book to write as climate change and its impacts permeate many areas of study, and the scope of the subject was hard to narrow down to just one medium-length text. We know the book has many failings, for which we take the blame. By necessity, the book provides a limited introduction to topics, gives guidance for further research, and then moves on. It is an introductory text to the problem, and as such, we sought to consider "the big questions" and to provide exercises at the conclusion of each chapter that allow students to further and actively engage with the material.

Despite its shortcomings, it is an ambitious book. We seek to explain most dimensions of this Gordian knot of climate change causes and responses and to convey to dedicated and idealistic students (like you) that the problem does have solutions—but also that they are partial, far-ranging, and will require sacrifice. We were tired of the climate denial of the 2016 presidential campaign, which only worsened under President Donald Trump (although we are greatly heartened by the election of President Biden and the thought and energy our Gen Z students give the environment).

This book was written as a joint project between a natural scientist and a social scientist in recognition that the greatest barriers to enacting sweeping climate change policy in the 21st century are political as well as scientific. Evidence of climate change is being amassed at an ever-increasing rate, but acceptance of the political challenges ahead lags way behind. As fathers as well as teachers and researchers, we believe that climate change can be contained within acceptable levels—and we desperately want to believe this. It is going to take sacrifice by each of us, however, as politicians are finally honing this message after President Biden's win in the presidential elections in 2020.

The economic fallout from the COVID-19 pandemic may help us all learn to work harder to resolve societal inequalities and tackle large problems by harnessing public resources. To solve climate change, we will have to alter our individual habits as well as our group behavior if we are to pass on to successive generations a planet resembling the one we inherited. We want to strike the balance for college students reading this text between presenting a realistic awareness of the scope and dimensions of the problem and the sense that there is still time, but only if we act quickly, deliberately, and effectively. As we write in Chapter 1, knowledge is power, and this book seeks to give you the knowledge needed for your generation to make better choices.

We have a long list of people to thank for their help by funding research, patiently answering questions and providing information, inspiring us to action, and creating an atmosphere propitious to the conduct of our work. Todd Eisenstadt's interest in these issues started during dinner conversations with his brother, Keegan Eisenstadt, a climate activist for decades (and proprietor of an emissions trading and development company: https://www.clearskyclimatesolutions.com) and his parents, retired New Mexico state legislator Pauline Eisenstadt and the late Mel Eisenstadt, a solar power advocate even in the 1970s. More directly related to this text, Eisenstadt thanks the Council on Foreign Relations for a 2018–2019 International Affairs Fellowship (IAF) for Tenured International Relations Scholars, which gave him time to reflect on the material and conduct research at the World Bank's Development Economics Research Group, where Michael Toman was a spectacular supervisor and colleague. Toman was a practical manager who helped Eisenstadt navigate the Bank, but also an intellectual and editor, who kept writing honest with helpful edits and good questions. At the Council on Foreign Relations, Victoria Harlan and Janine Hill were extremely encouraging.

At the World Bank, Eisenstadt also thanks Joseph Dickman, Mafalda Duarte, and Neha Sharma at the Climate Investment Funds (CIF), where Eisenstadt asked a lot of questions and learned about climate lending priorities. Also at the World Bank, he thanks Harsh Anuj, Mary Elinor Boyer, Carter Brandon (now of the World Resources Institute), Raul Alfaro Belico,

Meskerem Brhane, Nicholas James Callender, Viviane Clement, Genevieve Connors, Kerri Dionne Cox, Nathan Engel, Mirtha Escobar, Vivien Foster, Rashmin Gunasekera, Stéphane Hallegatte, Stephen Hammer, Daniel Lederman, Kanta Kumari, Neha Muki, Lizardo Narvaez, Neeraj Prasad, Loreta Rufo, Mark Sadler, and Luis Tineo. The World Bank sent Eisenstadt to Bangladesh, Barbados, Costa Rica, Dominica, and Ethiopia in 2019 and early 2020, allowing him to gather invaluable interviews, including those with Oluwatoyin Adejonwo, Cecil Charles, Rosamund Edwards, Tawfique Haque, Chris Jack, Brenda Mwalukanga, Janet Ngombalu, and Careen Prevost, all of which are cited in this text. Ifeoluwa Olawole of American University, Jie Lu of Renmin University of China, and Matthew Wright of the University of British Columbia collaborated in a survey of the climate vulnerable in Bangladesh, which was ably conducted by Professor Tawfique Haque and the Department of Public Administration at the North-South University, whose collaboration was enriching and fruitful.

At American University, the Center for Environmental Policy (CEP) and Director Dan Fiorino offered an institutional home and excellent feedback. We also received excellent comments and advice from Adam Auerbach, Ken Conca, Miles Kahler, Simon Nicholson, John Reeder, Danielle Wagner, and Paul Wapner at American University. Trustee Jack Cassell and Leslie Wong, assistant secretary to the Board of Trustees, provided information for the Chapter 7 case study on the university fossil fuel divestiture movement, as experienced by American University.

Former American University graduate students Laurel Wolf (Chapters 4 and 8), Fernando Hernandez (Chapter 3), and Gabriella Siegfried (Chapter 9), and undergraduates Dan Barrow (Chapter 5), and Lillian Frame (Chapter 7) authored didactic case study role-playing exercises accompanying the text. Dan Barrow also did a wonderful job harmonizing the manuscript's style, and Ifeoluwa Olawole and Gabriella Siegfried and Karl VonZabern helped with proofreading, editing, and indexing. Elizabeth Magill, also an American University student, was invaluable in obtaining permissions for all the republished figures that appear in the text. Then–College of Arts and Sciences Dean Peter Starr supported MacAvoy with funds for student assistance. School of Public Affairs Dean Vicky Wilkins and associate deans Allison Jacknowitz and Saul Newman helped fund Eisenstadt to attend the United Nations Conference of the Parties meetings in Lima (2014) and Bonn (2017), as well as the historic Paris meeting (2015). They also helped us commission students to write the case studies. Former Government Department Chair Candice Nelson also helped support Eisenstadt's trips to United Nations climate negotiations and Government Department staff member James Helms helped us get student assistance on a range of tasks.

At the Oxford University Press, we benefitted from the wisdom and experience of editor Jennifer Carpenter, who helped guide our project to completion even during COVID disruptions. At Oxford, we also thank Editorial Assistant Ishan Desai-Geller and Production Editor Patricia Berube, and the truly outstanding copy-editing job by Wesley Morrison. We would also like to thank the following group of external reviewers:

- Ross Astoria, University of Wisconsin–Parkside
- Plaxedes Chitiyo, Duquesne University
- Marjorie Hershey, Indiana University
- Peter J. Jacques, University of Central Florida
- Younsung Kim, George Mason University
- Sara Moats, Florida International University
- Simon Nicholson, American University
- Ryan W. Taylor, State University of New York–Purchase

The errors are ours alone. But we hope the benefits of reading and mulling over the contents of this book are widely shared by readers over dinner tables, on buses, in seminars, and at marches and demonstrations. Please send any suggestions or comments to help us improve subsequent editions of this book to either or both authors at climate.change.sacrifice@gmail.com.

1

How Humans Tried to Take Control of Nature

From Cuddly Teddy Bears to the Anthropogenic Tempest of Climate Change

When US President Theodore Roosevelt refused to fire on an injured and mangy black bear during a 1902 hunting trip in Mississippi, the president's valiant pity toward one of nature's proudest beasts was publicized by a political cartoonist in Washington, DC, giving rise to the Teddy Bear.[1] It was an apocryphal moment in the relationship between humanity and nature. The Teddy Bear, now a familiar fixture in children's bedrooms the world over, may still be the best representation of the exact moment when humans started viewing nature as a domesticated force at our service, rather than as a wild and untamable force bigger than us.

Well over a century later, the role of nature in the Anthropocene, the human-led environment, seems tamer than ever. However, humans may have overestimated our dominion. We have limited the ranges and populations of bears and many other large wild beasts, although we have not yet conquered much smaller threats, like disease-spreading mosquitos (i.e., *Plasmodium* protozoa [malaria] and the Zika virus) or, most notoriously, the COVID-19 pandemic, which ravaged the earth beginning in 2020. The pandemic caused by that virus showed us, just as much as climate change, that humanity's domination of our environment threatens to backfire. Even after the public health and economic fallout from the pandemic, we will still have to address perhaps the most complex and ominous remaining threat facing humanity in the 21st century: climate change. Ironically, this crisis was caused by our own pollution as we conquered the planet through industrialization and economic development, which lifted people out of poverty and improved peoples' standards of living over the past couple of centuries.

The smokestacks of the 19th century's coal-burning Industrial Revolution emitted pollutants like carbon dioxide (CO_2), called "greenhouse gases" (GHGs) because they had the effect of trapping heat inside the atmosphere, the same way that a greenhouse traps heat to grow "hothouse" tomatoes.

Evidence abounds that the climate is warming faster now than it was before the Industrial Revolution. The evidence includes an acceleration of melting glaciers and ice caps, coral reef bleaching, increasing rates of sea level rise, and changes to the ranges and migration of flora and fauna. These natural changes will be documented extensively in this book. Over 30 years of research, spearheaded by the world's most prominent international commission of thousands of scientists and technicians, the Intergovernmental Panel on Climate Change (IPCC), winner of the 2007 Nobel Peace Prize, has rendered overwhelming proof of the existence of climate change.

We do not have to use projections or our imaginations to understand the impacts of climate change. Based on thousands of measurements from satellites, land/ocean-based stations, and ice cores, we know that the planet has warmed 1°C since preindustrial times (IPCC 2018). Temperature is not shifting at the same rate everywhere, however, and many areas are experiencing the "warming" as increasingly extreme weather. As areas experience more extreme conditions with increased frequency, the term "climate weirdness" has been coined. Among the "weird" phenomena are epic droughts in the southwestern United States (especially California, where it has contributed to wildfires of unprecedented magnitude), heat waves in Europe that kill thousands, and increased frequency of large hurricanes worldwide since 1970 (Webster et al. 2005), just to name a few (IPCC 2018). Shifts in biological and physical systems are already apparent and well documented. For example, between March and November of 2016, 1,000 km of the Great Barrier Reef in Australia experienced a 40 to 60 percent coral decline because of warming water (Hughes et al. 2018), and marine species are shifting northward by 72 km a decade, fundamentally disrupting food webs (Scheffers et al. 2016). In fact, a landmark paper in 2016 showed that of 94 separate ecological processes (migration, species distribution, productivity, etc.) worldwide, 82 percent were already being affected by climate change (Scheffers et al. 2016).

However, even in the face of these scientifically verified truths, politicians, such as US President Donald Trump and Brazilian President Jair Bolsonaro, continue to deny the very existence of human-induced climate change, and for what can only be political reasons. Indeed, as the World Bank, headquartered some three blocks from the White House, mainstreams climate change considerations into its entire portfolio of lending for economic development around the world, as the European Union (EU) pours resources into saving forests to sequester CO_2, and as oil companies—climate change deniers for decades—acquiesce to the existence of climate change and seek to profit by improving the efficiency of renewable energy sources like solar, wind, and geothermal power, Donald Trump, former leader of the largest economy at the time of the writing, brashly denies the problem, confronting each new presentation of evidence with a more boisterous and daring denial. In what some have called the "Post-Truth Era," so-named because of policymakers' sometime insistence on political positions even in the face of contrary facts, climate change may be the quintessential "Post-Truth" issue.

A central question here is whether humanity is just another part of nature, as used to be more commonly believed, or if we stand "above" nature and hold a dominion decreed by God, as interpreted by some passages in the Book of Genesis in the Judeo-Christian tradition. Not long after "Teddy" Roosevelt's iconic bear hunt, the debate between conservation (putting natural lands at the disposition of human economic needs) and preservation (the imperative of maintaining lands in pristine shape) crystalized between the conservationist Gifford Pinchot and the preservationist John Muir over the proper use of the Hetch Hetchy Valley in Yosemite. This was a central debate in 20th century environmental studies, but as fewer and fewer natural lands remain, the debate has shifted. And in the 21st century, it is one of how much should humanity control what remains of the natural world. As we have attained ever higher levels of dominion, the "natural world" (as it used to be called in PBS documentaries) has become the "Anthropocene" (the human habitat).

As this book was going to press in early 2021, those concerned with climate change were expressing relief, and even optimism, that Trump's climate denial was giving way to incoming President Joe Biden's strong pledges to reassume leadership in fighting climate change. The issue had received "front and center" treatment during the US presidential debates, for the first time ever, and the only question, as the Biden team prepared to take power, was whether fighting climate change was going to be a primary engine of economic growth, infrastructure development, and "green" industry support, as advocated by the progressive wing of Biden's Democratic party, and in the "Green New Deal" proposal of liberal democrats Bernie Sanders and Alexandria Ocasio-Cortez, or whether Biden might adopt slightly less ambitious policies, which would still be the most ambitious ever in the US.

Humanity Has Impacted Climate, Ushering in the Anthropocene Epoch

Climate change is at the center of the debate over managing the Anthropocene Epoch, formally defined as geological "period of time during which human activities have had an environmental impact on the Earth regarded as constituting a distinct geological age" (Merriam Webster Dictionary). Yes, that's right. We live in an epoch in which humans have dramatically changed the land's surface with crops and cities, but also changed the atmosphere's chemistry with artificial compounds, inorganic acids, and, yes, vast amounts of CO_2. These changes are recorded in the ice sheets, deep ocean sediments, and the earth's surface itself. The impacts of these changes on our world's climate are empirically visible today, and have been through most of the 19th and 20th centuries, as our own strata of geological time. These strata, commonly seen as layers of rock, are clear in places such as the western United States' beautiful eroded canyons and mesas, like Arizona's Painted Desert, and the "tent rocks" of Bandelier National Monument in New Mexico. However, our contribution to the story these layers tell is not a proud one. We have caused a remarkable amount of damage in a very short time.

Human-generated climate change is what makes the Anthropocene Epoch distinct. And this begs the question: Given that humans created not just climate change but also our own geological epoch characterized by this phenomenon, how can humans undo the damage? The hard but responsible answer is that a World War II–type sacrifice will be needed just to get climate change under control; we have given up trying to revert the environment back to preindustrial temperatures. Some argue, with evidence we will present later in this book, that directed changes to the atmosphere to reduce incoming radiation, usually referred to as atmospheric geoengineering, are the answer. As we will show, however, those technologies are fraught with risk. Controlling climate change through disciplined cuts in GHG emissions as well as increased efforts to draw down what is in the atmosphere already is a much more sustainable and less risky course of action. It will take time, however, and we need to rapidly expand our current efforts.

We wrote this book after tiring of the "climate denial" politicization of the 2016 US presidential election that left our students so demoralized several of them came to class and lied down on the floor in exhausted despair. We seek to explain most dimensions of this Gordian knot and to convey to dedicated and idealistic readers (like you) that the problem does have solutions, but that they are partial, far-ranging, and will require sacrifice. As this book went to press, we were somewhat encouraged by the sense of sacrifice brought about by the COVID-19 pandemic lockdowns around the world, and peoples' willingness—by and large—to self-quarantine for the public good. Solving climate change will require such sacrifice by most of us, combined with a redistribution of wealth from fossil fuel–intensive industries to "clean" and "green" ones, along with a greater role for government in planning the economy, which was at least partly rejected during the COVID-19 scare, at least by the United States and a few other actors.

At the international level, the Paris Agreement was a United Nations Framework Convention on Climate Change nonbinding agreement drafted in 2015 to encourage the world's nations to tackle climate change by reducing emissions, sharing technology, and adapting infrastructure. It was signed by 195 countries, including the United States and China, the world's two largest economies and CO_2 emitters. Although it was nonbinding, the Paris Agreement was a true first step toward limiting our carbon output so that temperature would not rise 2°C above the 1950–1980 global mean temperature. While it has been criticized for not going far enough, every journey starts with a single step, and it is important not to minimize the good in pursuit of the perfect. The agreement was all the more remarkable because the United States had historically been hostile to measures combating climate change, as politicians were either skeptical about human-induced climate change or felt that taking measures to combat climate change would hurt the economy. In 2019, however, the United States stepped back into its historical role as a climate curmudgeon and began to formally withdraw from the Paris Agreement. In so doing, the United States gracelessly abdicated environmental leadership to China and the EU while undermining its own national security, not to mention snubbing the peoples of the 194 other nations who had signed on. This was a deeply shameful act and colossally short-sighted, perhaps comparable to the rejection of the League of Nations after World War I.

This book was written as a joint project between a natural scientist and a social scientist in recognition that the greatest 21st-century barriers to enacting sweeping climate change policy are political as well as scientific. Evidence of climate change is being amassed at an ever-increasing rate, but acceptance of the political challenges ahead lags way behind. In the pages that follow, we describe, in a preliminary way, some of the natural evidence of climate change and the political challenges this generates. Then, we outline a few of the ethical and political dilemmas we will explore throughout this book and introduce the book's organization and main argument—namely, that we can likely still keep climate change within acceptable levels, but that it will take a concerted sacrifice by each of us, as we are going to have to alter our individual habits as well as our group behavior if we are going to pass a viable planet on to successive generations. We hope that readers of this text will become more aware of the scope and dimensions of the problem, as their political and material decisions will have an impact on it. But knowledge is power, and in the following pages, we will try to motivate you.

The Art and Science of Modeling Impacts of Climate Change on Natural Phenomena

Weather and climate are not the same thing. Weather is daily or weekly variation or fluctuation in the state of the atmosphere. Climate is the regional or large-scale trends in averages of atmospheric characteristics. While weather forecasters make very specific predictions about local atmospheric conditions, climatologists project trends in averages over much larger time scales, potentially over years or even decades.

It follows, then, that discussions of the climatological trends we are witnessing don't sound like the typical weather forecast. For example, spring is coming almost a week earlier in the United States than it did 30 years ago; average global temperature has increased by 1 to 1.5°C over the last 80 years;

the Atlantic Thermohaline Circulation (of which the Gulf Stream Current is a component) has been slowing down the last five years; and worldwide precipitation has increased since 1951 and is intensifying in some areas (these topics and more are explored in Chapter 2). The macrotrends for these global phenomena are not disputed among atmospheric scientists, although the details can be. It is a little like knowing Serena Williams can serve a tennis ball at about 122 miles per hour, but disagreeing about whether humidity or wind will cause the ball's actual speed to be 120 or 124 miles per hour.

The global climate models are always improving as we gain computing power and a more complete understanding of the nuances in land/atmosphere/ocean interactions. However, uncertainty in those early models does not invalidate their generalized projections, even if those projections can be improved. This is sort of like correctly predicting snowfall but being unable to tell you the characteristics of the snowflakes. It is worth mentioning here that *any* model describing future behavior of even a simple system will have uncertainty, be it economic, biological, astronomical, or environmental. Modeling is not like measuring (i.e., the length of X is Y). Measurements are static data, while modeling uses data to describe dynamic future behavior. Because it is impossible to know if your predictions of the future are accurate, climate scientists run their models backward to see how accurately they model past climate. This is the way a model is validated and uncertainty assessed. Testing the model's ability to predict past climate also allows scientists to determine what "natural" forcing does to climate (natural forcing variables are those that affect climate but exist without considering human civilization: hydrological cycle, solar intensity, continent position, ice cover, cloud cover, etc.). It is important to understand that uncertainty is "baked in" to all rigorous models, and that it is a sign of professionalism and precision on the part of scientists and not of doubt, as climate deniers claim.

MODELING VERSUS MEASURING NATURAL DRIVERS

If a model is very good at "predicting" the past, then it is considered a good predictor of the future. However, one feature of any climate model is deciding how much CO_2 will be released in the future, and since we don't know this, scientists run many models with different atmospheric CO_2 concentrations and fluxes (and other variables). Sometimes this confuses people, because they hear that "the models don't agree"—and indeed they don't, because they assume different human-controlled variables in the future. We should worry if they do agree!

Another feature of climate models that might confuse people is the idea of uncertainty in modeling. People want science that is "certain." All scientific consensus, however, is based on probabilities, not certainties. Most scientists consider a 95 percent chance that one variable is affecting another as a significant finding, but there is still a 5 percent chance that this one variable does not affect the other. If enough scientists find the same degree of certainty about how the two variables affect each other, then you have "scientific consensus." Of the thousands of scientists working on mechanisms controlling climate change as well as its effects, there is consensus that human activity is currently driving at least 60 percent of the current climate shifts (Hansen et al. 2013).

Climate modelers must use data that varies by location and season, such as global emissions, plant transpiration rate, air movement rates, etc. Thus, there is a range in the scope of consequences for perturbing a large, very complex system. Uncertainties don't mean the science is flawed. The system is just complex, and a range of probabilities are possible. Remember, if you model the future behavior of *anything* in nature, there will be uncertainty.

The models of the 1990s, the first undertaken on a large scale, projected trends in climate and impacts on the biosphere for early in the 21st century that are coming to pass as we exceed 1°C warming and approach 1.5°C. Here are a few trends that were projected and have come to pass:

1. *Widespread coral bleaching:* 60 percent of the northern Great Barrier Reef bleached and experienced a rapid species assemblage shift from warm surface waters in 2016 (Hughes et al. 2018). Collapse of the reef will cause collapse of the fish communities and loss of associated economic activity.
2. *Rapidly rising sea level:* Sea level has risen about 7 inches during the last 100 years, which is more than the previous 2,000 years combined. Loss of coastal infrastructure and saltwater intrusion to groundwater is occurring in many communities. Billions are being spent on adaptation to the sea level changes.
3. *Loss of sea ice:* Arctic sea ice extent is 10 to 15 percent less, and thickness 40 percent less, than in 1960 (National Oceanic and Atmospheric Administration [NOAA] n.d.). At current melting projections, 2040 will see an ice-free North Pole. While ice may have been decreasing since the last glacial maximum, the rate of change is unprecedented.

The major factors that have always controlled climate are still with us: solar irradiance, the elliptical orbit and planet wobble, the position of the continents relative to the equator or the poles, etc. We know, for example, that solar irradiance intensity follows 11-year sunspot cycles (with an overlaying 33-year cycle on top of it). These fluctuations, however, to not correlate with atmospheric temperature change in the modern epoch.

Even the GHGs we worry about so much (methane and carbon dioxide) fluctuate over the long term. When the continents were located near the poles, the planet even experienced an ice age when CO_2 concentrations were four times higher than they are today. Obviously, today only one continent is on a pole (Antarctica, on the South Pole), and the ice around the North Pole is rapidly shrinking. In the summers, the Northwest Passage (sought after in vain from the 1400s to 1800s by European explorers) is finally open, and it is now possible for commercial shipping to make its way through. Today, however, GHGs are accumulating in the atmosphere at a rate unprecedented in the glacial record going back 800,000 years, and the rate of global climate change is following the high rate of CO_2 increase. We know this "extra" CO_2 is from humans, because it carries the chemical fingerprint of fossil fuels (this will be explained in Chapter 2). The CO_2 increase is not from some other source within the carbon cycle. This increase, and the changes coming with it, can be pinned on us. And we are the only ones who can get ourselves—and all of the earth's other species—out of this predicament. We argue that the kind of sacrifice last called for by the industrial nations in the World War II era is what it will take: an international Manhattan Project for the environment.

Where Do We Get the Energy to Tackle Climate Change?

The root of the problem lies in the energy sources we use to drive our economies. As the human population grows and governments worldwide lift their populations out of poverty, our energy use will only increase. Diversifying our collective energy portfolio is wise, since it hedges against the failure of any single energy source (by scarcity or geopolitical shifts). Traditionally, fossil fuels had been cheaper than solar or wind sources. That has changed, however, as the efficiency of wind turbines and solar cells has increased and larger alternative energy markets have become available (allowing the benefits of economies of scale). For example, since 1975, the cost of photovoltaic[2] modules (better known as "solar cells,") for residential use or power grids has declined a remarkable 99 percent because of increased efficiency in converting light to electricity, investments in research, and scale economies (Kavlak et al. 2018). Wind power in the United States has also been increasing its slice of the energy market. It has expanded 224 percent since 1999 and has dropped from seven cents per kilowatt-hour in 2009 to two cents per kilowatt-hour in 2017. These changes are a result of larger turbines that increase efficiency, but even these improvements will be dwarfed when offshore wind farms are developed in the United States. Currently, the only offshore US wind farm is in Rhode Island; however, many coastal states are planning to construct offshore windfarms. In fact, in December 2018, the US Bureau of Ocean Energy Management leased three areas off the Massachusetts coast for a remarkable 405 million dollars (previous record was 42.5 million dollars)! What is even more remarkable is that the company winning the lease was an oil company!

The rise of alternative energy and the expansion of natural gas will be discussed in detail in Chapter 5. Renewable energy sources were also helped in

2020 by the fall in the price of oil as air and surface transportation ground to a halt in March and April during the COVID-19 lockdown. As oil tankers waited in ports without adequate storage for the backlog of oil around the world, the price of oil briefly went negative in April before rising again in May (although it remained roughly half its per barrel costs before the pandemic). At the same time, the demand for electricity dropped between 18 and 25 percent, further reducing the demand for fossil fuels. Existing renewable energy electricity generation, which is automatically patched into the grid, has increased during the COVID-19 pandemic as well. This is mostly because the costs of generating electricity from aging coal plants is high, and utilities are cutting back on coal production and substituting more solar, which has dropped 80 percent in cost since 2010, and wind, which has fallen 40 percent in that same time (Plumer 2020b). The effects of COVID-19 on the energy market as well as CO_2 emissions are more fully discussed in Chapter 5.

NATURAL GAS AS A BRIDGE FUEL

The price of natural gas has been decreasing since the early 2000s, and it has been cheaper than coal for over a decade. This has allowed natural gas to become a "bridge fuel" to a less carbon-rich energy economy, because natural gas releases less CO_2 than coal per unit of fuel mass burned. For example, generating 1 million British thermal units[3] of heat from coal releases between 228 and 215 pounds of CO_2 (depending on the quality of the coal), whereas the same quantity of natural gas releases only 117 pounds (Energy Information Administration [EIA] 2019).

Even before the COVID-19 pandemic diminished fossil fuel use, the transition to natural gas instead of coal in electrical utilities was driving US emission reductions. In 2016, natural gas had already surpassed coal as the dominant fuel for electrical power plants in the United States (33 percent natural gas vs. 32 percent coal) (EIA 2019). US CO_2 emissions in 2014 were nine percent less than they were in 2005, largely because of our use of natural gas for generating electricity (US Environmental Protection Agency 2016). Although the US public is largely unaware of it, the federal government has taken some steps to reduce carbon emissions through the Energy Policy Act of 2005. That legislation intended to promote cleaner energy, improve efficiency, reduce CO_2 emissions, and assist the transition to cleaner energy. Between 2009 and 2018, the United States has experienced impressive economic growth while reducing emissions, yet the public is seemingly under the impression that reducing emissions will harm us economically. Therefore, it is worth reminding the reader that US emissions reductions during those years took place during the largest economic expansion in the last 100 years.

As mentioned, natural gas is sometimes referred to as the bridge fuel, taking the industrial economies to a lower carbon energy future. But why is

Table 1.1 **Energy released when reduced C and H are oxidized to CO_2 and H_20**

Bond	*Bond Energy (kJ/mol)*
H-C	413
C-C	347

Natural gas has four H atoms bound to each C (4:1 ratio), so when combusted 1,652 kJ/mole is released. Oil averages a 2:1 ratio for H:C, and for coal, it is closer to 1:1.

that? Natural gas is still a fossil fuel, and it is still being used in massive quantities. In 2017 alone, the United States used 27 trillion cubic feet of natural gas (EIA 2018). The answer is that natural gas (methane, or CH_4) has more energy-rich H-C bonds relative to coal, which has C-C bonds and fewer H-C bonds per unit mass. When you burn coal, each C-C bond that is broken produces energy (but less energy than when breaking a H-C bond) and two CO_2 molecules (Table 1.1). When you burn natural gas, H-C bonds are broken, and you generate CO_2 and H_2O, which enters the hydrological cycle. Therefore, if burning pure methane natural gas (CH_4), four H-C bonds are broken, and one CO_2 molecule is generated, along with two H_2O molecules. Energy is derived from each molecular recombination. Therefore, natural gas releases less CO_2 per mass burned than coal, and the heat generated is a bit higher. This is why natural gas is considered a bridge fuel away from carbon-rich fossil fuels to those with less carbon and a potential segue to renewables (again, this concept will be fully discussed in Chapter 5).

Drivers of Natural Climate Change Are Real

Obviously, the climate changed in the past long before humans even appeared on the planet. However, to argue that because there was climate change before humans means that humans aren't responsible now is like saying that because there were forest fires before humans means that humans do not cause forest fires. Clearly, there are forest fires resulting from natural causes as well as human causes. This is also the case with modern climate change. Our activity has accelerated the change on top of the natural drivers. Additionally, the problem with modern climate change is not the change itself, it is the rate of change. No ecosystem on earth has evolved under the rates of change we are observing now, and we do not know how resilient they will be to the change. It is highly likely that plants, invertebrates with limited mobility (which is most of them), and alpine species will experience local extinctions and massive population declines as environmental zones they adapted to shift beyond their migration speed (this will be thoroughly explored in Chapter 2).

The major climate change events (e.g., ice ages) in our planet's history have had several contributing factors, which occasionally overlap. The largest

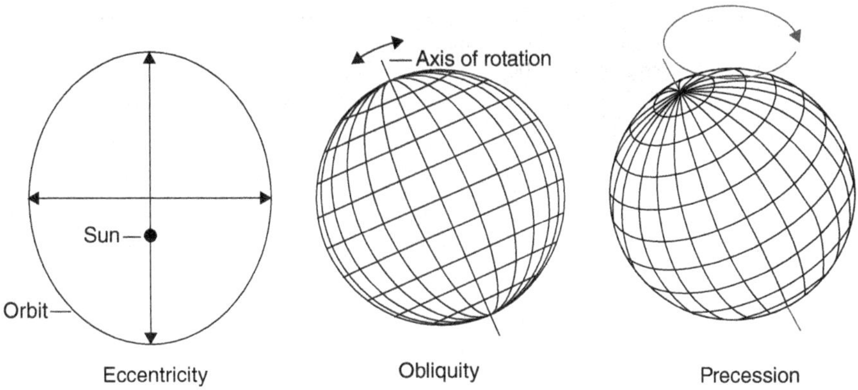

Figure 1.1 Features of Earth's movement relative to the Sun that contribute to the occurrence of ice ages. (a) Eccentricity of orbit, (b) precession of the axis of rotation, and (c) variation in axis tilt. Sources/notes: Simmon 2006.

scale, macro-driver appears to be regular and predictable changes in planetary motion and orbit (Figure 1.1). The first (in no particular order) is a change in the tilt of Earth's axis relative to the plan of its orbit. Every 41,000 years, the tilt gradually changes from 21.5 to 24.5 degrees. Today, Earth is tilted 23.5 degrees. (If the Earth was not tilted at all, it would be 0 degrees.) The second change is the "procession" or wobble of the Earth's axis of rotation. This is akin to the wobble of a spinning top. Earth completes a full wobble every 26,000 years. The third is the eccentricity of Earth's orbit around the Sun. Eccentricity is a measure of how circular the orbit is. The orbit cycles through periods of a very circular orbit to a more elongated orbit, and a full cycle takes approximately 100,000 years. These changes are collectively called the Milankovitch cycles (named after the Serbian scientist who discovered them), and they change the orientation of the planet relative to the Sun, which alters how much of the Sun's radiation is received but also how it is distributed. The cycles predict the majority of previous ice ages very well. There are a few discrepancies, however, because the Milankovitch cycles have not been the only driver of major climate change, even if they are the most important (at least in terms of ice ages).

The position of the continents relative to the equator or the poles has also influenced climate in the past. There have been two ice ages (250 million years ago and 2 million years ago) helped along by continents drifting to the poles (through plate tectonics). When continents are concentrated at the poles, snow and ice accumulate, because the land does not hold or distribute heat well. A few thousand years of snow that doesn't melt over a polar continent will generate continental glaciers.

Another major driver of past climate has been solar irradiance. The Sun's energy output has not been constant over geological time, and a 10 percent

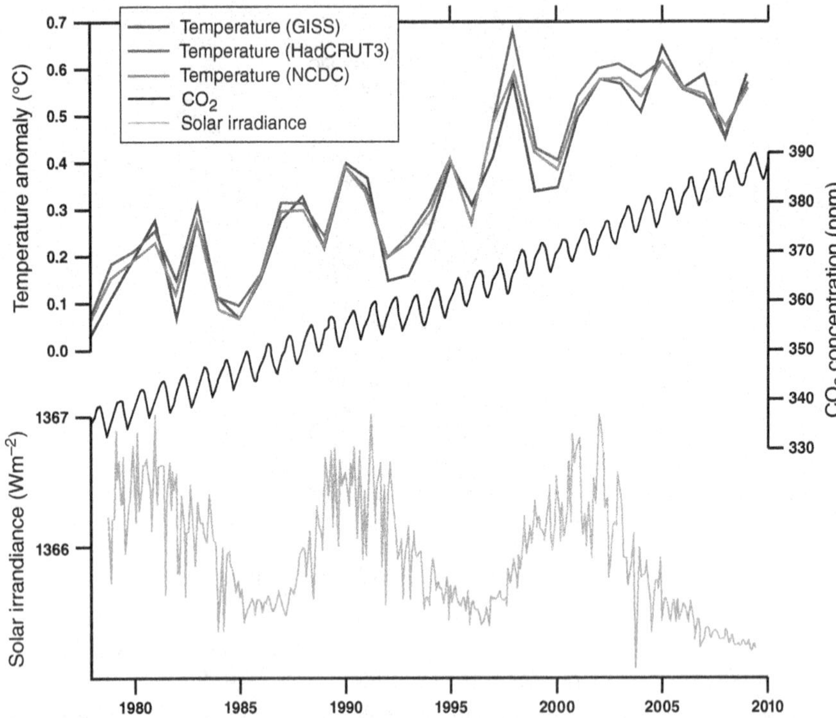

Figure 1.2 Temperature anomaly (defined as deviation from the 1950–1980 global average), shown in the upper part of the figure, is independent from the small changes in solar irradiance, shown at the bottom of the figure. Changes in atmospheric CO_2 are shown in the middle (the peaks and valleys reflect leaf decay in the fall and leaf growth in the spring, Northern Hemisphere). Sources/notes: Verheggen 2010.

change in solar irradiance could explain all temperature variation in our world's history (Hildore et al. 2010). Fluctuations in recent times have been very small, however. For example, between 1981 and 1984, irradiance dropped 0.07 percent. A drop of one to two percent could bring on a global cooling similar to that observed between 1600 and the early 1800s (sometimes called the Little Ice Age). A five percent drop could bring on an ice age. The current warming trends, however, do not correlate at all with changes in solar irradiance, which fluctuate with sunspot activity (Figure 1.2) and are simply too small to make any difference in global energy input. The difference in solar irradiance between high and low sunspot activity is 1,368 versus 1,365 kW/m^2.

The climate drivers outlined here are relatively slow-moving phenomena. Changes that can result in more rapid climate shifts are related to changes in ocean circulation or atmospheric gas composition. What the world is experiencing now is rapid change. Every year brings a new heat record for large swaths of the planet. The last 20 years have been the warmest on record, with the period from 2014 to 2018 being the warmest of these 20 years (US Global Change Research Program 2017). As the atmosphere warms, it can hold more water vapor, and as a result, rainfall intensity (centimeters per hour) has been increasing in many areas (this will be discussed in Chapter 2). Rapid warming has also caused the US Department of Agriculture to shift its "crop hardiness zones," as southern areas become too warm for traditional crops and warm-weather species distributions advance northward (Monier et al. 2016). All of these changes (and many more that will be discussed later in the chapter) are being observed year by year, which is a very short timeframe for a climatological shift.

Shifts this rapid must be due to a change in some feature of the planet that can be quantified. The fact that CO_2 has increased from 280 parts per thousand (ppt) around 1900 to 406 ppt in 2018 is indisputable. In terms of mass, the pre–Industrial Revolution amount of atmospheric carbon was 560 gigatons, and today, it is 750 gigatons (a gigaton is a billion tons). The fact that this increase in CO_2 is warming the planet is also indisputable. We know how much heat is trapped per unit CO_2; it is a matter of physics and is not mysterious. Just as increasing the heat under your tea kettle increases the vigor of the boiling water, so the increase in heat within the atmosphere makes the gases move faster. So the question then shifts to whether we can pin the extra CO_2 on human activity and what effects it will have on long-term climate patterns.

Over the past 50 years, scientists have learned the carbon cycle in a great deal of detail. It is known to plus or minus a few percent how much CO_2 goes into and out of the oceans, trees, soils, and rivers. The movement of CO_2 into and out of the various carbon pools are called "fluxes." Volcanoes emit CO_2, for example, and so each eruption causes a flux to the atmosphere and increases the atmospheric "pool." We know the flux from volcanoes is 0.05 to 0.1 gigatons of carbon (Schlesinger & Bernhardt 2013). Because fossil fuels are so expensive, their consumption is very well tracked. Worldwide C emissions from burning fossil fuels is 10.1 gigatons per year and rising (IPCC 2014). These fuels include coal, transportation fuel (gas and diesel), and natural gas. Although there are small year-to-year changes in C emissions from plants and soils, the fluxes are vanishingly small relative to the increased flux from fossil fuels, which we know very well.

THE SUESS EFFECT: A SMOKING GUN

No one can deny that the atmospheric concentration of CO_2 has increased by 31 percent since 1880, and we know that the increase correlates very strongly with fossil fuel combustion. However, another line of evidence that is less well known by the public also links the increasing CO_2 concentration to fossil fuels. The evidence is the atomic structure of the recently added atmospheric CO_2 itself.

Carbon has mass, and although one might think that all carbon has the same mass, it does not. Carbon emitted from burning fossil fuels is actually lighter (has less mass) than carbon emitted from the living biosphere. Carbon atoms with different masses are called isotopes, and they differ by a few neutrons. The most common carbon isotope is carbon-12 (^{12}C), so called because it has six protons and six neutrons in its nucleus. Carbon-13 (^{13}C) has seven neutrons, and carbon-14 (^{14}C) has eight neutrons. Fossil fuels have low abundances of ^{13}C relative to most living things and no ^{14}C at all.

The CO_2 generated from burning fossils is isotopically "light." It has more ^{12}C than CO_2 generated from the normal respiration of living things (decay included). The very large and "new" flux of fossil fuel–derived CO_2 has caused the total pool of atmospheric CO_2 to become isotopically "lighter." It is very easy to see that the CO_2 building up in the atmosphere has a greater abundance of "light" isotopes. This is called the Suess effect, named after the researcher who first found it (Suess 1953). Figure 1.3 shows the isotope change curve alongside the atmospheric CO_2 buildup curve. The change data in the figure are from Antarctic glaciers, but the Suess effect is observed in slow-growing corals, sponges, stalactites, plankton, tree rings, and other environmental markers that record CO_2 over time. The Suess effect is the smoking gun pointing to fossil fuels as drivers of the extra CO_2 we are currently enjoying.

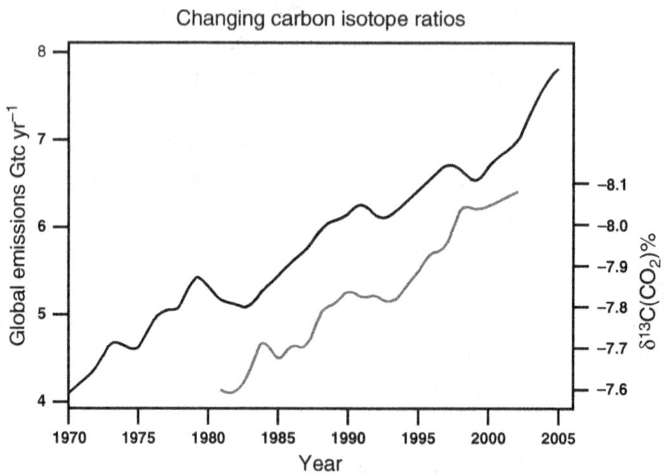

Figure 1.3 Atmospheric CO_2 concentration on left y-axis (black line) and mass of carbon-13/carbon-12 ($^{13}C/^{12}C$) in that CO_2 on the right y-axis (red line) over time. As ^{12}C increases relative to ^{13}C, the ratio becomes more negative. Fossil fuels are ^{13}C-depleted relative to CO_2 entering the atmosphere from the biosphere. The isotope ratio shift fingerprints the "extra" 120 parts per million (ppm) of CO_2 that has built up since 1900 as originating from fossil fuels. Sources/notes:IPCC 2007.

Technical and Scientific Means of Addressing Climate Change

From a technological and scientific perspective (without even considering the political and economic components), a solution to the problem of climate change will have several aspects. According to the IPCC, we need to become carbon neutral by 2050 to keep the warming to 2°C or less. Carbon neutrality does not mean zero emissions, but it does require an increase in carbon sinks to compensate for emissions. Energy use is linked to economic growth and the chance for a better quality of life for everyone. However, increased power supply does not mean that power has to come from fossil fuels; nor does it mean that we should deemphasize efficiency. As mentioned earlier, US emissions of CO_2 in 2017 dropped to 1990s levels without macroeconomic harm and without the general public noticing. Most of the reductions came from increased efficiency and increases in the supply of natural gas. Low-hanging fruit! Another recent important reduction in emissions has come very recently from the vast reduction in fossil fuel usage as a result of the COVID-19 pandemic. Experts have worried, however, that the only way to attain carbon neutrality will be for nations to "make concerted efforts to shift to cleaner energy as part of their recovery efforts" (Plumer 2020a). Indeed, reaching anything close to carbon neutrality over the decades ahead will require locking in some of the recent COVID "silver lining," and also the same pre-COVID substantial effort and innovation in many areas.

These areas will be discussed in detail in later chapters (particularly in Chapters 6, 7, and 8), but among them are:

1. Increased in efficiency and huge investment in low-carbon electricity, not only through solar and offshore wind but including batteries that can store the energy to reduce power fluctuation.
2. Investment in modernizing and expanding nuclear power. Although this energy source has its problems and is expensive, decarbonizing our electricity will have to include nuclear energy.
3. Carbon removal and storage from powerplants as it is emitted (in geological formations) plus scrubbing some CO_2 from what is currently in the atmosphere, perhaps by increasing weathering rates of silicate minerals, investing in massive reforestation, or some other approach.

It is important to note that the technology that will be used to decarbonize energy sources or sequester carbon will have to be shared with developing nations (IPCC 2018; see also Chapter 4). Whether this would be through a green fund administered by the United Nations or a more traditional development assistance approach, the developing world (its cities in particular) will need rapid and substantial assistance if they are to avoid the high-carbon modernization path the developed world took. There will need to be many other innovative approaches as well as we reach for carbon neutrality. There is no silver bullet, and we may fall short of our aggressive goals. However, we must make progress towards them and look for additional areas to improve.

Impacts of Climate Change on Human Populations

How does climate change affect people? At the most basic level, Lockwood (2016, 131–134) argues, using a slew of studies in psychology, that temperature change affects peoples' temperaments. Vector borne illnesses like dengue and malaria will have a much broader reach as climate change increases the range of these tropical illnesses carried by mosquitoes. Coastal sea level rise worldwide (with the worst impacts on small island nations), flooding (in South Asia and elsewhere), and drought (presently impacting much of Africa and the Middle East) will exacerbate human suffering, increase migration, and diminish food security for hundreds of millions of people. As we will show in the chapters that follow, concerns are increasingly being raised by analysts that climate change might also lead to group violence and nation-state instability.

While cost estimates for "fixing" climate change—discussed later in this book—vary depending on a range of scenarios, it is plausible that keeping warming from getting much worse (i.e., maintaining it within the 2.0°C targeted by the United Nations) could cost one to two percent of the world's gross domestic product (aggregate economic production measured annually). While the financial burden may fall more prominently on industrialized nations, it is important to realize that the developing nations are more vulnerable, by and large, as these are the countries receiving the brunt of climate-induced extreme weather events (cyclones, hurricanes, tsunamis, and droughts). The sacrifice is already being made by the people of Kiribati, who are preparing to relocate their entire population of some 120,000 people to Fiji, where they have a place to resettle if needed and as is growing increasingly likely.

No international legal figure for "climate migrants" exists, such as the definition used for political refugees. However, climate migration has already displaced millions of people across borders, and millions more within them. As stated by Kumari Rigaud et. al. (2018): "Climate change has emerged as a potent driver of internal migration, propelling increasing numbers of people to move from vulnerable to more viable areas of their countries to build new lives" (xix). Their study estimates that by 2050, some 143 million people in South Asia, sub-Saharan Africa, and Latin America alone will be forced to migrate within their countries. Internal migration is harder to measure than international migration, where border crossings are registered, but researchers are starting to estimate models.

Indeed, climate change has, in recent years, been considered by analysts as a rising independent source of political instability. Starting with a 2010 US Department of Defense review, military strategy has elevated climate change to an urgent threat, and this continues, even as US President Trump denied the existence of human-induced climate change. For example, elsewhere in the US international security community, a paper linking climate-related drought in Syria to the widespread internal crisis there has been circulating. This shows that policymakers in the US military have started to give climate impacts prominent consideration in planning for the future, even as political rhetoric continues to boisterously deny that it even exists. More broadly, a report from a politically centrist think tank on security reported in 2012 that "secure states do not automatically mean secure peoples and climate change is proving that," and furthermore, that "climate change is a risk to global security because it increases vulnerability in infrastructure, agriculture, energy, and other economic factors" (American Security Project).

In the longer term, no one will escape the impacts of climate change. However, in the shorter term (over say the next 20–50 years), most countries will "lose," but a few will actually "win." The differentiated effects of climate change may actually "warm" the permafrost in Russia and the tundra of Canada, moving maximal wheat production areas north. Additionally, businesses are starting to count on transportation through the previously impassible Arctic Ocean, meaning that hitherto unreachable areas will be accessible via waterway and shipping across the world will grow cheaper to many destinations through a new northern passage. This is despite the wide range of adverse impacts in most of the world, as headlined by Honduras, Haiti, Myanmar, Nicaragua, the Philippines, and Bangladesh, which between them have suffered almost 700 climate-related extreme weather events over the last 20 years, with nearly 200,000 deaths (Eckstein et al. 2019).

Climate Change Dilemma: How Do We Stop the "Denialist" Cross-Talk and Buckle Down?

Over the 30 years the IPCC has been researching climate change, the hundreds of scientists and technical experts involved have incrementally strengthened their language of the causal association of humans with climate change. Additionally, we know that climate change denial is more driven by politics than by science. Already in this short introduction we have considered the uncertainties "baked in" to rigorous modeling, which climate deniers have pointed to as justifications for critique.

Economic interests that have funded climate-denier campaigns for decades, however, have now given up that charade. As discussed further in Chapter 3, recent disclosures reveal that the leading fossil fuel company, ExxonMobil, which is responsible for some 10 percent of historical emissions (Heede 2013) and is still polluting the atmosphere (as these gases "hover" for about 150 years before dissipating), orchestrated a "climate denial" campaign that, for a time, included former US Secretary of State Rex Tillerson as "head denier." But ExxonMobil and other fossil fuel companies have acknowledged it is time to move on. The private sector has found, contrary to Kermit the Frog's famous song, that it is easy being green, and that in fact renewable energy sources are growing extremely lucrative.

Politicians, unfortunately, have not moved on. US Senator James Inhofe of Oklahoma, the chairman of the Senate Environment and Public Works Committee, famously brought a snowball to a US Senate meeting in 2015, presumably to show that the earth was not warming (Holthaus 2015).[4] President Trump's statement during the 2016 presidential campaign that climate change was a Chinese hoax was puzzling and unfounded, but it was his decision to remove the United States from the Paris Agreement after the first few months of his presidency that caused real harm (as mentioned earlier). The good news is that even before the COVID-19 pandemic, US emissions had decreased despite President Trump's strong support for burning coal, the most noxious GHG-emitting fossil fuel, and that China, the EU, and a slew of other climate champions are taking up world leadership on the issue.

How could climate denial have gotten so much traction in the United States as to generate lopsided public opinion polls showing that something like 30 percent of the public believes climate change does not exist? We do not want to dwell on this question, as we feel it is a distraction from addressing the scope of the problem and thus "feeds in" to the agenda of the climate deniers. However, it is important, from the standpoint of understanding how domestic politics derails climate action, to explain how scientists communicate findings differently from how politicians and political pundits do, and how climate deniers have exploited this gap.

To the science reporter Andrew Revkin (2014), the "tyranny of news" is at least partly responsible for the failure of the media to address climate change. Partly this is because climate change events are "so dispersed in time and geography that they will not constitute news as we know it" (237), partly because journalism "has long relied on the age-old method of finding a yea-sayer and a nay-sayer to frame any issue, from abortion to zoning" (239), and partly because the urgency of getting stories out leaves little time to cover complex issues full of nuance which take extensive space to explain" (243). Furthermore, Revkin adds, challenges exist in striking a proper balance in environmental stories between "the potent 'heat' generated by emotional content and the 'light' of science and statistics" (245).

To Revkin's reasons why the media oversimplifies and underreports climate change stories, we would add that the constant qualifications and ambiguities in scientists' reporting of their findings often confirm our reputations as nerds rather than rock stars. The main problem with climate change projections, in terms of reporting to the public, is that they are conditional and probabilistic rather than absolute. This is because they are projections about the future, and it is impossible to know the future with 100 percent certainty. Any responsible scientist therefore must convey results as a series of probabilities rather than empirical certainties. For example, if we continue to emit GHGs at our present rate, the atmosphere will warm between 2.0 and 3.5°C with a 90 percent probability. That is different from the kind of political statement a government official (or even a president) can make: "Climate change is a hoax . . ." Scientists' professional reputations rest on the precision with which they convey their findings; politicians' futures depend entirely on how many votes they get. Scientists must accurately convey information to other scientists using the scientific method; politicians are under no such restrictions. In fact, in the era of "fake news," political claims do not even have to be plausible. They just have to resonate with the right group of voters.

What would international organizations, national and subnational governments, industries and interest groups, and we as individuals have to do to really solve this problem? We do not yet have the precision in knowledge to answer with great specificity, although we do believe that international organizations would have to "hold the line" at a 1.5°C increase in global warming by the end of the 21st century (rather than the 2.0°C allowed under the current United Nations agreement),[5] try to get nations to volunteer genuine emissions reductions and then find ways to enforce this, facilitate emissions trading systems, offer assistance to develop new technologies, and help the world's most vulnerable adapt to climate change on the front lines. National and subnational governments are going to have to collect taxes to reckon with this problem (e.g., carbon taxes), regulate emissions in a fair and transparent way, and incentivize and require individual reductions in emissions. Individuals will have to diminish our use of fossil fuel–intensive transportation means, adopt more carbon-efficient diets (vegetarian diets, or at least less reliance on beef for protein), and find other ways of diminishing our carbon footprints.

Climate Change Dilemma: Can We Offset (Outsource) Some of this Sacrifice?

Led by a concerted misinformation campaign by oil and other interests, we have spent two precious decades debating the existence of climate change rather than working to stop it. Now we must stop asking whether we need to address this problem and instead calculate exactly how. Sacrifices have to be made in the short term to avoid greater ones in the longer term. And some believe that we do not need to diminish our own personal emissions if we balance these with efforts to diminish emissions by an equal amount elsewhere on the planet. Is that possible?

If we continue to consume, can we "offset" our carbon footprints by paying someone else to produce fewer emissions or absorb our emissions through carbon capture and storage via technological or natural means, such as planting and preserving more trees?[6] This is an important question, and—perhaps—a morally ambiguous one. For example, Nick Davis (2017), writing in *The Guardian* newspaper, concluded:

> The idea that we might cancel our own greenhouse gases by paying for projects that reduce the gases elsewhere was born in the early years of climate politics. It was adopted by the corporate lobby at the Kyoto Summit in 1997 and has grown into a large but deeply troubled adolescent—confused, unpredictable, and difficult to trust.

However, philosopher John Broome, in a systematic rebuke of objections to offsetting, argues that mathematically one can remove a ton of CO_2 from the atmosphere to offset a ton added. In this view, removing net CO_2 is the goal, and any methods that work should be used.

A net addition of zero can result, and offsets can be "remarkably cheap" (Broome, 90) as most offset projects are conducted in developing countries, where costs are generally lower and these projects can provide employment or other benefits to residents of the country doing the offsetting. While Broome does allow that the difficulties inherent in measuring CO_2 emissions, and thus offsetting them, are real and may lend themselves to "shenanigans of companies and nations" (91). But the moral question is not the same as this technical one, he argues, and he also acknowledges that individual emissions reductions will not be sufficient to solve the climate change problem in any event. "To improve the world, we shall have to adopt political means," he writes (96). "We shall have to work through our governments, because only governments can take action on the large scale that is required." While analysts increasingly turn to the private sector as public spending continues to diminish around the world,[7] we do nonetheless agree that governments are a foundational pillar in solving climate change.

Climate Change Dilemma: Equity and Climate Justice Across Countries

Part of the problem with offsets is that they are an option for the more affluent and would likely put the less affluent at their service. Aggregating this problem from individuals to communities and countries, we can begin to understand why it has been so difficult for nations to agree on a solution to the climate change problem. Rich nations, which tend to be concentrated north of the Tropic of Cancer, have more resources to "buy" their way out of sacrifice, but also benefit from two other strong advantages. First, these countries historically are the world's largest GHG emitters, having industrialized by burning carbon to an extent that is no longer going to be possible by "late industrializer" developing nations. Second, the Northern Hemisphere nations are suffering less directly from climate change than nations in the Global South (the equatorial zone and south).

The North-South Debate, so prominent in the Cold War between the United States and the Soviet Union (now split into Russia, Ukraine, and several other countries) in provoking geopolitical proclivities toward capitalism or communism, has reemerged, but with an environmentalist frame. As stated by Adil Najam (2015):

> Developing countries have consistently contextualized environmental issues as part of the larger complex of North-South concerns, particularly concerns about an iniquitous international order and their desire to bring about structure change in that order. This has become more poignant in recent years as environmental negotiations on issues such as climate change. (220–221)

More recently, Barbados Prime Minister Mia Mottley (2019) had a sharp critique for attendees at a 2019 conference, where she warned that the Caribbean had been ignored in the 1990s world trade liberalization and now risked being ignored regarding climate change:

> These countries remain on the front line of a war that we did not start, nor do we sustain . . . It is only when moral and ethical leadership is given, both at the national level and at the international levels, that we will summon the courage to fight these [climate] battles. Until such time, it [addressing the issue] is a form of idle entertainment for those who choose to watch.

This indictment of the countries that caused climate change, but do not have to suffer its immediate consequences as directly, was a strong invocation of the need for climate justice.

We will elaborate on this more extensively in Chapter 3, but here, we introduce the two main inequities in "climate justice" between North and South. Regarding the "penalty" on industrialization of developing countries, China and India led countries articulating this argument during climate negotiations in the 1990s (known as the Kyoto Protocol), winning exemptions from emissions reductions. While equity—and the United Nations—dictated that the industrial nations which had disproportionately caused the 150 years of GHG pollution settled in the atmosphere (think back to the coal-fired smokestack era of London depicted in Dickens' *Oliver Twist*) were the ones who had to cut emissions, and thus risk slowing economic growth, the world's worst emitter (China) and third-worst emitter (India) were not asked to reckon for their own poor recent records. Furthermore, these countries, other newer developers, and older industrial nations with large coal endowments—like Australia and Germany—are seeking to fuel industrialization with coal, the dirtiest of fossil fuels.

The other inequity is in climate victimization, which has already started to befall the Global South disproportionately. Droughts have beset California as well as Syria and Sudan. In the Middle East and Africa, however, these events have been life-threatening, whereas in California, they have just been inconvenient. Hurricanes have ripped through Louisiana, Florida, North Carolina, and New Jersey and caused a good deal of property damage, but few lives have been lost.[8] Yet compare this to hurricanes in Mexico, Honduras, Dominica, and Indonesia; cyclones and salinization in Bangladesh and India; and sea level rise in Tuvalu. Indeed, nine of the ten most climate-vulnerable countries are in the Global South (Eckstein et al. 2019), and the tenth is Puerto Rico, which becomes vulnerable for all the hurricane damages suffered when separated from the rest of the United States.

Climate Change Dilemma: Equity and Climate Justice Across Generations

Even more difficult to address than inequities of climate impacts across countries and regions is the inequitable distribution of climate impacts across time. One of the pioneer advocates for climate change, now-retired NOAA scientist James Hansen has argued that given the great inertias in the longevity of fossil fuel CO_2 in the atmosphere, "there is additional climate change 'in the pipeline' even without further change of atmospheric composition" (Hansen et al. 2013, 19). In what may even be an understatement, he further argues that "young people and future generations [are] inheriting a situation in which grave consequences are assured, practically out of their control, but not of their doing" (19–20).

Swedish high school student Greta Thunberg won global recognition in March 2019 for her weekly "school strikes for climate," propelling a movement of school strikers all over the world. Her protest started in August 2018, with weekly demonstrations outside the Swedish Parliament, and had morphed, some eight months later, into protests in 40 countries. The movement was largely publicized via social media, where Thunberg gained 250,000 followers on Twitter in time for the coordinated action. "I've said that I will continue to strike every Friday until Sweden is in line with the Paris Agreement," Thunberg said. "That may take a couple of years and I'll just have to try to be patient" (Filks 2019).

Time horizon may indeed be one of the most difficult aspects of galvanizing action to address climate change. As we will discuss briefly in Chapter 4, national politicians, who are elected and reelected to achieve immediate results, are not by nature professionally vested in what will happen 50 years hence, or even in 20 years. The urgency of our circumstances needs to be more clearly presented if the Generation Xers and Millennials are going to have a shot at fixing the climate before the burden passes even further down our lineage. And our inability to introduce temporal incentives for action is matched by our inability, thus far, to create incentives for international institutions and national governments.

Public Goods, the Tragedy of the Commons, and the Role of Interests

The late Mancur Olson (1965) argued that cooperation was more readily possible among small groups, as personal persuasion and reputation are more important in such groups, whereas in larger ones (unions, big businesses, governments, etc.), individuals could "free ride" by joining groups but not expending effort on their behalf. Even within large groups, smaller (or privileged) groups overcome the collective action problem, even when there are free riders who perceive benefits without paying the costs. Smaller groups (within the broader group) can be made to contribute to the cause through selective incentives, which Olson calls by-products. The problem for large groups with a common interest is that they are unable to deny anyone (even free riders) the benefits of any group successes. What are selective incentives? Coffee mugs and sweatshirts to identify Sierra Club donors, trade statistics and "Washington briefings" for trade association members, discounts on camping equipment to environmental club members (or two percent back from sporting goods stores, etc.). Political entrepreneurs help overcome barriers to collective action through charisma, appeals to beliefs, ideology, or just self-interest. We need a few more of those in the climate change space.

When collective goods provided by some application of Olson's theory are in the public interest (they are nonexcludable and jointly supplied—i.e., one person's use does not reduce the supply available to others), they are public goods. Most market goods are private goods, as their "owner" has full control over their use. Ronald Coase's (1974) lighthouse is a classic public good, but so is the provision of defense (whereas toothpaste and BMWs are private goods).

The commons is a special case, modeled on medieval town commons where sheep grazed. According to this model, discussed by Olson (1965), Hardin (1968), and others, each peasant gains by grazing his/her sheep on the village commons, but if they all graze their sheep, there is a depletion of the resource (the "Tragedy of the Commons" is an increasingly common reference point in environmental regulation). The broader point is that individual livestock owners, acting in their individual interests, act counter to the collective or public good by overgrazing their cows or sheep to the point of despoiling the commons and ruining it for all.

According to Nobel laureate Elinor Ostrom (1990), the resolution appears to be clearly defined boundaries, congruence between rules for using the commons and peoples' needs, individual rights for formulating and changing rules guiding the use of commons, monitoring arrangements (with graduated punishments) making the users responsible, low-cost means of resolving conflicts, and relative users of commons from outside government authorities. In other words, if the commons cannot be parceled into privately owned bundles (privatization), then a self-enforcing, self-restraining political arrangement must be found. The analogy to the climate problem is not overly taxing. Just consider the emissions overproduction by the industrial powers to the detriment of all, but especially the vulnerable nations, as stated by Barbados Prime Minister Mia Mottley. In reference to climate change, if we cannot ourselves reduce our emissions, we have to join together, Ostrom-like, and commission "third party" international institutions that will. But we also must collaborate and contribute as individuals and in smaller groups, like governments and social groups. How do we overcome the obstacles and make this so?

Where You Come In: Fostering Cooperation from Individuals to Nation-States to the Planet

We are confronting perhaps the most intractable scientific, technological, spiritual/ethical, political, economic, and legal problem facing our planet and its living beings for the foreseeable future. As professors, parents, and citizens, we felt it was time to provide a far-ranging text introducing the symptoms of the problem, but also to present outlines for its solution.

Information is the first step; a sober and comprehensive understanding of the full scope of the problem is essential. Journalistic shortcuts, climate deniers, and naysayers need to be understood as red herrings (and their habitats may also be shrinking). Credible and objective sources, like the IPCC reports, need to be judiciously considered even when they are messengers of bad news. We need to make scientists' jobs publicizing information easier, accepting that their reports may come with conditions and caveats and that scientists do not like to report anything without qualification. But that is authentic and should not need to be tweeted by Taylor Swift to be broadly accepted.

Furthermore, as the book will convey extensively in the chapters that follow, there is no sweeping solution to this crisis; there is only a series of partial and incremental steps toward resolution. As US climate movement leader and 350.org founder Bill McKibben (2006) wrote, "There are no silver bullets, only silver buckshot."[9] Indeed, proponents of one solution or another have emerged, somewhat like "snake oil salesmen" in the 19th-century Wild West. Be wary of them. For example, geoengineering, which as will be discussed in later in the book must understandably be considered as a Plan B, should not be viewed as a substitute for the whole menu of partial measures we need to adopt to mitigate emissions. The adoption of geoengineering would be a solemn moment, and better avoided if at all possible, although we must research and consider the possibility.

We will be much better off if we can solve climate change through a disciplined regimen of incremental measures. Like most worthwhile endeavors, solving this problem is going to be hard. And frankly, our job has only been made more difficult by the COVID-19 pandemic, which drained government coffers for years to come (although it does have the emissions reduction "silver lining" if we can capitalize on it). We realize that two "once-in-a-lifetime" crises—climate change and the pandemic—are a lot for any of us to handle. But information is power, and we want to empower you. This book offers a modest start in that direction.

It is time to act. In his 1941 "Four Freedoms" speech, President Franklin D. Roosevelt laid out the imperative of shared personal sacrifice at a moment of national emergency:

> We must all prepare—all of us prepare—to make the sacrifices that the emergency—almost as serious as war itself—demands. Whatever stands in the way of speed and efficiency in defense—in defense preparations of any kind—must give way to the national need.[10]

In recent years, climate change denial has become a mainstay in the platforms of populists in far-ranging countries from Australia to Brazil to the United States, as governments willing to—literally—mortgage long-term "national interests" in favor of short-term reelection strategies have questioned the need to adapt to climate change, or even to mitigate emissions. Luckily, most of the rest of the world has gotten on board.

Agronomists are drought-proofing rice seeds so that Bangladeshi farmers have crops to sell, shipping companies are planning new transcontinental routes through the melting Arctic ice, and oil companies—once among the loudest climate deniers—are helping research wind and solar renewable energy solutions. Insurance companies, military strategists, and fossil fuel companies are heavily vested in adaptation to and mitigation of the current and future climate realities. The Pope and Leonardo Di Caprio, as well as the ubiquitous Nobel Peace Prize–winning former Vice President Al Gore, are strong proponents of facing up to climate change to keep global warming to a minimum. We all need to get on board.

And regardless of how majestic black bears can be in the wild, we humans, having reduced them to Teddy Bears in the bedtime hug of millions of kids, are the only ones who can reclaim nature from our own simultaneously loving and lethal clutches. As already mentioned, President Franklin Roosevelt (a cousin of the Teddy Bear's inspiration) called for shared sacrifice with an urgency not uttered since as the Allies faced Hitler and the Nazis in World War II. The Reverend Martin Luther King cited the "fierce urgency of now" to energize action on civil rights in 1963. Both speeches have been used in seeking to motivate action on climate change.[11] The magnitude of the challenge is immense, but the effort that young people have to offer is high. It may be a copout for us Generation Xers (younger than the "Okay Boomers," but not by much) to pass the GHG-emitting torch to Generation Z. Understand that we do it with affection, respect, and a little bit of desperation.

Exercise 1.1 Making Coffee; Getting to Cooperation

OVERVIEW

A great ice breaker on the first day of class, the coffee game will help participants become familiar with the concepts of skills evaluation and resource endowments, now-versus-later costs and benefits, and concentration of resources.[12] Students will be split into two teams, each given an object or task. Students will both observe and participate in competition with and between two groups (phases 1 and 2, respectively) that are trying to achieve a common goal. The objective of the game is for all the component parts of making coffee to be given to one person so that they can make the coffee. The professor should walk into class and say, "Anyone else want some coffee?" and then pull out the materials and start dividing the class into groups. Don't discuss the exercise; just start it. The students will be a little surprised and not understand what is going on for five or ten minutes, but if you lead the exercise firmly, you will keep control of the class, they will fondly recall the chaos you created on their arrival, and they will remember how cooperation can be fostered or discouraged by different sets of incentives.

MATERIALS

- 1 coffee pot (moderately disassembled so that parts can be distributed to different players)
- 1 espresso machine (moderately disassembled so that parts can be distributed to different players)
- 2 bags of coffee
- Milk or creamer
- Water
- Stirrers or spoons

- Sugar
- Cups and napkins
- Coffee grinder
- Blue, red, and white chips or tokens
- Task sheets (coffee grinder, coffee pourer, coffee mixer—as many as are needed to match the number of students)

SETTING UP THE GAME

1. Split the students into two even groups (Team Coffee and Team Espresso).
2. Give the coffee maker (minus most of its parts) to Team Coffee and the espresso maker (minus most of its parts) to Team Espresso.
3. Distribute the coffee maker's parts to people in Team Espresso (i.e., the filter goes to one participant, the actual pot to another) and the espresso maker's parts to Team Coffee.
4. Distribute the remaining materials and task sheets to participants (water goes to one, stirrers and spoons to another, tokens to people on each team, etc.).
5. Write the point system on the board:
   a. Coffee pot = 25 points
   b. Coffee maker (person who eventually assembles all the component parts) = 50 points
   c. Each coffee or espresso component = 5 points
   d. Each task = 10 points
   e. Blue chips = 50 points each
   f. Red Chips = 25 points each
   g. White chips= 1 point each (for Team Espresso, white chips are worth 1 point each unless a consensus/cooperation is achieved when making the espresso, at which point they are worth 50 points each and the points are equally distributed to all)
6. Write the phase 1 objectives on the board:
   a. Team Coffee: One person makes coffee using all the needed components. The accumulated points only go to the coffee maker.
   b. Team Espresso: One person makes the espresso using all the needed components. The accumulated points are evenly distributed if consensus/cooperation is achieved.

PLAYING THE GAME

1. Allow the students to trade and bargain for 15 to 20 minutes until some outcome is achieved. There will be confusion, as there are no real rules besides the point system, but this will reinforce the complexity of the process and is a part of the game.

Continued

2. Phase 2: One team must make coffee, and winners are those of this team only—winning team gets coffee and the adoration of their colleagues.

Instructors only, please see in Appendix Instructor Guide to Exercise 1.1: Incentivizing Divergent Outcomes (to contrast cooperation and discord).

NOTES

1. According to Mooallem (2013, 62): "When Roosevelt saw the pitiful animal lashed to the tree, he refused to fire at it, or to have anyone else shoot it, either; he felt it went against his code as a sportsman. Instead, he asked a hunting companion to put the bear out of its misery with a knife. But that detail of the story would quickly get lost."
2. "Photovoltaic" refers to the production of electrical current between two semiconductors in the presence of light.
3. A British thermal unit (BTU) is the amount of heat required to raise one pound of water by 1°F. It is from the English system of measurements (predating the metric system).
4. Of course, had Senator Inhofe looked at weather patterns to discern how temperatures have increased every year for decades (presumably affecting the number of "snowball days" outside the Senate), he would have drawn different conclusions.
5. For comparison, 1.5°C is equal to 1.8°F, and 2.0°C is equal to 3.6°F. Most recently, a report by the Intergovernmental Panel on Climate Change addressed the importance of seeking climate change of not more than 1.5°C (see ICPP 2018).
6. The international Reducing Emissions from Deforestation and Forest Degradation (REDD+) initiative seeks to protect forests, especially tropical ones, as a means of diminishing CO_2, which is absorbed by trees in photosynthesis (emitting carbon and oxygen, which, when separated, do not promote climate change).
7. Public budgets as a percent of gross national product over time.
8. The exception is Puerto Rico, a US territory where in 2017 Hurricane Maria was estimated to have killed 2,975 people, after an original estimate that some 64 people had died. Puerto Rico's climate vulnerability has been largely ignored by the US "mainland." President Trump threw towels at people there but did not release much emergency assistance (Nakamura & Parker. 2018).
9. "Buckshot" refers to hundreds of BB-sized metal balls encased into a shotgun shell, which strike the target with many small hits rather than piercing it with a singular penetration.
10. See http://voicesofdemocracy.umd.edu/fdr-the-four-freedoms-speech-text/ (accessed on October 23, 2018).
11. On a World War II analogy to the retooling, solidarity, and sacrifice needed to confront climate change, see https://newrepublic.com/article/135684/declare-war-climate-change-mobilize-wwii (accessed on November 21, 2020). On an interfaith worship and demonstration of "urgency" for climate change, see https://www.interfaithmoralactiononclimate.org/uploads/7/8/5/4/7854310/imac_3-21-13program.pdf (accessed on November 21, 2020).
12. The inspiration for this exercise came from a 2013 talk at American University by University of New Haven Dean of Arts and Sciences Elizabeth Beaulieu, who mentioned that a colleague at Champlain University had used a similar exercise to teach cooperation.

REFERENCES

American Security Project. 2012. *CSR—Part One: Climate Change & Security*. Washington, DC: American Security Project. Accessed December 11, 2020 at https://www.americansecurityproject.org/csr-part-one-climate-change-security/

Broome, John. 2012. *Climate Matters: Ethics in a Warming World*. New York: W. W. Norton & Company.

Coase, Ronald. 1974. "The lighthouse in economics." *The Journal of Law & Economics* 17 (2): 357–376.

Davis, Nick. 2007. "The inconvenient truth about the carbon offset industry." *The Guardian*, June 16. Accessed October 23, 2018, at https://www.theguardian.com/environment/2007/jun/16/climatechange.climatechange

Eckstein, David, Marie-Lena Hutfils, and Maik Winges. 2019. "Global Climate Risk Index 2019." *Bonn: Germanwatch*. Accessed July 1, 2019, at https://www.germanwatch.org/en/16046

Energy Information Administration (EIA). 2018. "US natural gas total consumption." Accessed June 22, 2020, at https://www.eia.gov/dnav/ng/hist/n9140us2A.htm

Energy Information Administration (EIA). 2019. "Frequently asked questions (FAQs)." Accessed November 24, 2020, at https://www.eia.gov/tools/faqs/faq.php?id=73&t=11

Filks, Ilze. 2019. "Swedish student Greta's climate 'school strike' goes global." Reuters. Accessed July 1, 2019, at https://www.reuters.com/article/us-climate-change-youth-greta/swedish-student-gretas-climate-school-strike-goes-global-idUSKBN1QS1O0

Hansen, James, Pushker Kharecha, Makiko Sato, Valerie Masson-Delmotte, Frank Ackerman, David J. Beerling, Paul J. Hearty, Ove Hoegh-Guldberg, Shi-Ling Hsu, Camille Parmesan, Johan Rockstrom, Eelco J. Rohling, Jeffrey Sachs, Pete Smith, Konrad Steffen, Lise Van Susteren, Karina von Schuckmann, and James C. Zachos. 2013. "Assessing 'dangerous climate change': Required reduction of carbon emissions to protect young people, future generations and nature." *PLOS ONE* 8(12 December 2013):e81648

Hardin, Russell G. 1968. "The Tragedy of the Commons." *Science* 162(3859):1243–1248.

Heede, Richard. 2013. *Carbon Majors: Accounting for Carbon and Methane Emissions 1854–2010: Methods & Results Report*. Climate Justice Programme, Sydney, Australia. Accessed July 1, 2019, at http://climateaccountability.org/carbonmajors.html

Hidore, John J., John E. Oliver, Mary Snow, Richard Snow. 2010. Climatology: an atmospheric science. 3rd ed Pearson. 0321602056

Holthaus, Eric. 2015. "Watch Inhofe throw a snowball on the Senate floor to disprove global warming." *Slate*, February 26. Accessed October 24, 2018, at https://slate.com/technology/2015/02/inhofe-throws-a-snowball-on-senate-floor-to-disprove-global-warming-video.html

Hughes, Terry P., James T. Kerry, Andrew H. Baird, Sean R. Connelly, Andreas Dietzel, Mark C. Eakin, Scott F. Heron, Andrew S. Hoey, Mia O. Hoogenboom, Gang Liu, Michael J. McWilliam, Rachel J. Pears, Morgan S. Pratchett, William J. Skirving, Jessica S. Stella, and Gergely Torda. 2018. "Global warming transforms coral reef assemblages." *Nature* 556:492–496.

Intergovernmental Panel on Climate Change (IPCC). 2007. *Climate Change 2007: The Physical Science Basis. Contribution of Working Group I to the Fourth Assessment Report of the Intergovernmental Panel on Climate Change*, edited by S. Solomon, D. Qin, M. Manning, Z. Chen, M. Marquis, K. B. Averyt, M. Tignor, and H. L. Miller. Cambridge University Press, Cambridge, UK.

Intergovernmental Panel on Climate Change (IPCC). 2014. *Climate Change 2014: Synthesis Report. Contribution of Working Groups I, II and III to the Fifth Assessment Report of the Intergovernmental Panel on Climate Change*, edited by R, K. Pachauri and LA. L. A. Meer IPCC, Geneva, Switzerland.

Intergovernmental Panel on Climate Change (IPCC). 2018. Global Warming of 1.5°C. An IPCC Special Report on the impacts of global warming of 1.5°C above pre-industrial levels and related global greenhouse gas emission pathways, in the context of strengthening the global response to the threat of climate change, sustainable development, and efforts to eradicate poverty [Masson-Delmotte, V., P. Zhai, H.-O. Pörtner, D. Roberts, J. Skea, P.R. Shukla, A. Pirani, W. Moufouma-Okia, C. Péan, R. Pidcock, S. Connors, J.B.R. Matthews, Y. Chen, X. Zhou, M.I. Gomis, E. Lonnoy, T. Maycock, M. Tignor, and T. Waterfield (eds.)]. IPCC, Geneva, Switzerland.

Kavlak, G., J. McNerney, and J. E. Trancik. 2018. "Evaluating the causes of cost reduction in photovoltaic modules." *Energy Policy* 123:700–710.

Rigaud Kanta Kumari; Alex de Sherbinin, Bryan Jones, Jonas Bergmann, Viviane Clement, Kayly Ober, Jacob Schewe, Susana Adamo, Brent McCusker, Silke Heuser, and Amelia Midgley, Amelia. 2018. *Groundswell: Preparing for Internal Climate Migration*. World Bank,

Washington, DC. Accessed February 18, 2021 at https://openknowledge.worldbank.org/handle/10986/29461

Lockwood, Alan H. 2016. *Heat Advisory: Protecting Health on a Warming Planet*. Cambridge, MA: Massachusetts Institute of Technology Press.

McKibben, Bill. 2006. "Welcome to the climate crisis." *Washington Post*, May 27. Accessed October 25, 2018, at http://www.washingtonpost.com/wp-dyn/content/article/2006/05/26/AR2006052601549.html

Monier Erwan, Liyi Xu, and Richard Snyder. 2016. "Uncertainty in future agro-climate projections in the United States and benefits of greenhouse gas mitigation." *Environmental Research Letters* 11:055001.

Mooallem, Jon. 2013. *Wild Ones: A Sometimes Dismaying, Weirdly Reassuring Story about Looking at People Looking at Animals in America*. Penguin Books: New York.

Mottley, Mia. 2019. Speech to inaugurate the "Understanding Risk (UR) Caribbean Conference," May 27, Bridgetown, Barbados. Accessed July 1, 2019, at https://www.youtube.com/watch?v=-ZGp9qohcu8

Najam, Adil. 2015. "The view from the South: Developing countries in global environmental politics." In *The Global Environment: Institutions, Law, and Policy*, edited by Regina Axelrod and Stacy VanDeveer. CQ Press, Los Angeles.

Nakamura, David, and Ashley Parker. 2018. "'It totally belittled the moment': Many look back in dismay at Trump's tossing of paper towels in Puerto Rico." *Washington Post*, September 13. Accessed July 1, 2019, at https://www.washingtonpost.com/politics/it-totally-belittled-the-moment-many-look-back-in-anger-at-trumps-tossing-of-paper-towels-in-puerto-rico/2018/09/13/8a3647d2-b77e-11e8-a2c5-3187f427e253_story.html?noredirect=on&utm_term=.80657cd9a60a

National Oceanic and Atmospheric Administration (NOAA). n.d. "Arctic change." Accessed June 22, 2020, at https://www.pmel.noaa.gov/arctic-zone/detect/ice-seaice.shtml

Olson, Mancur. 1965. *The Logic of Collective Action: Public Goods and the Theory of Groups*. Harvard University Press, Cambridge, MA.

Ostrom, Elinor. 1990. *Governing the Commons: The Evolution of Institutions for Collective Action*. Cambridge University Press, New York.

Plumer, Brad. 2020a. "Emissions declines will set records this year. But it's not good news." *New York Times*, April 30. Accessed June 3, 2020, at https://www.nytimes.com/2020/04/30/climate/global-emissions-decline.html

Plumer, Brad. 2020b. "In a first, renewable energy is poised to eclipse coal in US." *New York Times*, May 3. Accessed June 2, 2020, at https://www.nytimes.com/2020/05/13/climate/coronavirus-coal-electricity-renewables.html

Revkin, Andrew C. 2014. "Climate change as news: Challenges in communicating environmental science." In *Climate Change: What It Means for Us, Our Children, and Our Grandchildren*, edited by Joseph F. DiMento and Pamela Doughman. Massachusetts Institute of Technology Press: Cambridge, MA, pp. 227–256.

Scheffers Brett R., Luc De Meester, Tom C. Bridge, Ary A. Hoffmann, John M. Pandolfi, Richard T. Corlett, Stuart H. M. Brutchart, Paul Pearce-Kelly, Kitt M. Kovacs, David Dudgeon, Michela Pacifici, Carlo Rondinini, Wendy B. Fodne, Tara G. Martin, Mora Camilo, David, Brickford, and James E. M. Watson. 2016. "The broad footprint of climate change from genes to biomes to people." *Science* 354. DOI: 10.1126/science.aaf7671

Schlesinger, William H., and Emily S. Bernhardt. 2013. *Biogeochemistry: An Analysis of Global Change*. 3rd ed. Academic Press: London UK.

Simmon Roger. 2006. "Paleoclimatology: Explaining the evidence." Accessed June 22, 202,0 at https://earthobservatory.nasa.gov/features/Paleoclimatology_Evidencew

Steffen, Will, Paul J. Crutzen, John R. McNeill. 2007. "The anthropocene: are humans now overwhelming the great forces of nature," *AMBIO: A Journal of the Human Environment*, 36(8), 614–621.

Suess, H.E. 1953. *Natural radiocarbon and the rate of exchange of carbon dioxide between the atmosphere and the sea. Proceedings of the Conference on Nuclear Processes in Geologic Settings*, pp. 52–56. National Academy of Sciences, National Research Council, Washington, D.C.

US Environmental Protection Agency. 2016. "US Greenhouse gas inventory report: 1990–2014." Accessed June 22, 2020, at https://www.epa.gov/ghgemissions/us-greenhouse-gas-inventory-report-1990-2014.

US Global Change Rsearch Program. 2017. *Climate Science Special Report: Fourth National Climate Assessment, Vol. I*, edited by D. J. Wubbles, D. W. Fahey. K. A. Hibbard, D. J. Dokken, B. C. Stewart, and T. K. Maycock. US Global Change Research Program, Washington, DC., DOI: 10.7930/J0J964J6

Verheggen, Bart. 2010. "Recent changes in the sun, CO_2 and global average temperature." Accessed June 23, 2020, at https://ourchangingclimate.wordpress.com/2010/04/11/recent-changes-in-the-sun-co2-and-global-average-temperature-little-ice-age-onwards/

Webster P.J., G.J. Holland, J.A. Curry, H.R. Chang. 2005. "Changes in tropical cyclone number, duration, and intensity in a warming environment." *Science* 309 1844–1846.

2

Learning to Listen to the Earth

The Evidence and Scope of the Problem

In decades past, projections of environmental shifts resulting from climate change seemed abstract, because they were not being experienced in a noticeable way by people at the time. One of these characteristics of projecting the effects of climate change has always been the time frame: "by 2040, we will lose the alpine glaciers," or "by 2050 with 2°C of warming, tropical reefs will be largely gone," and so on. Humans tend to weigh future risk differently than immediate risk, and because climate change is, by definition, a long-term and large-scale phenomenon, the absence of immediate impact fostered a feeling of complacency.

While no one disputes that atmospheric CO_2 concentration is 30 percent higher than it was in the late 1800s, there are those who say we are probably not the cause. The general argument is that since CO_2 has risen and fallen in the atmosphere long before humans were even around, it is presumptive to say that we are doing this ourselves. To quote Senator Jim Inhofe from the Senate floor in 2015, "The hoax is that there are some people who are so arrogant to think they are so powerful they can change climate. Man can't change climate. . . . God is still up there. The arrogance of people to think that we, human beings, would be able to change what He is doing in the climate is to me outrageous" (Bruenig 2015). In other words, if nature causes CO_2 to increase in the atmosphere on its own, the current increase should not be pinned on us—or so the argument goes. It is, of course, a fact that CO_2 concentration has risen rapidly in the geological past, well before humans arrived on the scene. However, to say that this means we cannot cause CO_2 to rise ourselves is the same as saying: "We have forest fires today. There were forest fires before we were around. Therefore, we do not cause forest fires." The example is a bit silly, but so is the argument that we cannot be responsible for CO_2 increases now, and for the associated warming now, because it also happened in the past.

There are, of course, natural fluctuations of CO_2 that arise with seasonal growth, fires, volcanic activity, and land-use change, just to name a few sources. However, we can look at CO_2 fluxes from those and see that the anomalous 30 percent increase over the last 120 years has been independent of season or fires or volcanic activity and now are driven largely by anthropogenic fossil fuel combustion (and, to a lesser extent, by land-use change). Not only is there a compelling correlation between CO_2 and fossil fuel burning, but the 30 percent extra CO_2 carries the ^{13}C-depleted isotope signature of fossil fuels. Recall, as discussed in Chapter 1, that the carbon in CO_2 derived from fossil fuels has a higher percentage of ^{12}C ("light" carbon with six neutrons) versus ^{13}C ("heavy" carbon, with seven neutrons) than most modern, photosynthetically fixed carbon. The carbon in atmospheric CO_2 has been getting "lighter" since the Industrial Revolution because of fossil fuel CO_2 emissions (the Suess effect; again, please see Chapter 1 for more details).

In more recent years, it has become more apparent to people that strange and alarming climate-driven environmental change is occurring. Although people might not know how much of these climatic shifts are caused by humans, scientists are less comfortable with dismissing the shifts as normal variations.

The reality is that we no longer need models to tell us that climate change is coming. It is happening right now, all around us. It manifests in different ways and with different degrees of severity. The changes are not only measured by physical parameters (temperature, glaciers retreating, rainfall intensity, etc.) but also biological parameters. In general, less is written about the biological changes that are occurring compared to the physical changes, but the entire biosphere (with the possible exception of the deep ocean) is responding to the rapid increase in atmospheric energy, which manifests as climate change. The changes in timing of naturally recurring events such as ripening of fruit, leaf unfolding or coloring, migration, nest building, and mating are termed "phenological." This chapter explores some of those changes, with the hope that readers will appreciate how something as abstract and large scale as climate change can impact the planet's biological function at an accessible level. We also review recent changes in the outlook for precipitation patterns, rising seas, and cyclones (hurricanes/typhoons), since those changes are impacting human communities right now.

Seasons and Phenological Change

We begin with changes in seasons—specifically, spring. It is now well established that spring is occurring earlier in the Northern Hemisphere. In the United States, it is arriving almost a week earlier than it did 30 years ago. Globally, the mean (average) advancement of spring is 2.3 days per decade, but it tends to come sooner the higher in latitude you go (Parmesan & Yohe 2003). A very visible sign of the shift is seen in flowering trees, which are very sensitive to temperature change. For example, the blooming of cherry trees is an iconic event in the temperate world, particularly in Asia and in the eastern United States, but the timing of the blooms has been changing. In Japan, a recent study examined 732 years of recorded blooming dates and temperatures. It was found that blooming strongly correlated with temperature and late blooming with prolonged times of sunspot minimums (when solar irradiance declined and the earth cooled; i.e., the Little Ice Age) (Aono & Kazui 2008). Over the last 30 years in Japan, however, the time of blooming has been occurring earlier every year regardless of the solar irradiance or sunspot activity (the effect of sunspots on solar irradiance was discussed in Chapter 1). And all the way across the planet, in Washington, DC, the cherry trees are now blooming five days earlier than they did in the 1920s (Waxman 2017). This is just one example with a single species, but the fact that the trend is occurring in two places with wide geographical separation indicates that warming is global.

SHIFTS IN MIGRATION PATTERNS

Another sign of spring has to do with bird migration patterns. People living in temperate climates understand the arrival of migrating birds is a sign that spring is coming. Now they can see some of these birds arriving a week earlier than they did 30 years ago. In 2015, researchers in Norway examined 30 years of tracking data for the pied flycatcher (*Ficedula hypoleuca*) and found that they were reaching spring breeding grounds five days earlier than they did in the 1980s (Cadahia et al. 2017). What is driving them north is the early warming of their overwintering grounds in western Africa. This seemingly innocuous change is not only impacting the birds (obviously) but other species as well. The birds migrate to Norway to breed, which means they need an ample food supply for their soon-to-be-hatched chicks. For the flycatcher, the food is a winter-hatching moth caterpillar. However, the moth caterpillars are not operating on the same climate-driven clock as the birds, since the insects are local and experiencing local climate effects, not the conditions in western Africa. The winter moth caterpillars are actually hatching 15 days earlier.

So while the breeding cycle has been accelerated for the birds, that acceleration is less than the rapid shift in the moth's life cycle. Therefore, when the flycatcher chicks hatch, the caterpillar larvae have already started to pupate and are on their way to becoming moths—that is, they are unavailable for food. The immediate effect of this is a decline in flycatchers and a boom in moth populations.

In another example, a 47-year study of the great tit (*Parus major*, a common bird in Europe), showed the egg laying has been occurring two weeks earlier than it used to. In this case, however, the shift in egg laying appears to have been driven by the earlier hatching of its caterpillar prey (winter moth caterpillar), and both species seem to be healthy (Charmantier et al. 2008). A recent study however, suggests that the current synchrony, maintaining 13 days between egg hatching and peak caterpillar abundance, will break down within 50 years if CO_2 emissions continue on their present trend (Simmonds et al, 2020).

This type of shifting migration/breeding patterns has been observed in many places. Marine fish that spawn in freshwater (salmon, herring, etc.) have been arriving earlier in recent years. Their return to freshwater to spawn is linked to estuary temperature. As coastal streams reach spawning temperatures earlier, the fish have "stepped up" their travel plans. Similar to what was observed with the flycatcher, the rate of adaptation to the climate shift has been different for different species. The species that prey on the spawning fish have not changed their behavior in the same way as the fish have (Scheffers et al. 2016). While these changes do not necessarily mean the species will not adapt (as they appear to have done in the case of the great tit in the United Kingdom [UK]), one cannot predict the ramifications of all the changes happening at once. Unforeseen consequences are likely.

SHIFTS IN GENETICS

Other shifts in biological systems have not been as obvious as the timing of returning migratory birds. Genetic shifts in species with rapid life cycles have been observed in response to rising temperatures. Water fleas (an important food for fish) in the UK have seen an increased frequency of genes used for coping with higher temperatures. Oceanic plankton in the Gulf of Cariaco have adjusted their thermal tolerance by 0.45°C in response to a 0.75°C change in the area (Scheffers et al. 2016). Note that the temperature adaptation is lower than the temperature change. The rate of adaptation therefore is not keeping pace with the rate of temperature shift. This is what primarily worries biologists. Even species with rapid life cycles, which should be best equipped to adapt to environmental changes, may not be able to meet the rapid rate of current and future climate shifts. If a plankton that produces several generations in a single season cannot keep up, how will species that have one offspring every few years? And how will plant species that cannot migrate, nor distribute seeds widely, fare?

Shifting Distribution of Marine Species

Changes in the distribution of ocean biota have also been observed. All organisms attempt to stay in the temperature zones for which they are physiologically adapted. The warming rate of ocean surface waters is one-third that of the land (0.07°C/decade vs. 0.24°C/decade, respectively) because of the massive heat-redistribution effects of ocean currents (Burrows et al. 2011). Even so, 75 percent of marine species are shifting toward the poles (up to a 1,000-km shift in range). The rate of this single directional movement is between 1.4 and 28 km per decade, with range shifts between 1.5 and 5.0 times faster than those observed on land (Burrows et al. 2011). This means that cold-water ocean organisms will begin drifting poleward as surface waters warm, or they will disappear if they cannot migrate quickly enough. For example, Antarctic krill (*Euphausia superba*) populations, upon which Antarctic ocean food webs depend, have retreated toward that continent over the last 90 years as surface waters have warmed. Their populations also have dropped approximately 70 percent relative to the 1976–1996 average (Atkinson et al. 2019), which will dramatically change how energy flows through oceanic food webs around Antarctica (up to and including whales and penguins).

Changes in equatorial marine systems are more variable and not as dramatic as what is observed in other areas; however, some zones of high productivity and diversity (coasts, coral reefs) are seeing species migrating (if possible) or experiencing harmful effects of warming waters (e.g., the 29 percent loss of corals in the northern section of the Great Barrier Reef discussed later in this chapter). It is worth noting that as tropical species follow warming waters toward the poles, there are no species available to move into the warm waters they vacated.

Other very visible changes are also occurring among charismatic Arctic fauna. The plight of polar bears (*Ursinus maritimus*) has received widespread attention as they have been tracked making epic swims (more than a week in the water without resting or eating) in search of sea ice large enough for them to haul out (moving from water to ice) and hunt. Arctic sea ice, however, is shrinking more and more every year, losing 14 percent of its surface area per decade. The bears need large expanses of ice to hunt their main prey, seals (Pagano et al. 2019), so the ice loss is an existential problem for them.

PLIGHT OF THE PACIFIC WALRUS

Less well publicized, but equally distressing, is the frailty of the Pacific walrus (*Odobenus rosmarus*). In the summers, female and young walruses feed on shallow benthic invertebrates to build up fat stores. In between feeding dives, they haul out on ice flows to rest. Those ice flows, however, are retreating to ever deeper areas of the Arctic, so the walruses now haul out on the coast of northwestern Alaska and the northern coast of the Chukotka Peninsula in Russia. The problem is that there isn't room for all of them. Walruses have terrible eye sight, and large numbers of young are trampled to death as the adults desperately try to find unoccupied patches on the beaches (Taylor et al. 2018). Some walruses are actually driven more than 100 m up the cliffs behind the beaches. Walruses weigh thousands of kilograms and are not very mobile, so try to imagine how desperate they are to find a place to lie down if they are climbing up cliffs! Tragically, when those on the cliffs hear their fellow walruses beginning to surge into the water (they do this in crowds), they try to join the migration and leap off the cliffs to their deaths. Such emotionally wrenching scenes appear in the 2019 Netflix series *Our Planet* created by Alistair Fothergill and provide a gut punch reminder that climate change results in visceral suffering by living things that have no control over it, yet are bearing the brunt of it.

Shifting Terrestrial Climate Zones

Since 1975, terrestrial isotherms (lines of equal temperature) have been moving poleward at 100 km per decade. This basically means that warmer average temperature is pressing poleward. Isotherms have also been moving up mountains, squeezing cold-adapted alpine communities into smaller and smaller areas. Intensification of the large-scale atmospheric circulation that ascends at the equator and descends at 30°N and 30°S (known as the Hadley cell) has created the rapid monodirectional isotherm shift. This has made the North American southwest dryer along with the Mediterranean, South America, southern Africa, southern Australia, and Madagascar. The increasingly dry conditions have also dried forests and consequently increased the heat of wildfires. Fifty percent of terrestrial species have shifted their ranges poleward by 600 km at an average rate of 6.1 km per decade. In temperate mountains, species ranges have shifted upwards by 400 m (Hansen et al. 2013). The distribution of life on the planet is changing, particularly in the subtropics. It always has, of course, but the rate of change is much higher than any ecosystem currently on the planet has ever experienced, even during the onset or recession of ice ages.

Effects on Biodiversity

The greatest threat to biodiversity on a local or regional scale is habitat destruction, which is highly visible, has many causes, and can be shaped by local, regional, or national policy. The effects of warmer years, year after year, is slower and more insidious. Widespread declines in frogs and other amphibians came as a surprise to researchers, since the losses were frequently in landscapes that were not developed (nor developing). The cause for the disruption therefore was not obvious. In many cases, it was an increase in fungal disease that reduced amphibian populations (and, in some mountain communities, wiped them out).

It is now understood that climate change is what allowed the fungal infections and other pathogens to spread (Hansen et al. 2013). We should think of these amphibians as "canaries in the coal mine" for climate change and threats to biodiversity. (Canaries were brought into coal mines because they were very sensitive to toxic gas that might be in the tunnels and shafts, so their distress was the first warning to miners.) Species (all species, not just animals) that are geographically constrained, such as populations on mountain tops or at high latitudes, will have nowhere to go as they seek to remain in their thermal environmental window. Additionally, species that are not resilient to rapid warming stress will experience losses. Unfortunately, the climate resilience for the majority of species is completely unknown. Scientists are trying to determine resilience for a number of species, particularly the endangered ones. For most species, however, we will learn about their resilience by watching their populations fluctuate as climate change overtakes them.

As mentioned earlier, many species are shifting in response to the movement of their climate zone, but even if they can keep pace with the shift, they could encounter obstacles such as mountain ranges or bodies of water that will block them. Sea level rise will also threaten coastal communities, and not just from those communities getting inundated. Increasing saltwater intrusion into freshwater habitat will stress communities and impact organisms sensitive to salts, such as invertebrates, fish, and many plants. For communities that are on low-lying islands (Bermuda, Barbados, Aldahabra, etc.), the challenge will be similar to that of the polar or mountain species—namely, there will be nowhere to go.

Although it is difficult to pin climate change as the exclusive cause for any of the extinctions happening currently (except the frogs mentioned earlier), if one looks at the geological record we see that mass extinctions frequently co-occur with rapid climate change. This does not necessarily indicate that climate change was the only cause of those extinctions, but it suggests that we should be mindful of the likely threat to diversity represented by rapid climate change in today's world. Losing diversity will impact ecosystem function and services (removing pollutants from air and water, pollination rates, nutrient cycling, pest mitigation, etc.) as well as the genetic diversity of the biosphere on which we have traditionally depended for new medicines, chemical compounds, and food (Hansen et al. 2013).

Effects on Cyclones (Hurricanes)

With the appearance of large, damaging cyclones (known as hurricanes in the United States) in the North Atlantic, more people are examining the possibility that climate change is involved in these storms. In 2012, Superstorm Sandy (an odd combination of a nor'easter[1] and a hurricane) did approximately 65 billion dollars of damage along the US eastern seaboard. Hurricane Maria devastated Puerto Rico and the Virgin Islands in 2017, causing unprecedented infrastructure failure and costing upward of 90 billion dollars. Hurricane Michael stalled over the North Carolina coast in 2018 and flooded the entire eastern half of the state, causing 25 billion dollars in damage (Federal Emergency Management Agency 2018).

The proximate cause of hurricanes is not well understood. It is clear that an ocean surface temperature of at least 26°C is required, but ocean waters of this temperature or higher do not necessarily cause hurricanes (e.g., the North

Indian Ocean never falls below 29°C). A reasonable working hypothesis has been that warming of the oceans and atmosphere would result in stronger, more frequent cyclones. However, doubling of CO_2 (relative to preindustrial concentrations) in models has led to variable results in terms of frequency, even if they do tend to project an increase in intensity (Bhatia et al. 2019, Webster et al. 2005).

In an effort to understand trends in cyclones worldwide, a study in 2005 examined all storms between 1970 to 2004 for any change in cyclone frequency or strength associated with rising sea surface temperature (Webster et al. 2005). This was done for all the world's oceans. The study found that in the North Atlantic hurricane frequency is increasing (99 percent confidence level); however, in other ocean basins, this was not the case. Globally, no relationship was found between increasing sea surface temperature and cyclone frequency. However, there was a significant shift in the strength of the storms. Category 4 and 5 cyclones (on the Saffir-Simpson Hurricane Wind Scale of 0–5) increased from 20 percent of the total in 1970 to 35 percent of the total by 2004.

This scale of increase in the frequency of category 4 and 5 storms occurred in all ocean basins worldwide. In an unrelated study, researchers tracked the rate of wind speed increase versus time (rate of intensification) for cyclones between 1982 and 2009, focusing on the storm transitions from category 1 to category 5 (Bhatia et al. 2019). These researchers found that the rate of intensification has also been increasing. So the storms are getting stronger faster. They also found that storms that experience rapid intensification (95th percentile of the 24-hour intensity changes) tend to become major hurricanes (categories 3–5), and these are the hurricane types that do the most damage. Needless to say, this was an alarming finding, and if it continues, areas of the world that experience cyclones can expect more damaging impacts from these storms. (We address these human costs of increased hurricane frequency and ferocity in Chapter 8.)

The Atlantic Thermohaline Circulation

The poleward movement of warm equatorial water toward the poles and the return of deep cold waters to the surface is generally called the Global Heat Conveyer (hereafter called the Conveyer) because so much energy is being redistributed planetwide from the process. Obviously, the Conveyer has a large impact on climate, since it is redistributing heat. However, some studies have suggested climate change could impact parts of the Conveyer mechanics, and it deserves a closer look.

Each ocean basin is a portion of the Conveyer, but we are going to focus for the moment on the Atlantic Thermohaline Circulation (ATC). Some recent findings have shown that it has slowed. In fact, the ATC is moving 15 percent less water through the currents than it did in 1950 (Caesar et al. 2018). The ATC is a series of currents that include the Gulf Stream, which moves heat energy from the Caribbean along the coast of North America and then shoots across the Atlantic to the UK and Europe. The amount of energy the ATC moves is mindboggling. Estimates are that one petawatt (one petawatt is equal to 10^{15} watts) of energy is moved from equator towards the UK. That is more than the energy used by 100 earth civilizations at any given moment (Schlesinger et al. 2006)! The amount of heat delivered by the ATC is also staggering. The ATC warms parts of Europe by 8°C (or 46°F). That huge amount of warming is the reason why the UK is relatively warm compared to other areas of equal latitude in North America. If you follow a line of latitude from southern England west across the Atlantic, you will run into Ontario's Polar Bear Provincial Park! The Scilly Isles (50°N) in the southernmost UK are almost subtropical (palm trees grow there!), which is certainly not the chilly climate you would experience if you crossed the Atlantic at the same latitude.

Such a massive heat delivery is possible because the warm Caribbean water is less dense than the colder North Atlantic water it encounters as it swings east, away from North America. The warm, dense water does not mix with the colder surrounding water, so the warm water stays on the surface in a sort of hot-water ribbon. Since both the warm and cooler water are saline, temperature becomes a barrier to mixing. Once the water reaches the UK, it cools and sinks, only to be replaced by new warmer water from the ATC.

So what does this have to do with climate change? Well, it turns out that at the end of the last ice age (Pleistocene, approximately 15,000 years ago), the ATC shut down. It stopped delivering heat to Europe, and the ice age there lasted 1,500 years longer than it did in North America. This prolonged glaciation in Europe is called the "Younger Dryas," named after a cold-weather plant that could be found in Europe during that period. It is thought that what shut down the ATC was a massive freshwater flood caused by a melting ice dam that held back a glacial lake in Canada (Lake Agassiz). That cold freshwater entered the Atlantic Ocean by either the St. Lawrence Seaway or the Mackenzie River, but either way, it deposited a thick, cold, freshwater lens over the Atlantic. Freshwater is less dense than saltwater, and the warm Gulf Stream portion of the ATC was stopped and forced under the freshwater. Unlike the "normal" situation described in the paragraph above (where temperature is a barrier to mixing), in this case the barrier is the salinity difference. The heat delivery to Europe was interrupted, and the heat energy was delivered to the deeper ocean instead, where it was absorbed by the vast volume of the Atlantic. The consequences of an ATC shutdown were probably global then, as they would be today. Models suggest that heat, which would otherwise warm Europe, would instead become redistributed across the planet. The actual effects will depend on the duration of the shutdown and also on the overlying climate change resulting from the other global disrupters. Models suggest that after 10 years of shutdown, and with continued high levels of CO_2 emission (the IS92A storyline), the ocean around the UK and Scandinavia will cool between 1.0 and 12.5°C, while the rest of the world's oceans will warm by between 0.1 and 2.0°C (Wood et al. 2006).

These days, we do not have glacial lakes of the sort that existed at the end of the Pleistocene, but Greenland is in the North Atlantic and is melting at an accelerated rate. Also, most climate models project that precipitation will increase in the North Atlantic as a consequence of climate change. These two possible large freshwater sources have attracted a lot of attention by European modelers and politicians, particularly in the UK (with good reason) (Schlesinger et al. 2006). An 11 percent primary productivity loss is projected to result worldwide from a shutdown of the ATC (Wood et al. 2006). Thankfully, current projections suggest that the ATC is unlikely to shut down this century, but there is a 50 percent chance of a shutdown by 2205 if global warming by that time reaches 3°C (Schlesinger et al. 2006). This sort of event is classified as a low-probability, major-impact event, and we need to be aware of tipping points (e.g., a meltwater volume threshold) that could shut the ATC down.

Changing Precipitation Patterns

It is widely accepted that a warmer planet will result in more evaporation and therefore a larger volume of water moving through the freshwater cycle. The extra water will result in greater overall rainfall; however, different regions will experience increases or decreases. In general, the subtropics will experience decreasing rainfall, while the higher latitudes will see increasing rainfall (Intergovernmental Panel on Climate Change [IPCC]) 2014. As models have predicted, we are observing polar movement of storm tracks and subtropical dry zones plus an expanding tropical region (Marvel & Bonfils 2013). These changes are altering our planet's biosphere in dramatic fashion.

Precipitation is the most important predictor of plant productivity, and shifting patterns will clearly impact terrestrial ecosystems. Increasing water availability will lead to more CO_2 being drawn out of the atmosphere as plants in areas with sufficient nutrient availability increase productivity. Because of this, the biosphere, including soil, is predicted to increase carbon storage through the midcentury (Wu et al. 2011). Because the precipitation patterns are shifting, however, some areas that are currently productive farmland could become dryer, which has obvious implications for food security (and national security, for that matter). One of the ways precipitation patterns are being altered is through changing jet streams, which are high-altitude streams of air that move weather systems around, either moving or trapping high- and low-pressure systems in the mid latitudes. Models have correctly predicted that the North American southwest would become dryer and that California would experience extremes of dry and wet seasons. In all areas of the United States, except Hawaii and the Southwest, precipitation has increased substantially since 1991(relative to the 1901–1960 average) (Figure 2.1).

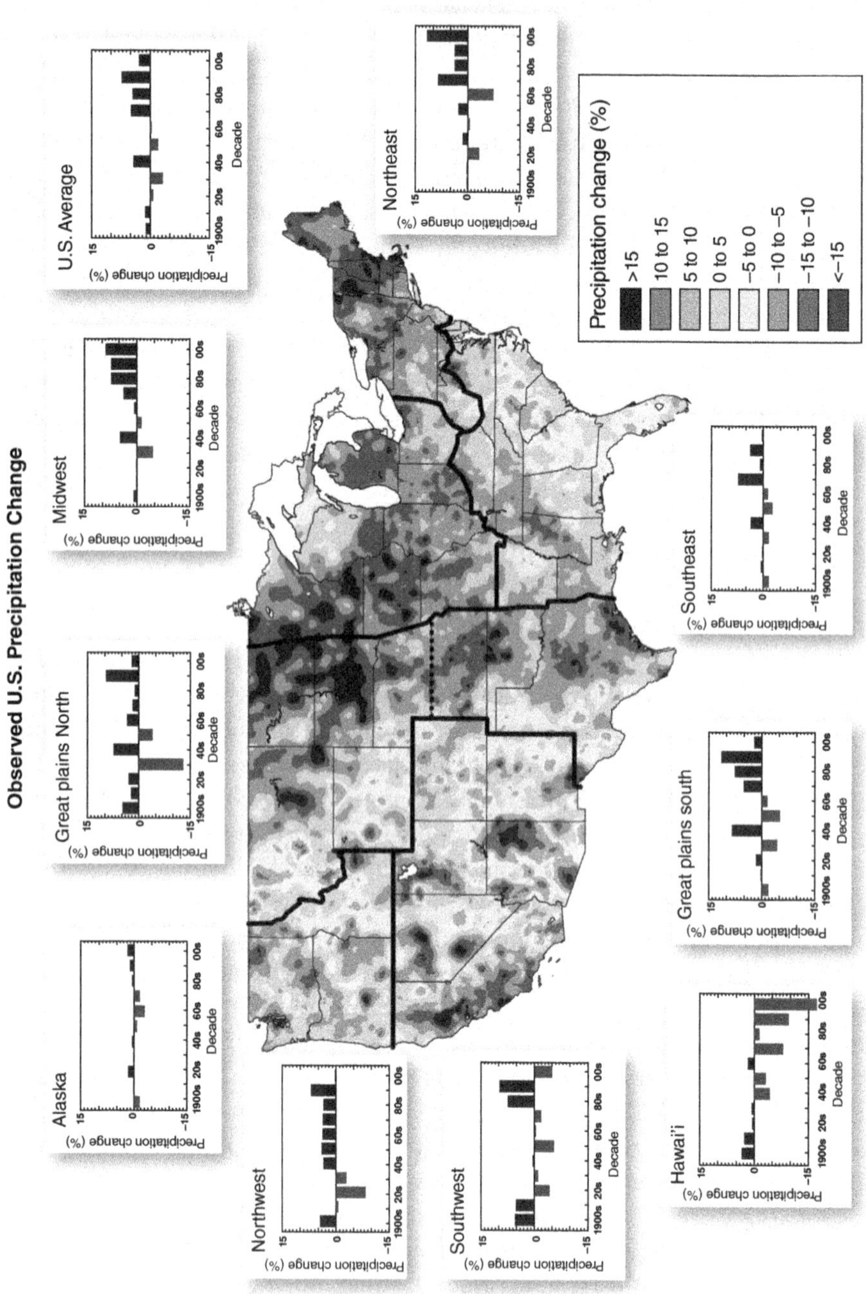

Figure 2.1 Shades of gray show percent precipitation changes for 1991– 2012 compared to the 1901–1960 average (that average is the 0 in each box) (Walsh et al. 2014, adapted from Peterson et al. 2013). © American Meteorological Society. Used with permission.

Alterations of precipitation patterns have profound implications for managing agricultural water supplies, but they also impact wildfires. For example, southern California has seen an increase in wildfire intensity and burn area recently as a direct consequence of years of drought brought on by a shift in the northern subtropical jet stream. Wildfires in southern California are projected to increase by 64 percent on average in the period from 2041 to 2060 relative to the period from 1981 to 2000 (Figure 2.2).

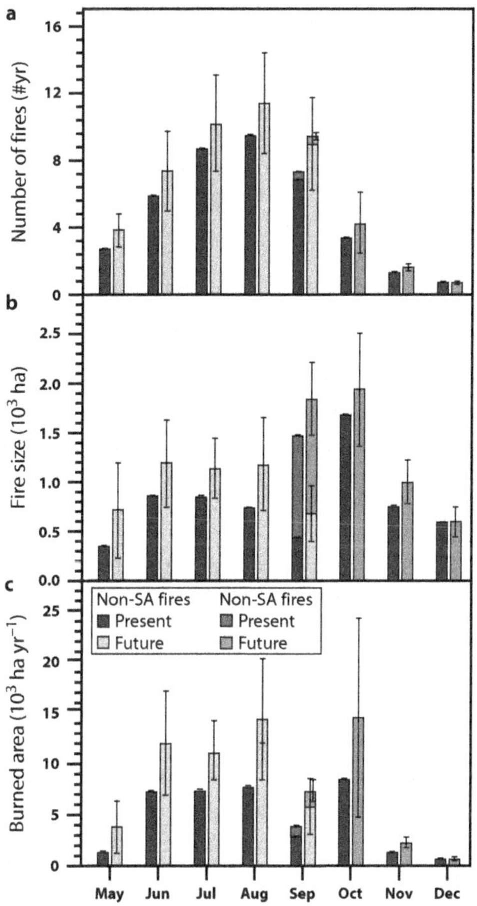

Figure 2.2 Shifts in fire number (a), size (b), and burn area (c) in Southern California expected by 2060. Scenarios were modeled using five global climate model simulations with the Weather Research and Forecasting model; fire parameter data were from the California Department of Forestry and Fire Protection and Resource Assessment Program (Jin et al. 2015). "SA fires" are associated with the Santa Ana winds which can be seen starting during the month of September in the figure. All fires In October through December are Santa Anna fires.

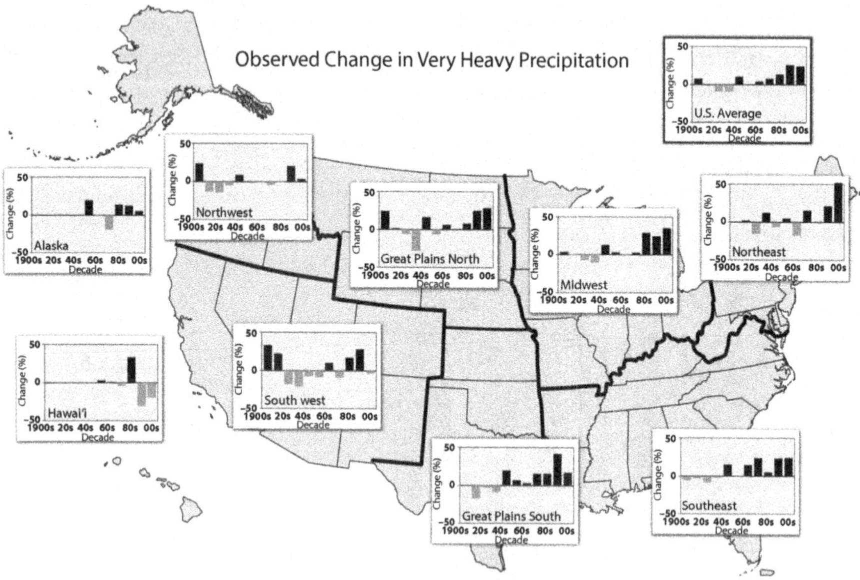

Figure 2.3 The percent change in annual precipitation falling in very heavy events for different regions in the United States. A very heavy event is the greatest one percent of all daily events between 1901 and 2012 for each region in the figure. The change is measured relative to the 1901–1960 average everywhere except Alaska and Hawaii, which are relative to the 1951–1980 average. (Walsh et al. 2014, adapted from Peterson et al. 2013). © American Meteorological Society. Used with permission

Although the shifting precipitation patterns were expected from the general circulation models (Wu et al. 2011), what has been a bit unexpected was the dramatic increase in rainfall intensity. Rainfall in centimeters per hour has been rising in many areas, especially during thunderstorms. The increase in rainfall is occurring in almost all areas of the United States, except for Hawaii and the Southwest (Figure 2.3).

Increasing intensity of rainfall is having real-time impacts. In the northeastern United States, heavy rain events increased 71 percent between 1958 and 2012. All this "extra" water has strained infrastructure in cities and developed areas as more water tries to leave landscapes characterized by impervious surfaces or floodplain development. Simply put, the drainage systems were not built for the extraordinary pulses of runoff occurring ever more frequently. What used to be 500- or 1,000-year floods are now occurring more frequently. For example, Baton Rouge, Louisiana, experienced a three-day storm event that dropped 25.5 inches of rain, causing extreme flooding and damaging over 60,000 homes (van der Wiel et al. 2017). And Ellicott City, Maryland, has seen two 1,000-year floods in only two years (Barton 2018)!

Sea Level Rise

In the last 100 years, we have observed a sea level rise of 18 cm (about 7 inches). This is more than the rise over the past 2,000 years *combined*. One hundred years ago, global sea level was rising at one mm per year. Today, the rate has it doubled. The rate of sea level change is variable depending on which region you examine, of course, because there are other contributors aside from climate change. For example, coastal Louisiana is experiencing sea level rise at four to five times the global average because the land is subsiding (the mass of development plus compaction of coastal sediments) (Figure 2.4).

Another example is the mid-Atlantic, where in some areas sea level is decreasing as the land continues to rebound in the postglacial period (particularly near Hudson Bay, Canada, and parts of Alaska) (Figure 2.4). Sea level rise is impacting coastal cities with more frequent flooding and consequent

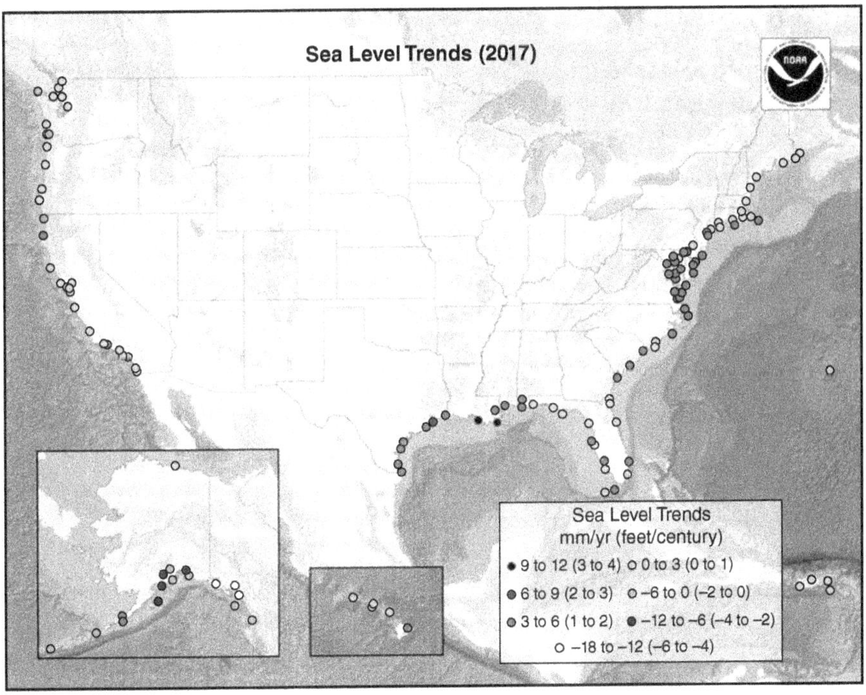

Figure 2.4 Sea level trends for different areas of the US coast based on tide stations (minimum of 30-year record). Most of the coastline is experiencing rising sea level. Louisiana is experiencing subsidence from development, sediment delivery losses, postglacial rebound, or other exacerbating factors. The southern coast line of Alaska and the Yukon/British Columbia, are seeing decreasing rising sea level because of tectonic uplift associated with the subduction zone located off-shore (National Oceanic and Atmospheric Administration/National Ocean Service 2018).

damage to infrastructure. In some areas, there is a combination of sea level rise and subsidence from development. Municipalities and military facilities are making plans to adapt against the inevitable, even if some US states are loath to actually admit the sea level rise has anything to do with climate change. In 2012, the North Carolina state legislature passed a bill forbidding state agencies from taking global warming into account when projecting sea level rise for coastal communities (Charlotte Observer 2012). Making the use of science illegal in planning for the future is probably not a wise choice, but recall that climate deniers persist even today.

Both glacial retreat and sea level rise are obvious consequences of a warming climate. Of the planet's more than 5,400 terrestrial glaciers, most are retreating, and the speed of that retreat is accelerating. According to the US Geological Survey (2019), all of the glaciers in Glacier National Park have been reduced in size since 1966, and the average reduction over the last 50 years has been 39 percent. Similar patterns of loss are being experienced from Chile to Pakistan, indicating that the cause of the melting is global and not a decline in regional snowfall.

All this ice loss is adding to the total "active" water moving through the hydrological cycle, which essentially means it is entering the oceans. Many human populations rely on the accumulated freshwater storage that glacial and mountain ice represent. Therefore, the loss of these glaciers will severely impact dry-season water supplies in large sections of South Asia adjacent to the Himalayas and in California adjacent to the Sierra Nevada, just to name two. Both the Greenland and Antarctic ice sheets have been losing mass since the 1990s, with the largest mass loss occurring in the years following 2002. Arctic (North Pole) sea ice has decreased every year since 1979 (IPCC 2015).

Ice that is floating in water (ice sheets and icebergs) will not add to sea level rise by melting, just as the ice in your drink does not cause it to overflow when it melts. The mass of the ice is displaced as soon as the ice hits the water, not after it melts. The main driver of sea level rise, however, is not melting ice from Greenland or Antarctica; it is thermal expansion of water. Warming water expands in a more or less linear direction once it is above 4°C. The rate of expansion is about 0.002 ml/cm^3, which doesn't sound like much until you consider the vast volumes of water in the world's oceans. The average depth of the ocean is 3,740 m, and a 3,000-m column of water expands 60 cm per degree warming. Of course, the entire ocean would take many centuries to rise in temperature (right now only the surface waters are warming), but the effects of rising sea levels are apparent already.

The US Commerce Department is responsible for providing estimates of coastal/waterway threats that might interfere with commercial activity. Consistent with rising sea level, they report a linear increase in high-tide flood events in one-third of areas with tide gauges, mostly along the Atlantic Coast and Gulf of Mexico (Sweet et al. 2018). As noted earlier, the global trend has increased from one mm per year in the early 1900s to three mm per year today, although the rate of rise is different for different areas (Figure 2.1). By 2100, high-tide flooding will occur 182 days per year for the US East Coast and Gulf of Mexico—in other words, "Today's flood will be tomorrow's high tide" (Sweet et al. 2018). Only 0.3 to 0.7 m separates frequent damaging flooding from major destructive flooding for half the days of the year.

Damage to infrastructure and flooding coastal cities are not the only impacts we are seeing as a consequence of sea level rise. The rising waters are forcing saltwater into coastal aquifers. The "intrusion" arises when the pressure of seawater mass forces groundwater to retreat away from the coast. Unless groundwater levels somehow rise in conjunction with the sea level, saltwater intrusion will accelerate as sea level rise accelerates (Nhung et al. 2019). Obviously, this will impact the freshwater supplies of coastal communities that rely on local well water. Effects may be mitigated by developed countries that either have or can deploy piped water from interior sources, but for the developing countries with limited water-delivery infrastructure mitigation may be more challenging.

Effects on Coral Reefs

As climate change increases the frequency and duration of extreme weather events, it is important to remember that these events can affect the oceans as well as terrestrial systems and that their responses will not be similar. Marine species, as with terrestrial species, will respond to climate change differently depending on their resilience to extremes. Some will adapt, but many will not.

The widescale ecosystems shifts being observed in terrestrial environments will also have corollaries in marine systems including reefs. Scientists have known for decades that corals are sensitive to warming waters and that prolonged exposure can kill large swaths of a coral reef. The symbiotic, photosynthetic zooxanthellae (algae), which live inside individual coral polyps and provide the coral with food, exit the polyps when under heat stress. This results in "bleaching," where the corals turn white when the colorful zooxanthellae disappear. Bleached corals will die if the zooxanthellae are not reestablished. It is possible, however, that if the conditions that precipitated the exodus are reversed quickly (as in a transitory marine heat wave), the algae will return.

Marine biologists have long warned that widespread coral reef collapse would commence in an environment with greater than 1°C warming above the 1950–1980 average. In 2016, a prolonged marine heat wave struck the northern section of the Great Barrier Reef off the coast of Australia. Twenty-nine percent of the 3,863 individual reefs that make up the Great Barrier Reef experienced a catastrophic die-off and loss of both physical structure and ecological function. Over a period of two to three weeks, millions of corals were exposed to water 3 to 4°C above the bleaching threshold (Hughes et al. 2018). The bleaching events covered 1,000 km south of Papua New Guinea, almost reaching Rockhampton, Australia, and resulted in an unprecedented ecological collapse of the whole region (Figure 2.5). The remaining living reefs shifted away from the fast-growing, wide-branching, and habitat-forming species and favored a "depauperate" collection of slow-growing, flattened species that are associated with lower diversity communities (Hughes et al. 2018).

The 2016 heat wave was a harbinger of what is to come with increasing frequency. Corals throughout the tropics will continue to degrade in the coming

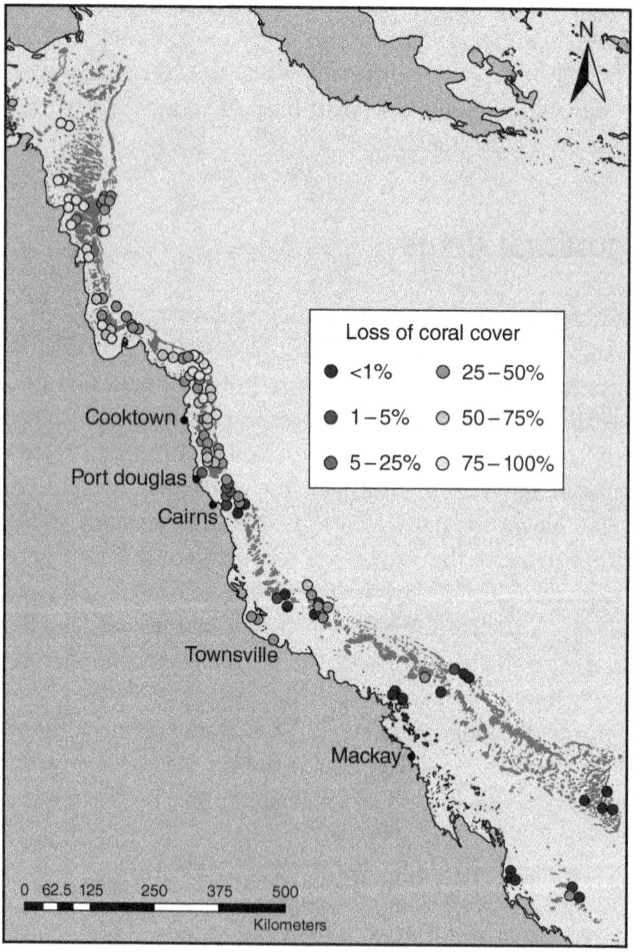

Figure 2.5 Percent loss of coral along the Great Barrier Reef in 2016. Losses were measured on 110 reefs between March and November that year. (Figure is reproduced from Hughes et al. 2018).

decades until some sort of climate equilibrium is established. Future tropical reefs likely will be composed of new, heat-tolerant species arrangements, with unknown impacts on diversity and fish communities that depend on reefs. As with most ecosystem responses to climate change, the unprecedented environmental shifts mean that, while the biosphere will certainly change, it is unclear how, or even if, the diversity of species will adapt. Our best hope is that these systems will adapt to a warmer, more extreme climate and achieve functional stability (i.e., maintenance of ecosystem service functions like promoting productivity at the base of the food web or removing pollutants from water, just to name two).

Effects on Biosphere Vegetation and Soils

It has long been known that plants will increase photosynthetic rate (productivity) and growth rate if they are kept in greenhouses with elevated CO_2. If given ample water as well as optimum light and temperature along with elevated CO_2 (up to 600 parts per million [ppm]), productivity can increase from 32 to 41 percent (Schlesinger & Bernhardt 2013). This increase in productivity is called the "Beta factor," and there is great interest in assessing increased primary production due to elevated CO_2 in the natural environment as opposed to just greenhouses. Imagine how convenient it would be if the plants just absorbed the extra carbon we are emitting to the atmosphere! Currently, the land is absorbing approximately 25 percent of anthropogenic CO_2 emissions (Canadell et al. 2007, IPCC 2014) because photosynthesis, which draws CO_2 out of the atmosphere, is outpacing respiration, which releases it.

So if plants will respond to extra CO_2 by becoming more productive and therefore fixing more CO_2 into sugars and woody tissue, then the biosphere might become a larger sink for CO_2 as the concentration of the gas increases. On the other hand, warming environments could increase soil respiration rates, causing them to release more CO_2. Warming environments could also lead to regional water scarcity and reduce photosynthetic rate despite the "extra" CO_2. Many studies have shown that water-stressed plants reduce photosynthesis even if plenty of CO_2 is available (Cox et al. 2006).

From a plant's perspective, water availability trumps all, and extra CO_2 or nutrients will not spur additional tissue growth if there is a hint of water scarcity. Field-based studies looking for a "CO_2 fertilization" effect in adult trees by measuring tree-ring width have not found the productivity increase observed in controlled experimental or greenhouse settings. In fact, models have suggested that, while the biosphere right now is a sink for CO_2, there might be a temperature threshold where a "sink to source" transition occurs (Cox et al. 2006). It is absolutely vital to know if this transition will occur and, if so, when and under what conditions. We can realistically decrease emissions from fossil fuels we are burning, but if the biosphere becomes a net CO_2 source, there may not be very much we can do about that.

In an effort to project whether plants and soils will continue to act as sinks for CO_2, models have been developed that combine (1) plant respiration response to temperature, (2) biosphere sensitivity to CO_2 increase *and* temperature, (3) projected temperature change, and (4) projected atmospheric CO_2 increase. These models are usually called "Global Dynamic Vegetation Models," and they are our best way to examine the biosphere's response to rising temperature and CO_2 without being able visit the future ourselves. So you can see how these models work, one has been included as an exercise at the end of this chapter.

1.5 vs 2.0°C Warming: The Meaning of 0.5°C Change

In this section, we examine the changes in risk of a 1.5°C (optimal) versus a 2.0°C (United Nations goal) global temperature increase, as this distinction was the basis of an important 2018 publication by the IPCC. The current consensus is that we have experienced 0.87°C of warming relative to a 1850–1900 baseline. Holding warming to 1.5°C would avoid most of the tipping points, where there is risk of either nonlinear positive warming feedback loops (e.g., melting ice caps decreasing the Earth's albedo, leading to more warming and greater melting etc.) or irreversible changes (e.g., ecosystem collapse). There is high confidence, however, that 1.5°C will result in the warming of large regions, which will also change in precipitation intensity or amount in some areas. As dramatic as the 1.5°C changes will be, we stand a greater chance of adapting to them or mitigating the effects than we do at 2.0°C. Although 0.5°C does not sound like a lot, when you take the volume of the atmosphere and all the gases in it and then plow 0.5°C into all if that, this is a vast amount of heat energy.

It is worth remembering that the an increase in global mean surface temperature of 1.5 or 2.0°C represents the global average. Some regions will experience relatively little temperature increase (tropical areas generally), while other areas will experience a substantially larger temperature change (e.g., the mid latitudes). Also, the differences in regional extreme temperature between the 1.5 and 2.0°C warming scenarios are substantial. In areas that will experience the most warming under the 1.5°C scenario, models show that their extremes will be two to three times worse under the 2.0°C warming scenarios (Hoegh-Guldberg et al. 2018). There are large and substantial differences in regional outcomes between 1.5 and 2.0°C warming, which is why a large group of scientists strongly believe we should aim to limit warming to 1.5°C. This section of the chapter examines the expected differences, and it shows that the risks to the stability of regional ecosystems and the human population at 2.0°C are large enough that limiting warming to 1.5°C should be our goal.

Assessing the differential impacts of 1.5 and 2.0°C is not only a matter of examining the regional extent of an altered climate and a difference in the percentage of ecosystem function affected, it is also an assessment of risk—that is, how many systems move from being at moderate risk to being at high risk? For ecosystems currently judged to be unique or threatened, the transition from 1.5 to 2.0°C greatly increases their risk of collapse, but entire regions (the Arctic), classes of organisms (corals), and risks to biodiversity move from moderate risk to high risk. For example, coral reefs will be reduced 70 to 90 percent by 2050 under a 1.5°C warming but risk total loss at 2°C. For the Arctic tundra, permafrost losses increase by about 10 percent under a 2.0°C scenario versus 1.5°C, with likely methane and additional CO_2 release. We also know that the prairie pothole ecosystems (hugely important for waterfowl) found in Montana, Minnesota, and the Dakotas as well as in Saskatchewan and Manitoba would survive under 1.5°C but not at 2.0°C (IPCC 2018). Risks from extreme weather also shift from moderate to high as one moves from 1.5 to 2.0°C, which will result in larger economic losses, infrastructure damage, and insurance costs (not to mention risk to human life).

As discussed earlier, we are already seeing increased risk from fire, which all scenarios show will only get worse. Fire risk between 1.5 and 2.0°C is exacerbated by increased risk in already drought-prone areas, such as the Mediterranean region and southern Africa. Most of the increased fire activity has been in environments that traditionally experience fires (e.g., chaparral [Mediterranean] ecosystems); however, the percentage of land surface that will be under increased fire risk is substantial. If warming goes to 1.2°C, 37.8 percent of the land surface will experience greater fire frequency. And if warming goes to 3.5°C, 67 percent of all terrestrial environments will experience more fires (IPCC 2018).

Recent work has shown that the risk of food shortages in the Mediterranean region, central Europe, Africa (western, southern, and Sahel regions), and the Amazon all move from moderate to high if warming surpasses 1.5°C and proceeds to 2.0°C (IPCC 2018). Global maize production will drop 8 to 14 percent due to reduced rainfall. It is possible that more drought-resilient and heat-tolerant strains of crops will be developed or that farmers will switch to crops that are more suitable for their altered climate (e.g., moving from citrus to peach production in Florida); however, that involves cultural change, investment in new plants, and abandoning the old (it takes a long time to grow a fruit tree).

Economic models show that warming will impact aggregate economic growth, and the difference between 1.5 and 2.0°C is substantial. While economic damage may not be dramatic as we increase to 1.5°C above baseline (again, there will be regions that do better and some that do worse regardless), moving to 2.0°C could cost 1.5 to 3.0 percent of gross world production (GWP) by 2050 and 3.5 percent by 2100 (Burk et al. 2018). These productivity decreases do not take in to account sea level rise and subsequent impacts on infrastructure, water supply, and economic activity in coastal areas. The projected GWP losses therefore are underestimates.

CHANGE AT 3°C

If the United Nations efforts fail to hold warming to less than 2°C and we reach 3°C (the current projection without further CO_2 reduction efforts), the planetary change will be dramatic and affect all of the earth's ecosystems, with the possible exception of the deep ocean. At 3°C warming, the Arctic will be ice-free in the summer, with massive habitat loss for polar bears, walruses, seals, and sea birds. Warming in the Arctic will be well above the 3°C global average, and a shift of more than 8°C relative to that average should be expected. This would result in large permafrost losses, collapsing infrastructure, and the release of large amounts of methane as well as additional CO_2. For soils with large amounts of organic carbon (e.g., those in the Arctic), there is a positive correlation between temperature and CO_2 release. At 3°C warming, Southeast Asia will experience increased risk of flooding from sea level rise and large increases in precipitation. Additionally, there will be reduced crop yields in the region and elsewhere. The Mediterranean region and southern Africa will experience less rainfall and increased risk of drought, also resulting in lower crop yields. Likewise, yields will decrease in West Africa, the Sahel, and all tropical regions. Global maize production will drop by 20 percent (IPCC 2018).

Conclusion

There is no doubt that the accelerated rate of climate change is impacting the biosphere. Although all systems change naturally, the current rate of change is greater than the rates of change we have observed in the past, whether directly or with various chemical or geological proxies. It is highly likely that rates of change and shifting climate will accelerate, but how most species will respond is not known. Some will adapt, some will not, but either way, ecosystems are shifting in unprecedented ways. We are conducting a planetwide experiment on ourselves and every surface ecosystem.

Having inventoried some of the most severe projected and already-occurring impacts of climate change, we will start addressing human efforts to mitigate climate change in the next chapter. As demonstrated, the difference between 1.5 and 2.0°C increases in temperature (by the end of the 21st century) may be determinant in our ability to adapt to the world's new climate. We will address an array of diplomatic, political, economic, and technological "partial fixes," but in this chapter, we conclude with two exercises to enable readers to understand some of the trade-offs inherent in climate modeling. While we address relatively self-contained systems of reforestation and carbon storage (addressed further in Chapter 6), the form the analysis takes is not unlike that used when modeling large and open systems, as we have suggested in this chapter.

Exercises

This section contains exercises where readers may examine policy for managing biosphere carbon sequestration or use models to examine the biosphere's response to CO_2 increase and associated warming. Our hope is that it will give the reader an opportunity to take what was discussed in the chapter and apply it. By evaluating mitigation through tree planting and by manipulating vegetation growth/soil carbon models, readers should see that it is possible to quantitatively assess policy goals and plan for the future. Tree planting as a carbon sequestration strategy will throw some light on the idea that growing more trees will solve the problem. By using the model, readers may gain an appreciation for the very different outcomes one gets depending on the initial conditions used. These exercises are meant to be stand-alone sections of the chapter that complement the material covered, but are not directly dependent on the chapter contents.

Exercise 2.1 Will Reforestation Efforts Make a Difference?

The biosphere is currently taking up approximately 25 percent of anthropogenic carbon emissions, and while this carbon sink is helpful, it is clearly not keeping pace with emissions as atmospheric CO_2 concentration continues to rise. Reducing emissions is going to be absolutely necessary if humans are to keep warming to less than 2°C, but enhancements of the biosphere sequestration would be helpful if feasible.

Right now, reforestation (or afforestation[2]) is increasing terrestrial wood carbon storage by about one gigaton of carbon (GtC) per year (emissions are approximately 10 GtC/year). Both China and India have embarked on massive tree-planting campaigns aimed at stopping desertification. By 2003, the expanding Gobi Desert in China was within 150 miles of Beijing, and seasonal sand storms were causing economic and health crises (Ratcliff 2003). The afforestation effort called for planting 9 million acres (3,642,200 hectares). As of 2019, however, 13 million hectares have been planted, and today's rate of new planting is 3 million hectares per year (Sun & Liu 2020).

Given that the tree-growing projects started in the late 1970s (so all the trees are less than 50 years old), find a tree growth curve on the web. Use it to find the biomass likely to be contained in the 13 million hectares. Be sure to write down any assumptions you have to make (there will be a few). Given the biomass you calculate, how much additional carbon is being captured by these planting projects? Find a tree carbon calculator on the web, and translate tree mass into carbon (since the trees are

Continued

not 100 percent carbon, of course). Given those amounts and the pressure on land use from development and farming/ranching, do you think tree planting projects might be part of a realistic solution to reducing CO_2 in the atmosphere? Support your assessment with data from .gov or .edu sources or from the peer-reviewed literature.

Exercise 2.2 Modeling Terrestrial Carbon Storage

This exercise gives readers a chance to utilize a model from the published literature using different parameters to project biosphere response to climate change. By completing this exercise, it will be clear how models measuring biosphere response to warming are extremely sensitive to the initial variables selected. Some of these variables are based on projections that depend on future human behavior, so there are a range of outcomes. This range does not mean the model is wrong. It just means that we do not know what the future will be, so it is prudent to consider a few possible futures. This allows us to ponder the range of outcomes that our current actions may create in the future.

In this exercise, we examine terrestrial soil/vegetation response to four possible future temperature changes and examine how carbon storage changes as atmospheric CO_2 moves from 190 to 1,100 ppm. To start, we have to recognize that respiration in all living organisms is sensitive to temperature. Respiration rate increases with increasing temperature, but the relationship is neither linear nor equivalent across groups of animals and plants. The variable that describes the relationship is called Q_{10} and is calculated using equation 2.1 below, where R_1 and R_2 are respiration rates at temperature T_1 and T_2 (which are 10°C apart).

$$Q_{10} = (R_2 / R_1)^{10/(T_2 - T_1)} \tag{2.1}$$

Cold-blooded reptiles, insects, mammals, plants, soil microbes, and so on all have different Q_{10} values. For the purposes of this exercise, we are going to assign a Q_{10} value of 2.5, which is between 0.5 and 1.0 times higher than commonly used for soils but may be more realistic for soils with higher moisture content (as expected in the future for some areas) (Meyer et al. 2018).

Another variable we need to establish is a measure of environmental sensitivity to the temperature change resulting from doubling the preindustrial atmospheric CO_2 concentrations, μ. Equation 2.2 shows that we need to assign a particular ΔT_{2xCO2} to find μ. ΔT_{2xCO2} is the temperature change expected from doubling CO_2 relative to preindustrial levels.

The IPCC estimates that if CO_2 is doubled, a temperature change of between 1.4 and 5.8°C will occur (IPCC 2014). In this exercise, you will pick four different ΔT values between these temperature extremes and solve equation 2.2, yielding four different μ values.

$$\mu = (\Delta T_{2xCO2} / 10) * (\log Q_{10} / \log 2) \tag{2.2}$$

Now that we have established some for ΔT values, we are ready for the terrestrial carbon storage model shown in equation 2.3:

$$C_{eq\,t} = II * (C_a / (C_a + C_{0.5})) * (C_{a(0)} / C_a)^{\mu} * (f(T) / r_0) \tag{2.3}$$

where $C_{eq\,t}$ is carbon storage in gigatons of carbon (a gigaton is equal to a billion tons), the output we are interested in. Note that note all C variables and II are going to be in GtC, which means that atmospheric CO_2 in ppm must be converted using 2.13 GtC for every 1.0 ppm. (As long as that conversion is used, you can just do the math without worrying about units, and the answer will be in GtC).

Also in equation 2.3, II is global primary productivity, which is 120 GtC; $C_{a(0)}$ is preindustrial carbon content (280 ppm = 596 GtC); $C_{0.5}$ is the mass of carbon in the atmosphere at 500 ppm; and C_a is carbon content in the atmosphere at any given time. Increase atmospheric carbon concentrations from 190 to 1,100 ppm in steps of 20 or 30 ppm at a time. You will compute the model for each of these carbon amounts as C_a.

The last term in equation 2.3 is a carbon turnover term (land atmosphere exchange), f(T) = 1. The r_0 term is related to residence time of carbon in soil, which will be set to 0.034.

If you use ΔT values of 2.0, 3.0, 3.5, and 4.5 and solve the model for concentrations of atmospheric CO_2 from 190 to 1,100ppm, you should generate curves that resemble those in Figure 2.6.

Try the following problems as well (you should turn in answers to the bullet point questions). The objective is to create a graph like Figure 2.6 and interpret it.

- Calculate μ ("*u*" in Figure 2.6) using different ΔT_{2xCO2} values between 1.4 and 5.8 (these are the temperatures changes in °C models project). What do you get? Use four different ΔT values to get four different values of μ.
- Plot $C_{eq\,t}$ storage values for your four μ values versus projected atmospheric CO_2 (ppm) and see what you get. Note that you must convert ppm to GtC for your calculations, but use ppm when you graph. Increase atmospheric carbon concentrations in steps of 20 or 30 ppm from 190 to 1,100 ppm.

Continued

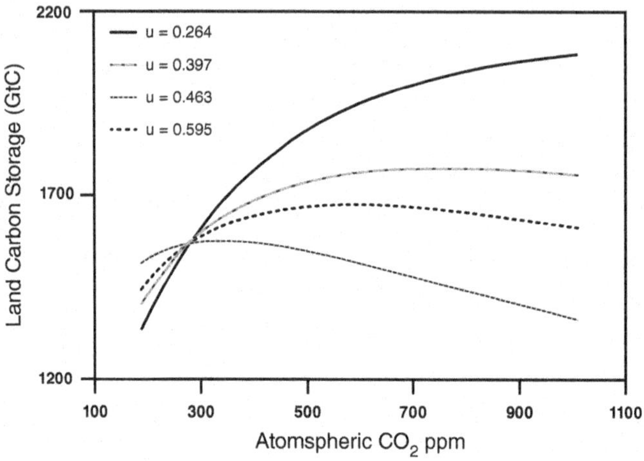

Figure 2.6 Land carbon storage as a function of atmospheric CO_2 using Q_{10} of 2.5, and ΔT values of 2.0, 3.0, 3.5, and 4.5°C.

- Interpret your results. What do high μ and low μ values project for carbon storage as atmospheric CO_2 increases? What do Q_{10} and μ measure in the first place? Are highs or lows in carbon storage realistic given the ΔT expected? (Assume a ΔT of 2.5 is "expected." Although such change will have drastic effects on ecosystems, it is quite likely we will experience it unless the world's governments take the problem seriously). If your results show an inflection point (when the slope changes from positive to negative, or vice versa), when is it projected to occur for your four different ΔT values?
- What implications does this have for our carbon emissions future? Can you think of an effective approach to inform a political leader or the public about the importance of avoiding the inflection point in land carbon storage?

NOTES

1. Nor'easters are large storms experienced in the New England area of the United States. Their winds come from the northeast, hence the name.
2. Afforestation is establishing a forest where one did not previously exist. Reforestation is reintroducing forests where they previously existed.

REFERENCES

Aono, Yasuyki, and Keiko Kazui. 2008. "Phenological data series of cherry tree flowering in Kyoto, Japan, and its application to reconstruction of springtime temperatures since the 9th century." *International Journal of Climatology* 28:905–914.

Atkinson, Angus, Simeon Hilll, Evgeny Pakhomov, Volker Siegel, Christian Reiss, Valerie Loeb, Deborah Steinberg, Katrin Schmidt, Geraint Tarling, Laura Gerrish, and Sévrine Sailley. 2019. "Krill (*Euphausia superba*) distribution contracts southward during rapid regional warming." *Nature Climate Change*, 9(2):142–147.

Barton, Mary Ann. 2018. Second '1,000-year" flood hits historic Md. city in Howard County" NACo Published May 29, 2018. https://www.naco.org/articles/second-%E2%80%981000-year%E2%80%99-flood-hits-historic-md-city-howard-county

Bhatia, Kieran T, Gabriel A. Vecchi, Thomas R. Knutson, Hiroyuki Murakami, James Kossin, Keith W. Dixon, Carolyn E. Whitlock. 2019. Recent increases in tropical cyclone intensification rates. *Nature Communications* https://doi.org/10.1038/s41467-019-08471-z

Bruenig, Elizabeth. 2015. "Why do evangelicals like James Inhofe believe that only God can cause climate change?" *New Republic* January 29. https://newrepublic.com/article/120889/evangelical-james-inhofe-says-only-god-can-cause-climate-change

Burke, M, WM Davis, NS Diffenbaugh. 2018. Large potential reduction in economic damages under UN mitigation targets. *Nature* 557:549-553.

Burrows, Michael T, David S. Schoeman, Lauren B Buckley, Pippa Moore, Elvira S Poloczanska, Keith M Brander, Chris Brown, John F Bruno, Carlos M Duarte, Benjamin S Halpern, Johnna Holding, Carrie V Kappel, Wolfgang Kiessling, Mary I O'Connor, John M Pandolfi, Camille Parmesan, Franklin B Schwing, William J Sydeman, Anthony J Richardson. 2011. The pace of shifting climate in marine and terrestrial ecosystems. *Science* 334:652-655

Cadahia, Luis, Antonieta Labra, Endre Knudsen, Anna Nilsson, Helena M. Lampe, Tore Slagsvold, and Nils Chr Stenseth. 2017. "Advancement of spring arrival in a long-term study of a passerine bird: Sex, age and environmental effects." *Oecologia* 184:917–929.

Caesar, L., S. Rahmstorf, A. Robinson, G. Feulner, and V. Saba. 2018. "Observed fingerprint of a weakening Atlantic Ocean overturning circulation." *Nature* 556:191–196.

Canadell JG, C Le Quere, MR Raupach, CB Field, ET Buitenhuis, P Ciais, TJ Conway, NP Gillett, RA Houghton, G Marland. 2007. Contributions to accelerating atmospheric CO2 growth from economic activity, carbon intensity, and efficiency of natural sinks. *Proceedings of the National Academy of Science* 104:18866-18870.

Charlotte Observer Replacement. Harish Alon. 2012 (August 2). New law in North Carolina Bans Latest Scientific Prediction of Sea-Level Rise. ABCnews. https://abcnews.go.com/US/north-carolina-bans-latest-science-rising-sea-level/story?id=16913782

Charmantier, Anne, Robin H. McCleery, Lionel R. Cole, Chris Perrins, Loeske E. B. Kruuk, and Ben C. Sheldon. 2008. "Adaptive phenotypic plasticity in response to climate change in a wild bird population." *Science* 320(5877):800–803.

Cox, Peter M., Chris Huntingford, and Chris D. Jones. 2006. "Conditions for source-to-sink transitions and runaway feedbacks from the land carbon cycle." In *Avoiding Dangerous Climate Change*, edited by Hans J. Schellnhuber, Wolfgang Cramer, Nebojsa Nakicenovic, Tom Wigley, and Gary Yohe. Cambridge University Press. Cambridge, UK, pp. 155–162.

Federal Emergency Management Agency. 2018. "Sandy, 5-years later." Accessed February 2, 2019, at https://www.fema.gov/disasters/historic

Hansen, Pushker Kharecha, Makiko Sato, Valerie Masson-Delmotte, Frank Ackerman, David J. Beerling, Paul J. Hearty, Ove Hoegh-Guldberg, Shi-Ling Hsu, Camille Parmesan, Johan Rockstrom, Eelco J. Rohling, Jeffrey Sachs, Pete Smith, Konrad Steffen, Lise Susteren, Karina, James Van von Schuckmann, and James C. Zachos. 2013. "Assessing 'dangerous climate change': Required reduction of carbon emissions to protect young people, future generations and nature." *PLOS ONE* 8(12):e81648

Hoegh-Guldberg, O., D. Jacob, M. Taylor, M. Bindi, S. Brown, I. Camilloni, A. Diedhiou, R. Djalante, K. Ebi, F. Engelbrecht, J. Guiot, Y. Hijioka, S. Mehrotra, A. Payne, S. I. Seneviratne, A. Thomas, R. Warren, and G. Zhou. 2018. "Impacts of 1.5°C global warming on natural and human systems." In Intergovernmental Panel on Climate Change, Geneva, Switzerland. Global Warming of 1.5°C. An IPCC Special Report on the impacts of global warming of 1.5°C above pre-industrial levels and related global greenhouse gas emission pathways, in the context of strengthening the global response to the threat of climate change, sustainable development, and efforts to eradicate poverty. Masson-Delmotte, V., P. Zhai, H.-O. Pörtner, D. Roberts, J. Skea, P.R. Shukla, A. Pirani, W. Moufouma-Okia, C. Péan, R. Pidcock, S. Connors, J.B.R. Matthews, Y. Chen, X. Zhou, M.I. Gomis, E. Lonnoy, T. Maycock, M. Tignor, and T. Waterfield (eds.)]. Geneva Switzerland.

Hughes, Terry P., James T. Kerry, Andrew H. Baird, Sean R. Connelly, Andreas Dietzel, Mark C. Eakin, Scott F. Heron, Andrew S. Hoey, Mia O. Hoogenboom, Gang Liu, Michael J. McWilliam, Rachel J. Pears, Morgan S. Pratchett, William J. Skirving, Jessica S. Stella, and Gergely Torda. 2018. "Global warming transforms coral reef assemblages." *Nature* 556:492–496.

Intergovernmental Panel on Climate Change (IPCC). 2014. *Climate Change 2014: Synthesis Report. Contribution of Working Groups I, II and III to the Fifth Assessment Report of the Intergovernmental Panel on Climate Change*. IPCC, Geneva, Switzerland.

Intergovernmental Panel on Climate Change (IPCC), 2018: Global Warming of 1.5°C.An IPCC Special Report on the impacts of global warming of 1.5°C above pre-industrial levels and related global greenhouse gas emission pathways, in the context of strengthening the global response to the threat of climate change, sustainable development, and efforts to eradicate poverty [Masson-Delmotte, V., P. Zhai, H.-O. Pörtner, D. Roberts, J. Skea, P.R. Shukla, A. Pirani, W. Moufouma-Okia, C. Péan, R. Pidcock, S. Connors, J.B.R. Matthews, Y. Chen, X. Zhou, M.I. Gomis, E. Lonnoy, T. Maycock, M. Tignor, and T. Waterfield (eds.)].

Jin, Yufang, Michael L. Goulden, Nicholas Faivre, Sande Veraverbeke, Fengpeng Sun, Alex Hall, Michael S. Hand, Simon Hook, and James R. Randerson. 2015. "Identification of two distinct fire regimes in Southern California: Implications for economic impact and future change." *Environmental Research Letters* 10(9):094005.

Marvel, Kate, and Celine Bonfils. 2013. "Identifying external influences on global precipitation." *Proceedings of the National Academy of Science* 110(48):19301–19306.

Meyer N, Welp G, Amelung W. 2018. "The temperature sensitivity (Q10) of soil respiration: controlling factors and spatial prediction at regional scale based on environmental soil classes." *Global Biogeochemical Cycles* 32:306–323.

National Oceanic and Atmospheric Association/National Ocean Service. 2018. "US sea level trend map." Center for Operational Oceanographic Products and Services. Accessed March 13, 2019, at https://tidesandcurrents.noaa.gov/sltrends/slrmap.html

Nhung, Tran Thi, Phu Le Vo, Vu Van Nghi, and Ho Quoc Bang. 2019. "Salt intrusion adaptation measures for sustainable agricultural development under climate change effects: A case of Ca Mau Peninsula, Vietnam." *Climate Risk Management* 23:88–100.

Pagano, A. M., G. M. Durner, K. D. Rode, T. C. Atwood, S. N. Atkinson, E. Peacock, D. P. Costa, M. A. Owen, and T. M. Williams. 2019. "High-energy, high-fat lifestyle challenges an Arctic apex predator, the polar bear." *Science* 359:568–572.

Parmesan, Camille, and Gary Yohe. 2003. "A globally coherent fingerprint of climate change impacts across natural systems." *Nature* 421:37–42.

Peterson, T. C., R. R. Heim, R. Hirsch, D. P. Kaiser, H. Brooks, N. S. Diffenbaugh, R. M. Dole, J. P. Giovannettone, K. Guirguis, T. R. Karl, R. W. Katz, K. Kunkel, D. Lettenmaier, G. J. McCabe, C. J. Paciorek, K. R. Ryberg, S. Schubert, V. B. S. Silva, B. C. Stewart, A. V. Vecchia, G. Villarini, R. S. Vose, J. Walsh, M. Wehner, D. Wolock, K. Wolter, C. A. Woodhouse, and D. Wuebbles. 2013. "Monitoring and understanding changes in heat waves, cold waves, floods and droughts in the United States: State of knowledge." *Bulletin of the American Meteorological Society* 94:821–834. DOI:10.1175/BAMS-D-12-00066.1.

Ratcliff, Evan. 2003. "The Great Green Wall of China." *Wired,* March 1, 2003. Accessed June 16, 2020, at https://www.wired.com/2033/04/greenwall

Scheffers, Brett R., Luc Scheffers, De Meester, Tom C. Bridge, Ary A. Hoffmann. John M. Pandolfi, Richard T. Corlett, Stuart H. M. Butchart, Paul Pearce-Kelly, Kitt M. Kovacs, David Dudgeon, Michela Pacifici, Carlo Rondinini, Wendy B. Fodne, Tara G. Martin, Mora Camilo, David Brickford, and Jame. E. M. Watson. 2016. "The broad footprint of climate change from genes to biomes to people." *Science* 354. DOI: 10.1126/science.aaf7671

Schlesinger, Michael E., Jianjun Yin, Gary Yohe, Natalia G. Andronova, and Li Bin. 2006. "Assessing the risk of a collapse of the Atlantic Thermohaline Circulation." In *Avoiding Dangerous Climate*

Change, edited by Hans J. Schellnhuber, Wolfgang Cramer, Nebojsa Nakicenovic, Tom Wigley, and Gary Yohe. Cambridge University Press, Cambridge, UK, pp. 37–47.

Schlesinger, William H, and Bernhardt Emily S. 2013. *Biogeochemistry*. 3rd ed. Elsevier, Oxford, UK.

Sun, Wanlong, and Xuehua Liu . 2020. "Review on carbon storage estimation of forest ecosystem and applications in China." *Forest Ecosystems* 7:4. Accessed November 22, 2020, at https://doi.org/10.1186/s40663-019-0210-2

Sweet, William V., Greg Duesk, Jayantha Obeysekera, John J. Marra. 2018. "Patterns and projections of high tide flooding along the US coastline using a common impact threshold." NOAA Technical Report NOS CO-OPS 086. https://tidesandcurrents.noaa.gov/publications/techrpt86_PaP_of_HTFlooding.pdf

Taylor, Rebecca L, Mark S. Udevitz, Jay V. Chadwick, John J. Citta, Lori T. Quakenbush, Patrick R. Lemons, and Jonathan A. Snyder. 2018. "Demography of the Pacific walrus (*Odobenus rosmarus divergens*) in a changing Arctic." *Marine Mammal Science* 43(1):54–86.

US Geological Survey. 2019. "Area of the named glaciers of Glacier National Park (GNP) and Flathead National Forest (FNF) including Little Ice Age extent, and years 1966, 1998, 2005, and 2015." Accessed March 5, 20149, at https://www.usgs.gov/data-tools/area-named-glaciers-glacier-national-park-gnp-and-flathead-national-forest-fnf-derived

van der Wiel, Karin, Sarah R. Kapnick, Geert van Oldenborgh, Kirien Whan, Philip Sjoukje, Gabriel A. Vecchi, Roop K. Singh, Julie Arrigh, and Heidi Cullen. 2017. "Rapid attribution of the August 2016 flood-inducing extreme precipitation in south Louisiana to climate change." *Hydrological and Earth System Science* 21:897–921.

Walsh, J., D. Wuebbles, K. Hayhoe, J. Kossin, K. Kunkel, G. Stephens, P. Thorne, R. Vose, M. Wehner, J. Willis, D. Anderson, S. Doney, R. Feely, P. Hennon, V. Kharin, T. Knutson, F. Landerer, T. Lenton, J. Kennedy, and R. Somerville. 2014. "Our Changing Climate." In *Climate Change Impacts in the United States: The Third National Climate Assessment*, edited by J. M. Melillo, Terese (T.C.) Richmond, and G. W. Yohe. US Global Change Research Program, pp. 19–67. Washington DC, USA doi:10.7930/J0KW5CXT.

Waxman, Olivia B. 2017. "Springtime's cherry blossoms have a dark side." *Time,* March 21. https://time.com/4698670/cherry-blossoms-climate-washington-japan/

Webster PJ, GJ Holland, JA Curry, HR Chang. Changes in tropical cyclone number, duration, and intensity in a warming environment. 2005. *Science* 309:1844–1846.

Wood, Richard, Matthew Collins, Jonathan Gregory, Glen Harris, and Michael Vellinga. 2006. "Towards a risk assessment for shutdown of the Atlantic Thermohaline Circulation." In *Avoiding Dangerous Climate Change,* edited by Hans J. Schellnhuber, Wolfgang Cramer, Nebojsa Nakicenovic, Tom Wigley, and Gary Yohe. Cambridge University Press, Cambridge, UK, pp. 49–54.

Wu, Zhuoting, Paul Dijkstra, George W. Koch, Josep Penuelas, and Bruce A. Hungate. 2011. "Responses of terrestrial ecosystems to temperature and precipitation change: A meta-analysis of experimental manipulation." *Global Change Biology* 17:927–942.

Zervas, C. 2009. "Sea level variations of the United States 1854–2006." NOAA Technical Report NOS CO-OPS 053. Accessed June 23, 2020, https://tidesandcurrents.noaa.gov/publications/Tech_rpt_53.pdf

3

International Relations and Global Climate Politics

By the end of this century, sea level rise of some five to six feet will likely submerge large portions of island nations, endangering the livelihoods of tens of thousands of people and, in the end, making them refugees when they have nowhere to live. "Who wants to believe that their home won't be here?" asked Anote Tong, then-president of the Pacific atoll nation of Kiribati, responding to news of his government's purchase in 2014 of some eight square miles on the Fijian island of Vanua Levu for over 7 million dollars (cited in Keating 2018). "People here don't even like to plan for next week. But we've got to be hardheaded about it."

Counting the atoll nation's "blue carbon" or oceanic area, Kiribati's 33 small islands cover more than 1.3 million square miles—huge in terms of total area, but tiny in terms of land mass. How will the nation's 120,000 people take to moving from a population density one-tenth that of Alaska's (considering Kiribati's total area, by land and sea) to one greater than that of Chicago? And the people of Kiribati may be lucky in that they have a place to go. This scenario will play out many times in the decades ahead, and most sinking island states have not yet demonstrated the foresight, or resources, of Tong and his government.

Is it fair that the people of Kiribati must relocate when in fact they did not cause the problem of climate change, the blame for which falls disproportionately on some 20 companies in the United States, Europe, and—very recently—China and India? Who should pay to relocate the people of Kiribati and force entire nations of people to migrate? For the first time ever, the terms "loss and damage" were included in the 2015 United Nations Framework Convention on Climate Change (UNFCCC) agreement (UNFCCC 2015). These terms represent climate change–induced harm, whether reparable or irreparable, to which countries cannot adapt. Since 1991, climate-vulnerable nations—usually in the developing world—have sought compensation and liability from developed countries, which are the largest greenhouse-gas (GHG) emitters. These political developments have occurred in the absence of much systematic research on how affected populations will react to these changes.

More specifically, what are the related changes in public opinion and political behavior? For instance, will people migrate in response to climate change? If so, which people are more likely to migrate, and where will they go and why? If not, how will they adapt? Who do they blame for their sufferings and losses? How does the fact of climate change shape the nature and scope of their political attempts to seek redress?

Even without answers, these questions will only grow in significance. According to the Internal Displacement Monitoring Center, the total number of disaster-induced displacements in the world in 2015 was already 15.23 million. This number could triple by the late 21st century. For example, migration from the Middle East and Africa has increased recently and likely will greatly accelerate in coming years, and security analysts worry that causes of such migration, including climate change, could increase the severity of other threats to governance, security, and well-being. North Atlantic Treaty Organization (NATO) Analyst Brent Oglesby (2019), said that "climate change hasn't thus far been seen as a direct cause of instability, but rather as a major threat multiplier." He added that increased migration from Africa and the Middle East to Europe helped prompt the creation of a NATO policy framework that, starting in 2014, would "focus not just on Russia to the east but also on security threats to the Alliance from the south."

Within the purview of "loss and damage," climate migration has been a controversial topic given the contentious issues of adjudicating responsibility for financing and receiving displaced populations. The United Nations (UN) has issued several recommendations, including one to ensure adequate resettlement opportunities for those who are temporarily and permanently displaced, and another to ensure that the least-developed countries are provided with technical assistance and financial resources to cope with climate-related displacement (Kälin 2010, Rolnik 2009). Still, states have full sovereignty to decide whether they choose to address the issue, or how to avoid it. And while pledges of international funding support have been made, they just have not—for the most part—been delivered to those in need, or in anywhere near the needed amounts.

A range of proposals has been presented, including one that says climate migrants should be provided permanent residency abroad, with recipient countries accommodating these migrants proportionate to the receiving nations' historical national emissions (Holthaus 2016). Another, in collaboration with the UN Development Program, proposed to negotiate the creation of a "new protected status applying specifically to environmental migrants" to protect rights of those forcibly displaced by slow-onset process impacts (Wentz & Burger 2015). Additionally, it proposed establishing a climate change displacement coordination facility as a step toward developing a legal framework to negotiate future resettlements.[1]

These legal debates about compensation and resettlement are transpiring with insufficient information from those who are the most affected and the potential program recipients: migrants themselves. The urgent need for systematically collecting valid and reliable data to measure climate migration, as this project proposes, was underscored at the UN Climate Change Conference in Bonn, Germany, in May 2016[2] (UN Newsroom). As the UN commences extensive debates to establish a system for cost-sharing loss and damage across member nations, no systematic source of information to understand comprehensive cross-regional attitudes across the range of levels of vulnerability has been reported anywhere.[3] Vulnerable nations have advocated for considering climate displacements as "loss and damage," acknowledging that there are climate impacts to which these nations cannot adapt (Karasapan 2015). The brunt of this migration will occur in the Global South. Indeed, where these migrants will go, and which countries will let them in, is a visualization of the broader question: Who is responsible for this problem, and who should "pay"?

The Broader Problem of Climate in International Relations: An Anarchic World or an Ordered One?

Two broad, major theoretical "schools" of thought exist in international relations. Realism is associated with the "*real politique*" of political calculations based on shows of strength by analysts like Henry Kissinger and—slightly earlier—Niccolò Machiavelli. This theory is attributed to authors like Hans Morgenthau (1978), Ken Waltz (1979), and the US diplomat George Kennan (1947), who conceived of "containment" as the way to limit the expansion of the Soviet Union during the Cold War (from about 1947 to 1989) by allying with bordering states wherever possible, "pinning in" the Soviets, and then outspending them in a nuclear arms race that forced the Soviet Union to essentially go bankrupt and then collapse into a dozen nations. Realism assumes that the world is anarchic, that nations must attend to their own needs (self-help), that human nature is essentially evil (or at best neutral), and that ethics are subordinate to politics. In realism, one nation gains at the expense of others (relative gains), states are unitary actors that overcome internal dissent, and military power is the ultimate (and to some the only) measure of state capability.

While this depiction is a simplified one, and entire books have been written to differentiate variants of realism, what is relevant to our story is that pure realists have little use for creating norms and standards of conduct through international organizations like the UN or international law, which they associate with liberalism, the other largest school of thought. Realists might be profoundly concerned with climate change inasmuch as these changes shift strategic advantages from some nations to others, such as by thawing the Arctic passage, thus giving Russia, Canada, and the United States new mobility vis-à-vis other nations. They also might be concerned that Chinese investment in green technologies, such as greatly improved solar photovoltaics that can be set up without need for a utility grid, might decentralize power generation. This could enhance China's geostrategic position in Africa and "power up" threatening nonstate actors, like some of the insurgencies that have been active in Nigeria, the Democratic Republic of Congo, Somalia, Kenya, and so on. But realists will not really concern themselves with efforts to fortify the UN, beyond possibly in the security-issue area, to solve climate change, such as by incorporating climate change into the dominion of the UN Security Council, where the five most powerful "permanent member" nations[4] and 10 rotating ones can take immediate actions, such as sending peacekeeping missions or issuing economic sanctions. Realists believe that power must reside in states and their security apparatus; international organizations have no real authority. Their interest in international institutions is in using these to project their military and economic power, which are the units that really matter.

Liberalism, pioneered by Keohane and Nye (1977) and in environmental politics by Peter Haas (1990), Oran Young (2010), and Elinor Ostrom (2015), among others, argues—again, in an oversimplified, bottom-line view to suit our purposes here—that human nature is perfectible, that harmony of interests among states or individuals is possible to achieve, and that beyond states, nonstate actors (like international institutions) can also matter. Furthermore, according to most liberal theorists, groups within nations participate directly in national foreign policy and can be more determinant even than executives or their militaries, and affairs of state transcend security interests to include issues also relating to the economy and environment.

The UNFCCC is a quintessential liberal institution in that it tries, mostly through carrots (as it has only small sticks of economic sanctions and sending "blue helmet" security forces that can defend civilians but not confront militaries) to mitigate GHG emissions. As indicated in Chapter 1, the adjudication of the climate change dilemma of equity and climate justice across nations has been left to the UN. This chapter shows that while the UN's science advisors (who are not financially compensated by the UN by the way) have done a stellar job pushing the boundaries of their research in a wave of reports by the Intergovernmental Panel on Climate Change (IPCC), the full-time UN diplomats have not achieved such a strong record.

The "Tragedy of the Commons" presents liberal types of responses to what has traditionally been depicted as a problem resolvable through liberal means. The sheep overgraze in the village public space, and the solution is to instate rules, require permits, rotate grazing, and so on. There is a central authority, the village mayor and government, who can resolve the problem and enforce the rules. The problem with law in international relations is that there really is not a reliable central enforcer. Furthermore, in the case of climate change, the villagers are not all equal, a tacit assumption in the scenarios of Ostrom, Hardin, and others. In climate change, the Global South "villagers" are much more numerous and relegated to a small part of the commons (say, 80 percent of the population given access to only 10 percent of the resources), whereas the Global North set up the system (i.e., nominated the mayor and village police), has many fewer sheep, and yet controls the lion's share of the commons as well as the most fertile part and the part with access to water.

Climate Change and Security: From Remote Perceived Link to Admitted Threat Magnifier

Energy security has been a high-level security issue ever since US President Jimmy Carter in 1979 called the energy crisis "the moral equivalent of war," in "The Crisis of Confidence" speech where he further argued that "the energy crisis is real. It is worldwide. It is a clear and present danger to our nation (Carter 1979)." The problem for climate change is that while fossil fuel shortages and peak oil may be readily viewed, as we discuss in Chapter 5, as a "clear and present danger," using the white-knuckles phrase popularized later in a Tom Clancy spy novel, the danger from GHG emissions is opaque and hard to measure rather than "clear," and the danger is mostly in the future and thus not manifestly "present." How can the language referencing climate change be "weaponized"—or at least "securitized"—to make us appreciate the scope of the problem, its attendant risks, and the urgency of addressing it?

The IPCC in the 1990s did make inroads to sounding alarms, and former Vice President Al Gore's 2006 book *An Inconvenient Truth: The Planetary Emergency of Global Warming and What We Can Do About It* also seems to have added a note of urgency. But perhaps it was only when climate change consequences started to draw the attention of the US military, downsized from its Cold War peak capacity but still the largest in the world, that realists began to see the clear and present danger. Indeed, in 2010, the US military Quadrennial Defense Review, the most important military planning document pinpointing security threats, started factoring climate change into its considerations, and this continued after 2016, even as the climate-denying President Donald Trump took office and continued to insist, as he had during the campaign, that climate change was a hoax. In 2016, the Stockholm International Peace Research Institute published an entire comprehensive report on the security risks posed by climate change (Mobjork et al. 2016). More specifically, the authors considered six risks posed by climate change: extreme weather events, food security, water security, sea level rise, climate migration, and climate-related violent conflict. While climate change has been viewed as a "threat magnifier" during much of the 21st century, it became viewed as a security threat unto itself starting around this time.

The 2010 US Department of Defense (DoD) Quadrennial Defense Review contextualized climate change as a vital security issue with its reportage that

> more than 30 US military installations were already facing elevated levels of risk from rising sea levels [in 2008]. DoD's operational readiness hinges on continued access to land, air, and sea training and test space. (Foley 2012)

Many US naval bases are at risk, but chief among these are the Indian Ocean and Middle East naval hub of Diego Garcia (tiny sinking island administered by Great Britain) and "floating bases" anchored in Bahrain, Guam, Florida (Eglin Air Force Base), and Virginia (Norfolk Naval Air Station).

Increasingly, climate change is viewed as a driver, or at least a complicating factor or threat multiplier, in security analysis. The climate-induced drought in Syria and elsewhere in the Middle East is being viewed as a cause of food riots there, which led, in part, to the Arab Spring political protests. Protestors in Tunisia, the most successful democratizer as a result of these protests, waved baguettes to protest the high price of basic staples in the wake of torrential rains in faraway Canada and drought in Argentina in 2010, both of which diminished wheat crops, causing world wheat prices to more than double in 2011 (Johnstone & Mazo. 2013, 16). Syria suffered more direct adverse weather conditions. From 2006 to 2011, some 60 percent of the nation's land experienced the worst drought ever, prompting a mass exodus of people whose hunger and discontent intertwined with the volatile geopolitics of the region. And lest we refer to these weather events as "one-off" occurrences, analysts see them as trends (as discussed in Chapter 2). These dire occurrences are climate changes, rather than just weather changes.

In addition to serving as background causes requiring consideration in the security realm, climate change is also a primary cause unto itself driving other emerging geopolitics. Perhaps the best example may be the race to control the new Arctic Passage. As sea-ice melts, leaving a sea lane shortcut from one side of the world to the other, Canada, Norway, Denmark, and the United States have challenged Russia's 2007 expedition to plant that nation's flag in the North Pole seabed. That transportation route may over time become as important to maritime transit as the Panama Canal and the Strait of Hormuz, and it may literally be only the tip of the iceberg of new security threats created by climate change, which we are only starting to consider. It is widely speculated that within a few decades, water will be as valuable as oil is now. Climate change will clearly alter natural endowments of this resource through flooding and drought, and the race is on to see who can most accurately predict the direction of changes and their impacts. In the economic realm too, a zero-sum race to develop technology for a smooth transition to "net zero" carbon usage is on. In preliminary salvos, a competition has emerged between Chinese companies (with government support) and US companies (with less direct government involvement), such as over the construction of more efficient solar photovoltaics and battery storage of energy (both of which are discussed in Chapter 5). While nations, companies, and individuals sometimes view these as zero-sum "races," many argue that the more appropriate way of viewing most climate-related issues is as a positive-sum cooperation. Inasmuch as there has been success devising international institutions to mitigate climate change, this success arises from liberal approaches to international relations and efforts in that tradition to overcome the "Tragedy of the Commons." We discuss the evolution of the international institutions to limit climate change in the sections that follow.

The History of Liberal Institutionalism and Climate Change: The IPCC, Science, and Policy

A remarkable understanding of climate change already existed over 50 years ago, when US President Lyndon Johnson's science advisory committee in 1965 sent him a report called *Restoring the Quality of Our Environment*, which stated the following in its introduction: "Pollutants have altered on a global scale the carbon dioxide content of the air and the lead concentrations in ocean waters and human populations" (cited in Nuccitelli 2015). This prescient report refuted a range of myths that climate debunkers still perpetrate, such as the fact that human carbon emissions pose climate dangers (i.e. climate change is caused by humans). The authors wrote:

> Only about one two-thousandth of the atmosphere and one ten-thousandth of the ocean are carbon dioxide. Yet to living creatures, these small fractions are of vital importance . . . Within a few short centuries, we are returning to the air a significant part of the carbon that was slowly extracted by plants and buried in the sediments during half a billion years/ (cited in Nuccitelli 2015)

The US government seems to have been the first on record to recognize the dangers of climate change and, it seems, the first to withdraw from the Paris Agreement, the UN's current agreement seeking to regulate and diminish worldwide emissions. The agreement was adopted in December 2015, and former US President Donald Trump announced in June 2017 that his government intended to withdraw from the accord, achieving that withdrawal in 2020, before President Joe Biden rejoined the Paris Agreement in January 2021. How do we explain this complete policy reversal?

While the domestic politics of the US climate policy turnabout will be addressed a bit more in Chapter 4, we address here the early decades of climate change research (conducted mostly by US-government funded university scholars and federal agency scientists), the emergence of a "liberal" regime of international institutions to address the problem (starting with the IPCC, launched in 1988), the negotiation of the first international agreement to reduce or "mitigate" carbon dioxide (CO_2) emissions (the Kyoto Protocol), and the rise of the current UN instrument, the Paris Agreement, to address the problem. These are clearly liberal institutions seeking to overcome the "Tragedy of the Commons" by giving nations, peoples, and economic entities motivation to cooperate and assume that political actors who know the stakes of climate change will rise to the challenge.

Many do argue, however, that international "liberal" institutions often lack pure motivations, and that the kind of instrumental behavior assumed by realists helps explain why liberals, relying on the persuasion of scientific explanations of phenomena with adverse effects still mostly in the distant future, have been unable to make headway. Has the liberal regime of institutions stopped climate change, or at least slowed it down? Has the UN and its institutions made any progress at all? This is an open question, and certainly we do not yet have enough evidence to pass judgment. But what do the preliminary data show? In the sections that follow, we discuss some of the successes and failures to date of the UN institutions to mitigate climate change.

Getting the Science Right: Importance of the Intergovernmental Panel on Climate Change

International institutions like those under the UN umbrella tend to run on liberal assumptions, as the basis for their existence is to arrive at cooperation. The problem is that realists tend to undervalue such institutions, or even dismiss them. And those who point out that the UN has never "won a war" are partly right. The Korean War, perhaps the most important war initiated by the UN, was a stalemate (and some argue that it was basically a US-waged war given political cover by the UN). And weapons inspectors from the UN needed US backing to gain access to Iraqi dictator Saddam Hussein's military installations, where the dictator was said to be developing nuclear weapons. In the security realm, the UN has not been very successful.

However, the UN has had some humanitarian successes. The World Health Organization (under the UN) has helped eradicate smallpox and other preventable diseases,[5] UNICEF has saved millions of children from starvation by delivering food during famines, and the UN has helped save world historic sites—such as the Egyptian pyramids—from degradation and destruction.

While the World Health Organization has made some important strides in medical research, the UN is not known primarily for its scientific prowess.

That said, one can readily understand the importance of "getting the science right," especially given all the scrutiny the politicization of the science has caused over the last few decades. In this regard, the unheralded heroes of the climate change movement may be the hundreds of scientists who have collaboratively drafted some six IPCC reports since the group's founding in 1988. (In 2007, however, these scientists shared the Nobel Peace Prize with former US Vice President Al Gore in 2007, for efforts to publicize awareness of human-caused climate change and help generate conditions for the international community to do something about it.)

The IPCC, consisting of some 400 natural and social scientists and economists from around the world, has drafted five reports, with each one casting human contributions to climate change in more certain language. In addition to publishing five comprehensive reports (with first releases from a sixth slated for 2021), the IPCC was important in the establishment of the body created to implement policies suggested by the science, the UNFCCC. Ratified at the 1992 so-called Rio "Earth Summit"[6] and driven by IPCC data, the UNFCCC called for nations to stabilize GHG emissions to prevent anthropogenic interference with the climate. The UNFCCC did not call for a specific country-by-country "budget" of reductions, or a system for making sure that emissions cuts were being enacted. The UNFCCC did call for regular meetings of countries belonging to it. These annual meetings become known by yet another acronym, the Conference of the Parties (COP). The third such meeting, at Kyoto in 1997, yielded finalization of the Kyoto Protocol (discussed later). But first we consider one of the UN system's unquestioned successes, the Montreal Protocol.

Ideas into Climate Action: The Montreal and Kyoto Protocols

The Montreal Protocol sought to diminish ozone-depleting chlorofluorocarbons (CFCs) in the atmosphere, which are only indirectly related to climate change but which may be similar enough to inform us about how the UN can achieve such additional necessary milestones. We will also consider the Kyoto Protocol, the worldwide agency's first—mostly failed—effort to regulate GHG emissions, before outlining key provisions of the Paris Agreement, the current international treaty to reduce emissions, which has been in effect since 2015.

The Montreal Protocol was a 1987 agreement to limit and then eliminate the emission of CFCs, common until the 1990s in "spray" cans of hair color and spray paint, which were known by that time to be depleting the atmosphere's ozone "protective shield" and thus exposing peoples' skin and eyes to greater sunlight and increased risks of skin cancer and cataracts. Unlike later agreements, which did not have the strong support of industry, the Montreal Protocol received overwhelming industry support from companies like DuPont, which was pleased to phase out the ozone-depleting chemicals as they were out of date and superseded by cheaper and less harmful ones (Beinecke 2012). The agreement was flexible, with phase-out periods that depended on individual countries and conditions, offered financial incentives to help developing countries initiate the "phase outs," and placed trade restrictions on nations that did not comply ("Saving the Ozone Layer"). In 2017, the Kigali Amendment, governing refrigerant and coolant chemicals and aimed at reducing GHG emissions in a related area, was approved. The Montreal Protocol is credited with reducing the ozone "hole" over the Antarctic by some 20 percent in the 30 years since it was enacted (Quan 2018), and comprehensive statistical analysis (Wu et al. 2013) has demonstrated that the detrimental effects of CFCs on the ozone layer are only half of what these would have been without the mitigation caused by the agreement and its effective implementation.

PROS AND CONS OF THE KYOTO PRECEDENT

The Kyoto Protocol, the first effort to comprehensively address the entire climate emissions problem (rather than just ozone depletion, which is a related problem but only a small part of the composite), met with less success. The agreement opened for signatures in 1997 but only entered into force in 2005, when 55 nations (known as Annex I countries) covering 55 percent of the 1990-level emissions agreed to cut worldwide emissions by about 5.2 percent within decades. Some nations, like Australia and the United States, failed to ratify the agreement, citing concerns that the costs of reducing emissions would reduce the competitiveness of industry. Other nations, like Canada and Japan, initially ratified and implemented the agreement and then withdrew later based, in part, on changes of authorities in power. Russia initially refused to participate but then joined the agreement upon realizing that the country's economic contraction in the 1990s had, on its own, shrunk GHG emissions to below the Kyoto-imposed cap. Given that Russia produced fewer emissions than its budgeted limit, the country joined the agreement and then sold "space under the cap" to European companies in nations that polluted more than their share. In essence, Russia turned the bad luck of economic contraction, which shrunk smokestack output to less than half of 1990 levels by 2000, into profits by letting polluters elsewhere buy Russia's unused emissions. We address carbon markets much more in Chapter 7.

Signatory compliance with the Kyoto Protocol was mandatory, although it only covered nations responsible for about 30 percent of global emissions by the time it expired in 2015. In addition to the aforementioned emitters like the United States (responsible in 2010 for 22 percent of the world's GHG emissions), which never managed to ratify the agreement (it failed to be approved by the US Senate), the world's other two largest polluters—China and India—were not held to emissions targets, as developing nations (dubbed Annex II countries) did not face the required reductions imposed on the Annex 1 signatories. Of the countries that were included in Annex I, the European Union climate pioneers (often led by Germany, Norway, and Sweden) participated in a market-based emissions trading system (ETS; see Chapter 7) as well as in direct domestic mitigation reduction efforts.

The worst GHG polluters either did not participate in the Kyoto agreement or were not mandated to reduce their emissions. Kyoto did not get renewed in 2012 despite several "false start" UN Conference of the Parties (COP) meetings, such as at Copenhagen in 2009. The Kyoto Protocol did set concrete emissions reductions by signatory nations and made a credible effort to enforce these cuts and monitor member nations' progress (within the limits of technology for measuring emissions, which are invisible to the naked eye). However, the agreement only covered about a third of the world's emissions, as the worst polluters were not forced to adhere to a "carbon budget" of reduced emissions.

In hindsight, the Kyoto Protocol did not diminish worldwide emissions in any meaningful way (Aichele & Felbermayr 2012, Aimer & Winkler 2017), and it cost us all valuable time. The UNFCCC spent several years trying to conceive of the basis for a new framework but was unsuccessful until 2015, when the Paris Agreement was reached. As shown in the next section, this agreement is now the cornerstone of international cooperation to mitigate climate change (i.e., reduce GHG emissions) and, as diplomats recognized for the first time starting in 2009, to adapt to inevitable changes in the climate, which humans have failed to stop.

ASSESSING THE LARGEST EMITTERS (CORPORATIONS AS WELL AS COUNTRIES)

Before discussing the current UN climate framework, stemming from the 2015 Paris Agreement, Table 3.1 lists the largest emitters in the world to demonstrate the percentage of global GHG emissions regulated by the Kyoto Protocol and the percentage not regulated. For purposes of comparison, we also list the percentage share of global emissions by the 20 largest fossil fuel companies.

The comparison is "apples to oranges," to be sure, as we are comparing a cross-section of global emissions by nation in one year, 2016, to shares of accumulated CO_2 over nearly 30 years. Still, the table does show how just a handful of companies have been responsible for the emissions build-up in our atmosphere, just like a handful of countries. Unfortunately, the success of the Montreal Protocol in getting industry on board has not been matched by the

Table 3.1 **Global GHG Emitters in by Percentage of Global Emissions (2016) and "Carbon Major" Companies and Their Share of Global Emissions (1988–2015)**

Emitting Countries	*Percentage of Global Emissions (2016)*	*Emitting Companies (Public or Private)*	*Percentage of Accumulated CO_2 (1988–2015)*
China	29.0	China National Coal Group	14.3
United States	16.0	Saudi Arabian Oil	4.5
India	7.0	Gazprom OAO	3.9
Russian Federation	5.0	National Iranian Oil Co.	2.3
Japan	4.0	ExxonMobil Corp.	2.0
Germany	2.0	Coal India	1.9
Korea	2.0	Petroleos Mexicanos (Pemex)	1.9
Islamic Republic of Iran	2.0	Russia coal	1.9
Canada	2.0	Royal Dutch Shell PLC	1.7
Saudi Arabia	2.0	China National Petroleum Corp (CNPC)	1.6
Indonesia	1.0	BP PLC	1.5
Mexico	1.0	Chevron Corp.	1.3
Brazil	1.0	Petroleos de Venezuela SA (PDVSA)	1.2
South Africa	1.0	Abu Dhabi National Oil Co.	1.2
Australia	1.0	Poland Coal	1.2
United Kingdom	1.0	Peabody Energy Corp.	1.2
Turkey	1.0	Sonatrach SPA	1.0
Italy	1.0	Kuwait Petroleum Corp.	1.0
Poland	1.0	Total SA	0.9
France	1.0	BHP Billiton Ltd	0.9
Rest of the world	19.0	Rest of the companies	21.9

Sources: https://www.ucsusa.org/resources/each-countrys-share-co2-emissions; https://6fefcbb86e61af1b2fc4-c70d8ead6ced550b4d987d7c03fcdd1d.ssl.cf3.rackcdn.com/cms/reports/documents/000/002/327/original/Carbon-Majors-Report-2017.pdf?1501833772, 14.

100 "carbon majors" responsible for 71 percent of all GHG emissions worldwide since 1877, the first year for which measurements are available. The comparison shows that without corporate cooperation, climate change will not be brought under control.

While laws and legal regimes are limited to nation-states, the international regulation of companies through greater vigilance might also be beneficial in the future. Chapter 4 further discusses the "carbon majors." Here, we only suggest that thinking in terms of emissions by nation-states and accounting for emissions that way may not be the only, or even the best, means of addressing the problem. But that is the way the UN is organized, and hence we return now to the nation-based approach to addressing climate change, as expressed in the Paris Agreement. (We consider fossil fuel company–centered views a bit more in Chapter 4.)

The Paris Agreement: Voluntary, Nationally Determined Contributions but Without Needed Ambition

The UNFCCC debated a successor regime to the Kyoto Protocol at several annual meetings, including at the 2009 Copenhagen COP, which came the closest before the Paris meeting six years later to establishing some sort of successor to Kyoto. The Paris Agreement doubled down on the Kyoto Protocol's "precautionary principle" from the 1992 Rio Treaty, which stated: "The Parties should protect the climate system for the benefit of present and future generations of humankind, on the basis of equity and in accordance with their common but differentiated responsibilities and respective capabilities" (UNFCCC founding document cited in Hermwille et al. 2017). The focus on mitigation (i.e., reducing emissions) was also kept in the Paris Agreement, which is a benefit, by concentrating "laser beam" attention on CO_2 pollution as the main component of the problem, and a liability, by not giving sufficient attention to other causes of climate change (e.g., the human behavior causing the emissions) or the emerging bugbear of climate change: adaptation. Adaptation refers to the need for us all to learn to live with climate change. Another liability of the Kyoto Protocol that was extended to the Paris Agreement was the consensus rule regarding decision-making; that is, decisions must be approved by all parties rather than just by a majority.

What does the Paris Agreement say? Basically, it offers a "bottom-up approach" to mitigation, allowing each nation to voluntarily declare a "pledge" in emissions reductions in accordance with its interests and capabilities. This system allowed negotiators to break down what former US negotiator Todd Stern refers to as the "firewall" between the Kyoto Annex I countries (the developed countries with emissions reductions) and the non–Annex (developing exempt from emissions reductions) countries. To Stern (2018), the Kyoto Protocol solved the problem of "common but differentiated responsibilities" among countries by assigning "legally binding targets and timetables for reducing emissions to developed countries, backed up by rigorous rules for accounting transparency, and compliance, while asking virtually nothing of developing countries" (6). Indeed, Paris asks all countries for something, even if that something can be very little indeed (at the discretion of each and every nation). However, Stern's firewall problem has given way to a new problem, documented by the IPCC in a special 2018 report: the "ambition gap" between each country's Nationally Determined Contributions (NDCs) and the actions needed to keep climate change from increasing average world temperatures by more than 2°C (about 3.9°F) by the end of the 21st century.

The world has heated—on average—about 1.0°C since the Industrial Revolution in the middle of the 19th century. The IPCC argued in its 2018 report, requested by the Alliance of Small Island States, several of which are losing surface area to rising tides every year, that the permanent damage wrought by a 2°C change is unacceptably severe, that diplomats need to tighten the Paris Agreement to further lower emissions so that warming is kept to 1.5°C, and that such an outcome is still possible, although improbable without more stringent ambition in emissions reductions (IPCC 2018; Roelfsema et al. 2020. Paris negotiators acknowledged that in some cases countries merely "repackaged" a range of programs already underway in sectors like energy and development as NDCs.[7] In other cases, the lack of NDC ambition was found to be a result of domestic politics. For example, Eisenstadt and Stevens (2017) found that nations with a large dependence on fossil fuels were less likely to propose ambitious cuts in their emissions.

Regardless of the "original sin" of letting nations propose their own emissions reductions bottom-up, rather than setting an international objective based on scientific research and then assigning, top-down, each nation's obligations, some top-down provisions do exist in the Paris Agreement. Individual NDCs may be weak, but they have to be updated periodically and also reviewed to make sure that pledged emissions reductions are indeed being achieved. Additionally, five-year "stock takes" are being held to ramp up NDCs and aggregate these into worldwide tallies from which progress can be measured. Countries also are called upon to adopt strategies over the longer term for "de-carbonization," which is an end to reliance on carbon-based energy sources (fossil fuels) and a transition to a postcarbon economy. The "Paris Rulebook" or implementing legislation was mostly agreed upon at COP 24 in Katowice, Poland, in 2018, although work remains on finalizing means of transparently reviewing nations' progress. This process was to be formally undertaken at COP 25, which was moved to Germany at the last minute and scaled back after social unrest engulfed the original location, Chile, and then it was slated for COP 26 in Scotland, which was postponed a year to 2021 due to the COVID-19 pandemic. This first Paris global stock take will be critical, as many researchers, including Kate Dooley (2018), have noted that "the national pledges adopted in Paris are still woefully inadequate to meet the 1.5°C or 2.0°C global warming goals."

Perhaps the most ambitious of the Paris provisions was Article 6.2, allowing Internationally Transferred Mitigation Outcomes, which are basically offsets between countries. Recall from Chapter 1 the controversy over offsets and the critique that these "cure" the symptom rather than the cause. Nevertheless, the Paris Agreement incorporated these in part as a means of creating an ETS and galvanizing private sector support. The strengths and weaknesses of ETS are discussed extensively in Chapter 7, but it is important to note here that this policy approach (as opposed to other means of raising public funds, like carbon taxes) has received a strong endorsement from the UN.

Overall, the Paris Agreement offered a clear improvement over the anchorless debates that had gone on for well over a decade as the UNFCCC delegates searched for a successor to the failed Kyoto Protocol. Still, after the scientific community managed to coordinate so smoothly in casting the dire state of climate change, disappointment with nations' lack of ambition in their NDCs, and the with UN's failure to push them further, has been palpable. Bodansky and Rajamani (2018) point to several unresolved issues from the Kyoto Protocol that remained as challenges in the Paris Agreement, mainly (1) how legally binding the Paris Agreement should be, (2) how broadly prescriptive (as opposed to letting nations make their own case) it should be, and (3) how to overcome the "common but differentiated" contradiction with policies that are either more common or more differentiated.

Regarding legality, Bodansky and Rajamani (2018) point out that, on the one hand, Paris creates binding procedural obligations regarding NDC preparation, communication, and accounting, but on the other hand, it does not "make the content of Parties' NDCs legally binding, nor, more generally, does it create new substantive obligations relating to mitigation or finance" (186). With regard to prescriptiveness, the Paris Agreement demonstrates much more flexibility than the Kyoto Protocol, but also less teeth. And on the issue of differentiation, Paris—unlike Kyoto—does not differentiate regarding NDC or emissions reduction obligations, although it is lax regarding all nations. It does differentiate in language, asking developed countries to immediately undertake economywide transformation, while asking developing countries to move in this direction. The developed nations are also held to tighter standards of transparency of reporting on NDC progress, as the developing countries are asked to report in accordance with their capacities (rather than being given a categorical treatment based on historical treatment). Because emissions reductions are at the core of mitigation, it is worth briefly considering the implications of different logics of emissions responsibility, and how these might impact actual national obligations.

Distributive Dilemma of Climate Change Solutions: Who Should Suffer and Who Should Pay?

The concept of "environmental justice"—an umbrella precursor to "climate justice"—was really born in the United States in 1982, when civil rights activists mobilized to stop the state of North Carolina from dumping 120 million pounds of highly contaminated soil in Warren County, which had the state's highest concentration of African Americans. Indeed, Mohai et al. (2009) cite dozens of studies of the hundreds showing that "ethnic minorities, indigenous persons, people of color, and low-income communities confront a higher burden of environmental exposure from air, water, and soil pollution from industrialization, militarization, and consumer practices" (406). In turn, "climate change justice" has become a term used by developing world nations to contest their disproportionate share of the climate burden as they suffer the consequences of Warren County–like dumping of emissions, over the last 150 years, by industrial nations into the atmosphere. The industrial nations, which started by lining the atmosphere with coal-generated soot—the worst of all—in the 1800s, have industrialized without "paying" for climate change. Now they are asking nations like China and India not to emit and, by extension, to limit their industrialization and otherwise absorb the loss of development revenues represented by the diminishment of GHG emissions. In the Paris Agreement, these countries did agree to voluntary emissions targets but India initially contested the agreement and only signed at the last minute.

A second type of climate injustice, and perhaps a more forward-looking variant, concerns the distribution of suffering and costs associated with climate change. It seems those costs are borne disproportionately by developing tropical countries, where the tempest of climate change torments more forcefully via hurricanes, drought, and other extreme weather events. Over the last 20 years, the nations most affected by climate-related weather have been (in descending order—based on deaths and property losses) Honduras, Haiti, Myanmar (formerly Burma), Nicaragua, the Philippines, Bangladesh, Pakistan, Vietnam, Thailand, and the Dominican Republic (Eckstein et al. 2018). The hurricanes and cyclones, which have caused up to some 8,000 deaths per year on average over a 20-year period (Myanmar) and up to two percent of gross domestic product (GDP) on average lost per year (Honduras), transcend national borders, as do their causes.

It seems that the Westphalian idea of national sovereignty (i.e., the notion that every nation had control over occurrences within their borders and thus the right to regulate these) may be obsolete when we consider causes and solutions to climate change, just as it is in other areas like economic flows (e.g., trade and investment have not known national boundaries since the internet was used for these starting in the 1990s) and governance (e.g., the European Union, with stops and starts, has shown that supranational governance is possible, the United Kingdom's "Brexit" aside[8]). As more collective consideration is needed of climate problems, the designation of nation-state boundaries will be increasingly seen as arbitrary and may in fact obstruct efforts to mitigate climate change.

Skillington (2017, 164–165) argues that as "dystopias of global resource wars and restricted border movements" will come into being if we cannot adopt collective solutions to climate change, it will be essential to take collective responsibility, defend the human rights of the tens of millions displaced by climate, and to create a transnational forum to debate how the rights of the displaced can best be exercised. While the UN seems like the organization best-suited for this labor, that body has thus far failed to recognize "climate migrants" as a category of refugees or to establish any mechanisms for addressing their needs. At the 2013 meeting of the UNFCCC, the Warsaw International Mechanism for Loss and Damage (or WIM, in UN parlance) was established to help distribute resources to climate change's biggest "losers." But the precise form the WIM takes is still being decided.

Meanwhile, the most severe form of this climate injustice may be the lack of preparations to ensure the continued right to self-determination of low-lying islands that will disappear with the continuation of sea level rise. Island-states like Kirabati and Tuvalu in the Pacific and the Maldives in the Indian Ocean face an extensive risk, which was popularized when former Maldives President Mohamed Nasheed held a cabinet meeting underwater in 2009.[9] Skillington (2017) argues that some 11 island countries face immediate risk of sinking, and that new legislation is needed focusing on "the collective rights of sovereign communities of people whose common way of life is being destroyed by a degraded environment" (193–194). New Zealand has offered legal residency to citizens of Fiji, Kiribati, Tonga, and Tuvalu since 2001, but this has been based on good will rather than any human rights declaration or new law.

What can be done to ease the suffering of those whose homelands are sinking and who will have to move their entire nations elsewhere? Do "homelands" define nation-states? Can Kirabati's culture, history, and identity be maintained—and prosper—even in Fiji or New Zealand? While such questions may define the ideal case of acculturation to a new environment, still larger questions loom before people even get such second chances. Who will help get people and their possessions off sinking islands? Who will help make sure they have a place where they can relocate? How can the legal and human rights of these climate refugees be guaranteed, even if they formally become stateless?

The costs of upholding the rights of climate refugees is another complicated element of the problem. The issue of who is responsible for climate-related "loss and damage" was taken up by the Paris Agreement, which, in Article 8, speaks to the need to support mechanisms for supporting loss and damage such as through the WIM. However, as noted by Gewirtzman et al. (2018), the Paris Agreement does not mention loss and damage anywhere in its financial article or in its call to enhance the transparency framework, and an administrative ruling has stated, furthermore, that the article "does not provide a basis for any liability or compensation." Who is responsible for climate change? The leading polluter nations (the United States, China, and India)? All nations? The "carbon major" GHG-emitting companies? This is highly debatable, and different arguments point to different culprits. Let's look at several.

Ways of Thinking About the Ethics of Who is Responsible and Who Should Pay

The dilemma of "who is to blame" was written right into the UNFCCC's founding document. The Convention opened for signatures in 1992 and entered into force in 1994, with 195 nations as "parties" at that time. The objective of the specialized UN group was to stabilize GHG emissions at 1990 levels, exchange information, and undertake technology transfer to enable all nations to mitigate emissions. Article 3 of the UNFCCC opened up the "climate justice" question, stating:

> The Parties should protect he climate system for the benefit of present and future generations of mankind on the basis of equity and in accordance with their common but differentiated responsibilities and respective capabilities. Accordingly, the developed country Parties should take the lead in combatting climate change and the adverse effects thereof. [bold emphasis added]

This is a powerful but ambiguous statement. First, does it not appear that "common" and "differentiated" are at odds with each other? Which nations should have common responsibilities, and which should have differentiated ones? And should their responsibilities be based on their capabilities? If so, what kind of capabilities? Their present capabilities? Their potential future ones? Let us consider five different scenarios, attributing blame to different actors, and their logics. As Table 3.2 illustrates, different measures of responsibility for emissions render different culprit nations. Why is there not more overlap among nations in Table 3.2?

FIVE LOGICS FOR ALLOCATING EMISSIONS BLAME (AND RESPONSIBILITY)

First, consider responsibility as a function of a country's role in historical emissions; the "longitudinal" approach, as posited by Singer (2016). Remember that, as mentioned in Chapter 1, the effects of GHGs stay in the atmosphere

Table 3.2 **Nation-State Responsibility for Emissions by Logic of According Blame**

LOGIC 1: Cumulative Emissions by Country (1850–2011)	*LOGIC 2: 2010 Cross-Sectional Emissions by Country*	*LOGIC 3: Emissions per Capita by Country (2011)*	*LOGIC 4: Ambition Level of Paris Pledges (Emissions Reductions by 2030)*
United States	China	Brunei	Dominica
European Union	United States	Kuwait	Iceland
China	India	United Arab Emirates	Switzerland
Russian Federation	Russian Federation	Oman	Solomon Islands
Japan	Japan	Australia	Luxembourg
India	Germany	Bahrain	Marshall Islands
Canada	South Korea	Luxembourg	Madagascar
Mexico	Canada	Saudi Arabia	Liechtenstein
Brazil	Indonesia	United States	Monaco
Indonesia	Saudi Arabia	Canada	Tuvalu

Sources: Mani et al. 2018; https://www.wri.org/blog/2014/11/6-graphs-explain-world-s-top-10-emitters; https://www.theguardian.com/environment/2011/apr/21/countries-responsible-climate-change

for some 150 years. This strongly implicates the industrial nations (and the "carbon major" oil companies, as discussed in Chapter 4) and most closely follows the North-South Debate and the Kyoto arrangement, whereby industrial countries like those of the European Union and the United States (which never entered the treaty) were responsible for the lion's share of emissions reductions, for compensating the developing countries, and for allowing these countries—led by China and India—to increase emissions in order to industrialize before reducing them as part of a UN-wide "net zero" emissions effort by the mid-21st century. The logic here was best stated by "The Pottery Barn Rule," made by former US Secretary of State Colin Powell in relation to another international crisis: "You break it; you bought it" (Safire 2004).[10] The claims by nations like India that they had a right to pollute, even given what we know about GHG emissions (which was not known 160 years ago when Great Britain industrialized, or even after World War II when smokestacks peaked in the United States and much of Europe), have thrown wrenches in many recent climate negotiations. In this approach, the United States accounts for fully 27 percent of cumulative emissions, while the European Union accounts for 25 percent, China 11 percent, and the fourth-place finisher, the Russian Federation, emitting only 8 percent.

Second, a less extreme variant of the pro-development argument made by some industrializing nations prompts a "cross-sectional" logic that freezes the status quo, meaning that nations' proportion of worldwide emissions will be frozen at 1995 or 2010 levels and that future emissions will abide by this same proportion. This distribution of responsibility still casts the blame on the industrial powers, but it discounts past contamination and only focuses on the present, hence reducing the percentage of emissions attributed to the few largest historical emitters, like the European Union and the United States, and instead distributing the burden more evenly with present high emitters, like China and India. Here, China tops the list with 29 percent, followed by the United States with 16 percent, India with 6 percent, and Russia with 5 percent.

Third, an "equal shares for all" approach disaggregates nations down to individual residents and charges governments according to per capita pollution. The logic here approximates the previous two in that nations that emit more GHGs are more responsible, but in this case, the data are generated at the level of the individual citizen rather than nationwide. Another way of saying this is that the emissions data are adjusted to the population. Still another way of doing this might be to adjust emissions data to the amount of economic activity per resident (e.g., emissions per each resident's annual average income). This disaggregated approach has not been widely used but holds high-emitting, low-population countries responsible, to the benefit of two of the highest emitters (China and India), which have large populations. Using the data of Mani et al. (2018), Brunei had the largest number of metric tons of CO_2 emissions per capita with 37, whereas Kuwait also had 37 and the United Arab Emirates came in third with 29. Lower in the Top 10 were the United States (23 metric tons per capita) and Canada (21 metric tons per capita).

Fourth, a Rawlsian moral argument favors those in the poorest and most vulnerable nations, like Bangladesh, Haiti, Pakistan, and perhaps nations like Kiribati in danger of extinction. The logic, not depicted in Table 3.2, is that given that any of us could have been born into the worst circumstances, we all collectively owe it to those worse off to help improve their lot. This is the argument most often made by the Alliance of Small Island States (AOSIS), led by nations like Barbados, Dominica, Fiji, Kiribati, the Maldives, Singapore, and Tuvalu.

Fifth, the logic followed by the Paris Agreement in allocating responsibility for climate change, and thus by implication the parties responsible for leading efforts to rectify it, simply allows nations to take responsibility themselves and offer whatever they want to solve the problem. Unfortunately, this approach leaves the onus on each nation to decide its own contribution, in a logic that can perhaps be best described as "charitable giving." Using the Mani et al. (2018) measures of NDC ambition, the nations reducing their emissions by the greatest percentage are those that have few emissions to begin with, such as Dominica, which pledged a 73 percent reduction by 2030 over 2015 levels, and low-emissions, affluent nations with little industry, such as Iceland, which pledged a 60 percent reduction, and Switzerland, which pledged a 58 percent reduction. This approach, depicted as Logic 4 in Table 3.2, does not really diminish emissions unless the major emitters make meaningful contributions. This may happen under increasing pressure from the UNFCCC.

INADEQUACY OF THE PARIS AGREEMENT TO KEEP WARMING TO 1.5°C (OR EVEN 2.0°C?)

Given these approaches to allocating blame and responsibility, and the "charitable giving" option under which the UNFCCC presently labors, we can comprehend how the IPCC recently released a report warning that present NDCs under the Paris process are inadequate for restricting temperature increase to the target of 1.5°C, or even 2.0°C, by the end of this century (Intergovernmental Panel on Climate Change (IPCC 2018). Indeed, the NDCs pledged in Paris, to be ratcheted up over the next few years, will only limit warming to 2.7°C by end of 21st century (with 50 percent probability) according to the New Climate Institute Action Tracker (2020). Climate Interactive (2020) estimates that the present level of pledges will only keep climate change within 3.3°C by the end of the century.

While national "winners and losers" from the Paris process may be difficult to assess, Mani et al. (2018) estimated the effects of Paris NDC pledges on national economies. Their work shows welfare losses of 1.0 to 1.5 percent among the European countries, Australia, New Zealand, and Mexico, while major emitters China and India did not pledge emissions reductions until 2030 and thus will not experience welfare effects over the coming decade. The economists concluded that the United States would witness a welfare decline of about 0.7 percent by 2030, whereas the effects of Paris Agreement commitments would have negligible welfare effects in Asia, Latin America, the Middle East, and North Africa (Mani et al. 2018, 23).

We'll Always Have Paris: Overcoming Limitations of the International Climate Change Regime

While climate change mostly demands a liberal approach to international relations, part of the problem with liberals in international relations is that the incentives they depict sometimes do not align with the motivations of actors in the real world. The regime of climate change institutions has evolved fairly quickly in just a few decades, although it is fair to say that the scientific knowledge base has advanced much more quickly than the political will to put forth the effort required. The IPCC scientists, organized more according to their specialization of expertise than according to whether they came from developed or developing countries, were able to conduct their work with a minimum of politics. Contrarily, it sometimes seems that the diplomats of the UN and, especially, the representatives of a few nations—such as the United States under Trump conducted their politics with a minimum of science. Citizens of climate-denier countries must continue to demand accountability from their leaders, and means must be found to add some "top-down" compliance mechanisms to bridge the Paris emissions ambition gap.

While the entire Paris process has at times succumbed to the domestic pandering of leaders in a few nations—like Australia, the United States, Saudi Arabia, and Brazil in recent years—domestic politics and pressures from binational bargaining (between two nations rather than among 195) might also be used to leverage climate change mitigation. In other words, as considered further in Chapter 8, it may behoove nations to negotiate bilaterally, as the United States under President Obama and China under President Xi Jinping did in 2014, pledging to cut emissions by over a third each by 2030. In addition to bilateral negotiations, "climate clubs" or cartels, composed of several important emitters (rather than all the world's nations), might also be able to strike deals. And if a UN Climate Directorate can be created at the UNFCCC, consisting of a dozen or so members, whether permanent or rotating, with executive authority, the inertia of the consensus approach to negotiating—meaning that every single nation must agree before language can be accepted—may be overcome. Moving to decisions based on majority rule, rather than requiring unanimity, has also been discussed but is unlikely to be popular.

One of the benefits of the Paris Agreement is its flexible duration and provisions for ratcheting up expectations. COP meetings since Paris in 2015 have for the most part not exploited this authority, but it is possible that top-down requirements could be imposed to prompt UN member nations to finance research and development via some sort of climate Marshall Plan;[11] address the oft-ignored but inevitable need to address adaptation as well as mitigation, which has received the vast majority of efforts to date; add a system for insuring vulnerable nations against climate risk, which takes increasing numbers of lives and property in extreme weather events; and create a legal category for climate refugees. More resources are needed to start creating the infrastructure and research to generate such achievements, and means must be found to collect from nations the 100 billion dollars pledged by developed nation donors since the 2009 Copenhagen meeting to solve this problem. As of 2020, only a tenth of the funding had actually been collected, and the Green Climate Fund, established by the UN as the new technical agency/development bank for climate lending, has been slow to get on track (see Chapter 7).

While the UNFCCC's record on climate change mitigation is not overly impressive, and while the Paris Agreement itself is highly perfectible, the treaty has served as an important point of convergence for civil society and the private sector, as well as national governments. Analysts have argued that incremental and partial "polycentric" solutions will be the only way to address the complexities of this wicked problem, and even criticizing the UN framework helps unify critics and allows us all to adopt similar terms and language. That is a key starting point, and although it is getting late in the game for starting points, nonstate actors as well as national and subnational governments are investing extensive and dynamic energies in the effort, and may ultimately offer our best chance. In Chapter 4, we discuss these domestic actors who are tailoring the Paris Agreement to their contexts.

Exercise 3.1 A Hypothetical United Nations Simulation to Specify Loss and Damage in the Paris Agreement

SCENARIO

At the hypothetical COP 30 in Rio, the Alliance of Small Island States (AOSIS members plus the Philippines) is proposing to rewrite Article 8 of the Paris Agreement and to create a concrete institution to administer loss and damage by climate-affected nations on behalf of a UN-formed institution (to be named in the proposal).

ASSIGNMENT

In this exercise, students represent individual countries. For an 18-student class, these countries can be Australia, Bangladesh, Brazil, Canada, China, Dominica, Ecuador, Ethiopia, Germany/European Union, Great Britain, India, Japan, Nicaragua, Nigeria, Russia, Saudi Arabia, Sweden, Tuvalu, and the United States. The largest emitters must be present (China, European Union [represented through Germany as leader], India, and the United States), as must some of the recalcitrant nations (Australia, Brazil, Russia, Saudi Arabia, and the United States were good examples in 2020). Some of the most vulnerable nations should also be part of the simulation (small island nations of the Pacific and the Caribbean, Sahel-area African nations, Bangladesh, Philippines, and the Central American nations).

The AOSIS + Philippines single team-written proposal (in no more than 2 pages of proposed text and 6 to 10 pages of commentary – constituting four or five of an 18-student group) constitutes the preparatory text of all members (in other words, this is a group writing assignment), and they will submit the text to all parties by a given deadline. This group drafting this proposal and commentary—the proposing Team A—will be led by the AOSIS countries and some other vulnerable nations (such as Tuvalu, Dominica, Ecuador, Bangladesh, Nigeria, and Ethiopia). The AOSIS member nations (Tuvalu and Dominica) should draft a two-page opening statement, the rest of the team should write a two-page text proposal, and each country must write an individual one-page commentary on the proposal before class. The rest of the class (Team B—United States, China, India, Sweden, Saudi Arabia, Great Britain, Russia, and Germany/European Union) must draft a two-page response (and one-page country positions) prior to class on the start date of the simulation exercise and be ready to respond when the AOSIS presents the proposal. That group, the responding team, led by China, India, Sweden, Great Britain, and Germany/European Union, should draft a statement or statements (depending on levels of agreement) to bring to class on a given deadline. Brazil and Russia (and Australia and the United States at times, depending on who governs) can be spoilers, acting in their own interests rather than those of the group. The two main groups may need to meet outside of class to prepare. The "response group" will meet during class while the "proposing group" listens to and evaluates the INDC simulation. Teams A and B will bring any revised agreements or new proposals to class on a given deadline.

PROCEEDINGS OF SIMULATION

The 150-minute negotiation will proceed as follows: On Day 1, three-minute statements by the two AOSIS nations are followed by four of the non-AOSIS nations (25 min). The UNFCCC chair (your instructor) will then convene a discussion section to go over the proposal in its entirety over the course of 60 minutes, with questions by the group of two minutes or fewer. That will end the first class.

On Day 2, 25 minutes will be allocated to further discussion, and then 45 minutes will be left for improvised floor statements and the taking of a vote on a revised proposal, after which closing statements will be made by two members of each group for three minutes each. Two three-minute responses from each group will be allowed, and then the nonparticipating students will cast votes on two issues: Which side won the debate based on their performances? Which side should have won from an equity standpoint (in other words, regardless of which side argued better, which side had a stronger position)? Team leaders will be responsible for ensuring that each participant on their team is assigned at least one three-minute role in the opening or closing statement.

Continued

The proposing team may present a formal amendment to the Paris Agreement text (recommended), or they may present their proposal as a "secondary source" text (not as a direct text to be inserted into the agreement). The responding team must reply in kind. Both groups are encouraged to present PowerPoint slides of each specific proposal that can be shown while the text in question is being considered.

POST SIMULATION DEBRIEF

Each participant will write a short memo (one page) to submit at the debrief, with a paragraph or two for each of the following questions: (1) Was the outcome that occurred the expected outcome or not? Were there opportunities for the negotiations to go differently? If so, what were they? (2) Do the AOSIS countries have any chance of actually getting a statement in the revised agreement about liability and compensation? (3) Was the best outcome attained (and tell the instructor your criteria for asserting what is "best")? If not, how might this optimal outcome be attained?

Climate Interactive (https://www.climateinteractive.org/tools/#simulations) offers a great resource for simulations. Specifically, see https://www.climateinteractive.org/programs/scoreboard/ for the best rendering of how national changes may (or may not) impact international emissions.

NOTES

1. Australia vehemently opposed the creation of this facility before and during the COP 21. See Milman 2015 and Karsapan 2015.
2. Although researchers have made some advances recently in measuring climate migration via surveys, these past endeavors have been either case-based without drawing broader implications (see, e.g., Bilsborrow 2009) or not representative (Warner et al. 2009). Moreover, these data need an urgent update and a systematic approach. A review of the past literature underscores the need for investigating "the effects of environmental factors on outmigration based on quantitative multivariate methods" (Laczko & Aghazarm, 2009, 31).
3. The only survey to date on small island-nation vulnerability was conducted by United Nations University. See http://i.unu.edu/media/ehs.unu.edu/news/11747/RZ_Pacific_EHS_ESCAP_151201.pdf
4. The five permanent members of the Security Council at the UN's San Francisco creation in 1948 were Taiwan, France, the Russian Federation, the United Kingdom, and the United States. In recognizing that geopolitical strength should matter more than who was a Western ally at the close of World War II (the original justification for the status), the UN in 1971 replaced Taiwan with the People's Republic of China. Over the years, several other large regional powers, like Brazil, India, Japan, Mexico, and Nigeria, have campaigned for permanent-member status.
5. In 2020, President Trump sought to remove the United States from the World Health Organization, claiming that it had not adequately predicted the COVID-19 pandemic. As of early 2021, President-elect Biden had pledged that the US would re-join the WHO.

6. As you can see, the cities where these meetings are held often become part of the names of agreements and/or reference points for meetings. They may seem unimportant but are central to how we refer to UN. Consider for example the Kyoto Protocol named for the city of its signing in 1997, and the Paris Agreement, signed there in 2015.
7. Todd Eisenstadt attended the 2015 Paris meeting where this practice of diplomatic "regifting" was acknowledged by multiple negotiators.
8. After decades in the European Union (EU), the United Kingdom (UK) voted via referendum in 2016 to withdraw from the supranational arrangement. After further deliberation, the UK, one of Europe's largest economies (along with Germany and France, which remained in the EU and took over leadership), commenced the formal, one-year process of withdrawing early in 2020. The withdrawal, the first among the nearly 30 members of the EU, is widely viewed as part of a move away from multilateral organizations (like the United Nations), and toward unilateralism. Such an international movement may undermine efforts to collectively solve climate change.
9. "The Island President" documentary dramatizes the difficulties faced by the Maldives, in part based on the challenges of climate change. See, for example, Hornaday 2012.
10. Powell, President George W. Bush's secretary of state, was referring to the US invasion of Iraq, in an ill-fated effort to quell sectarian violence there.
11. The Marshall Plan was a US initiative launched in 1948 to help Western Europe recover from World War II. The US government transferred billions of dollars in assistance to help spur Europe's economic recovery. To this day, the Marshall Plan exemplifies successful economic assistance and the globalist (as opposed to the isolationist) position of the US after World War II. The US assumed the mantle of leader of the Western powers (industrialized capitalist democracies in Europe and North America) against the Soviet Bloc (what is now Russia and other Soviet Union nations, as well as Eastern Europe's then-communist nations).

REFERENCES

Aichele, Rahel, and Gabriel Felbermayr. 2012. "Kyoto and the carbon footprint of nations." *Journal of Environmental Economics and Management* 63:336–354.

Aimer and Winkler. 2017. "Analyzing the effectiveness of international environmental policies: The case of the Kyoto Protocol," in *Journal of Environmental Economics and Management* (82: C): 125-151.

Beinecke, Frances. 2012. "Saving the Ozone Layer: Lessons for Fighting Climate Change," blog. Accessed December 16, 2020 at https://www.nrdc.org/experts/frances-beinecke/saving-ozone-layer-lessons-fighting-climate-change

Bilsborrow. 2009. "Collecting data on the migration-environment nexus," in Laczko, Frank and Christine Aghazarm eds. *Migration, Environment and Climate Change: Assessing the Evidence.* Geneva: International Organization for Migration. 113-196.

Bodansky, D., and L. RajamaniL. 2018. "The Issues that never die." *Carbon & Climate Law Review* 12(3):184–190.

Carter, J. 1979. "The Crisis of Confidence Speech." Accessed January 14, 2019, at https://www.americanrhetoric.com/speeches/jimmycartercrisisofconfidence.htm

Climate Interactive. 2020. "UN Climate Pledge Analysis" Webpage accessed December 24, 2020 at https://www.climateinteractive.org/ci-topics/climate-energy/scoreboard/

Dooley, K. 2018. "We finally have the rulebook for the Paris Agreement, but global climate action is still inadequate." Accessed January 14, 2019, at http://theconversation.com/we-finally-have-the-rulebook-for-the-paris-agreement-but-global-climate-action-is-still-inadequate-108918

Eckstein, D., V. Kunzel, and L. Schafer. 2018. "Global Climate Risk Index 2018: Who suffers most from extreme weather events?" Accessed November 5, 2018, at https://germanwatch.org/sites/germanwatch.org/files/publication/20432.pdf

Eisenstadt, T., and D. Stevens. 2017. "The predominant role of domestic interests in setting ambition of National Determined Contributions to mitigation efforts."

Foley, C. 2012. "Military basing and climate change." Accessed January 14, 2019, at https://www.americansecurityproject.org/wp-content/uploads/2012/11/Military-Basing-and-Climate-Change.pdf

Gewirtzman, J., S. Natson, J. Richards, V. Hoffmeister, A. Durand, R. Weikmans, S. Huq and J. T. Roberts 2018. "Financing loss and damage: reviewing options under the Warsaw International Mechanism." *Climate Policy* 18(8):1076–1086. DOI: 10.1080/14693062.2018.1450724

Gore, Al. 2006. *An Inconvenient Truth: The Planetary Emergency of Global Warming and What We Can Do About It.* Emmaus, PA: Rodale Books.

Haas, Peter M. 1990. *Saving the Mediterranean? The Politics of International Environmental Cooperation.* Columbia University Press, New York.

Hermwille, L., W. Obergassel, H. Ott, and C. Beuermann. 2017. *UNFCCC Before and After Paris: What's Necessary for an Effective Climate Regime?* Wuppertal Institut, Wuppertal, Germany. Accessed January 14, 2019, at https://epub.wupperinst.org/frontdoor/deliver/index/docId/6104/file/6104_Hermwille.pdf

Holthaus, Eric. 2016. "Rising tide of migration accompanies sea level rises, as predicted," blog accessed December 24, 2020 at https://phys.org/news/2016-05-tide-migration-accompanies-sea.html

Hornaday, A. 2012. "Mohamed Nasheed talks about 'The Island President,' keeping fragile democracy afloat." *The Washington Post*, April 13. Accessed November 5, 2018, at https://www.washingtonpost.com/entertainment/mohamed-nasheed-talks-about-the-island-president-keeping-fragile-democracy-afloat/2012/04/12/gIQAUzvCFT_story.html?utm_term=.4468f545eeb9

Intergovernmental Panel on Climate Change (IPCC). 2018. Global Warming of 1.5°C.An IPCC Special Report on the impacts of global warming of 1.5°C above pre-industrial levels and related global greenhouse gas emission pathways, in the context of strengthening the global response to the threat of climate change, sustainable development, and efforts to eradicate poverty [Masson-Delmotte, V., P. Zhai, H.-O. Pörtner, D. Roberts, J. Skea, P.R. Shukla, A. Pirani, W. Moufouma-Okia, C. Péan, R. Pidcock, S. Connors, J.B.R. Matthews, Y. Chen, X. Zhou, M.I. Gomis, E. Lonnoy, T. Maycock, M. Tignor, and T. Waterfield (eds.)].

Johnstone, S., and J. Mazo. 2013. "Global warming and the Arab Spring." In *The Arab Spring and Climate Change*, edited by Caitlin E. Werrell and Francesco Femia. The Center for Climate and Security, Washington, DC, pp. 15–22. Accessed January 14, 2019, at https://climateandsecurity.files.wordpress.com/2012/04/climatechangearabspring-ccs-cap-stimson.pdf

Kälin, Walter. 2010. "Report of the Representative of the Secretary-General on the human rights of internally displaced persons." UN Doc. A/HRC/13/21, January 5. Accessed December 24, 2020 at https://www.brookings.edu/research/report-of-the-representative-of-the-secretary-general-on-the-human-rights-of-internally-displaced-persons-2/

Karasapan, Omer. 2015. "Refugees: Displaced from the Paris climate change agreement?" Brookings Institution, December 7. Accessed July 26, 2016, at http://www.brookings.edu/blogs/future-development/posts/2015/12/07-refugees-climate-change-cop21-karasapan

Keating, J. 2018. "The sinking state: This is what happens when climate change forces an entire country to seek higher ground." *The Washington Post*, July 26. Accessed November 5, 2018, at https://www.washingtonpost.com/news/posteverything/wp/2018/07/26/feature/this-is-what-happens-when-climate-change-forces-an-entire-country-to-seek-higher-ground/?utm_term=.ebbd4acf666d

Kennan, George F. ("X"). 1947. "The sources of Soviet conduct." *Foreign Affairs* 25(4):566–582.

Keohane, Robert O., and Joseph S. Nye. 1977. *Power and Interdependence*. Harper Collins, New York.

Laczko, Frank and Christine Aghazarm eds. *Migration,Environment and Climate Change: Assessing the Evidence*. Geneva: International Organization for Migration. 31

Malin Mobjörk, Malin, Maria-Therese Gustafsson, Hannes Sonnsjö, Sebastian van Baalen, Lisa Maria Dellmuth and Niklas Bremberg. 016. "Climate-related security risks: Towards an integrated approach." Stockholm: Stockholm International Peace Research Institute. Accessed December 24, 2020 at https://www.sipri.org/publications/2016/climate-related-security-risks

Mani, M., Z. Hussein, B. Gopalakrishnan, and D. WadhwaD. 2018. *Paris Climate Agreement and the Global Economy Winners and Losers*. World Bank Group, Washington, DC.

Milman, O. 2015. "UN drops plan to help move climate-change affected people." *The Guardian*, October 7. Accessed August 18, 2016, at https://www.theguardian.com/environment/2015/oct/07/un-drops-plan-to-create-group-to-relocate-climate-change-affected-people

Mohai, P., D. Pellow, and J. Roberts. 2009. "Environmental justice." *Annual Review of Environmental Resources* 34:405–430.

Morgenthau, Hans J. 1978. *Politics Among Nations: The Struggle for Power and Peace*. 5th rev. ed. Knopf, New York.

New Climae Institute. 2020. "Climate Action Tracker." Website accessed December 24, 2020 at https://climateactiontracker.org/

Nuccitelli, D. 2015. "Scientists warned the US president about global warming 50 years ago today." *The Guardian*, November 5. Accessed June 23, 2020, at https://www.theguardian.com/environment/climate-consensus-97-per-cent/2015/nov/05/scientists-warned-the-president-about-global-warming-50-years-ago-today

Oglesby, Brent interview. October 9, 2019. African Risk Conference Addis Ababa, Ethiopia. North Atlantic Treaty Organization Socio-Cultural Analyst for the NATO Strategic Direction—South.

Ostrom, Elinor. 2015. *Governing the Commons: The Evolution of Institutions for Collective Action*. Cambridge University Press, New York.

Quan, D. 2018. "'Great success': Scientists salute Montreal Protocol for shrinking ozone hole." *National Post*, January 8. Accessed January 7, 2019, at https://nationalpost.com/news/canada/great-success-scientists-salute-montreal-protocol-for-shrinking-ozone-hole

Roelfsema, M., van Soest, H.L., Harmsen, M. *et al.* Taking stock of national climate policies to evaluate implementation of the Paris Agreement. *Nature Communication* 11, 2096 (2020). https://doi.org/10.1038/s41467-020-15414-6

Rolnik, R. 2009. "Report of the Special Rapporteur on adequate housing as a component of the right to an adequate standard of living, and on the right to non-discrimination in this context." UN General Assembly, New York. Accessed December 24, 2020 at https://documents-dds-ny.un.org/doc/UNDOC/GEN/N09/446/64/PDF/N0944664.pdf?OpenElement

Safire, W. 2004. "If you break it . . ." *New York Times*, October 17. Accessed November 16, 2018, at https://www.nytimes.com/2004/10/17/magazine/if-you-break-it.html

Singer, Peter. 2016. *One World Now: The Ethics of Globalization*. New Haven: Yale University Press.

Skillington, T. 2017. *Climate Justice and Human Rights*. Palgrave MacMillan, New York.

Stern, T. 2018. *The Paris Agreement and its future*. Brookings, Washington, DC. Retrieved January 14, 2019, at https://www.brookings.edu/research/the-paris-agreement-and-its-future/

UN Framework Convention on Climate Change (UNFCCC). 2015. *Adoption of the Paris Agreement*. Accessed July 25, 2016, at http://unfccc.int/resource/docs/2015/cop21/eng/l09r01.pdf

UN Newsroom. 2016. "New data brings new answers on climate migration." Accessed December 20, 2020, at https://climatechange-theneweconomy.com/new-data-brings-new-answers-on-climate-migration/

Waltz, Kenneth N. 1979. *Theory of International Politics*. McGraw-Hill, New York.

Warner, Koko, Tamer Afifi, Marc Stal, and Olivia Dun. 2009. "Researching environmental change and migration: evaluation of EACH-FOR methodology and application in 23 case studies worldwide," in Laczko, Frank and Christine Aghazarm eds. *Migration,Environment and Climate Change: Assessing the Evidence*. Geneva: International Organization for Migration. 197-243.

Wentz, Jessica, and Michael Burger. 2015. "Designing a climate change displacement coordination facility: Key issues for COP 21." Sabin Center for Environmental Studies, Columbia Law School. Accessed July 28, 2016, at https://web.law.columbia.edu/sites/default/files/microsites/climate-change/unfccc_climate_change_displacement_coordination_facility.pdf

Wu, Y., L. Polvani, and R. Seager. 2013. "The Importance of the Montreal Protocol in protecting Earth's hydroclimate." *Journal of Climate* 26(12):4049–4068.

Young, Oran. 2010. *Institutional Dynamics: Emergent Patterns in International Environmental Governance*. Massachusetts Institute of Technology Press, Cambridge, MA.

Case Study 3.1

FOREST CONSERVATION VERSUS "LIVING OFF THE LAND"

TODD A. EISENSTADT AND FERNANDO HERNANDEZ[1]

REDD+ and the Socio Bosque Program in Ecuador

On April 26th, 2019, the Waorani tribe of Ecuador celebrated a rare victory. In a courtroom in Puyo, a small Ecuadorian village in the western Ecuadorian Amazon, Judge Esperanza Pilar del Araujo sided with the Waorani and ordered Ecuador's Minister of Hydrocarbons, Carlos Perez, to implement proper community consultations to include the rights and traditions of the Waorani when deciding to extract petroleum and minerals from their ancestral lands. The Ministry of Hydrocarbons, in charge of mining and nonrenewable energy, as well as the government of Ecuador have been aggressively pushing for more extractive projects in the Amazon, which has led to some of the world's highest deforestation rates.

The oil project also meant the destruction of the Waorani way of life, setting a precedent for other Amazonian tribes in Ecuador to protect their lands and keep those ecosystems free from extraction and the resulting deforestation and destruction of ecosystems. Most Amazonian tribes are small and live off the land they have occupied for centuries. The Waorani tribe, for instance, is a group of 3,000 people, many of whom hunt and gather their food across some 30,000 hectares constituting the Waorani Ethnic Reserve (an area the size of Maryland and Delaware combined). As hunter-gathers nestled in the confluence of the Ecuadorian Andes and the Amazon rainforest, the rainforest ecosystem and their way of life are intertwined. A small nomadic group related to the Waorani has no permanent address and is referred to as the "Uncontacted Peoples."

The Waorani never felt that their voices were heard. The tribe claimed that the consultation process had been conducted without the presence of the tribe's leaders, who were out hunting when the 2012 government process took place (Riederer 2019). In theory, these processes were inclusive and comprehensive citizen participation programs. Such citizen participation has often been required in nations with large indigenous populations like Ecuador since shortly after the UN-affiliated International Labor Organization (ILO) declared in the early 1990s that indigenous people around the world had rights to FPIC (free, prior, and informed consent) before their communal lands could be used for resource extraction. In practice, however, the Waorani's experience reflected other practices by national governments and oil companies, which often split tribal groups by "dividing and conquering," such as offering shares in the spoils to some while denying or even repressing others. Around Ecuador, prior consultations have sometimes been used by government leaders to prop up some tribal leaders favorable to their positions and to repress others (Eisenstadt & West 2019, 141–145).

Indigenous Communities, Prior Consultation, and Rainforest Preservation

The April 2019 decision by the Pastaza Provincial Court (and appealed by the Government of Ecuador) voided the 2012 consultation process in which the Waorani had given up land rights. Oil exploration would cease until a proper and inclusive consultation process could take place. Prior consultation was legally mandated in the 2008 Ecuadorian constitution. The implementation of such consultations was part of a trend across Latin America, which had become the site of nearly half the world's extractive conflicts by 2015. These conflicts were so severe that in some cases—like the Bagua, Peru, massacre of 2009—they claimed dozens of lives of indigenous peasants who tried to defend their lands against oil drilling and other outsider incursions (Eisenstadt & West 2019, 73).

The Waorani themselves were deeply divided over whether to support oil extraction, which at the very least results in employment for tribal members and a short-term economic boom, or to shun oil extraction and its associated environmental degradation and consequent effects on local hunting, farming, and ecotourism. Divisions within Amazonian tribes as to how to exploit the resources in the lands they control are rooted in this "divide-and-conquer technique," sometimes used by the Ecuadorian government to co-opt resistance to new extraction projects. On the one hand, many Waorani felt like Kemperi, a Waorani elder who said:

> Many companies want to enter, everywhere. But they do not help; they have come to damage the forest. Instead of going hunting, they cut down trees to make paths. Instead of caring for [the forest] they

> destroy. Where the company lives, it smells nasty, the animals hide, and when the rivers rise, the manioc and plantain in the low areas have problems. We respect the environment where we live. We like the tourists because they come and go away. When a company comes, it does not want to leave. Now the company is in the habit of offering many things, it says that it comes to do business, but then it makes itself the owner. Where the company has left its environment, we cannot return. It stays bad. (Kimerling 2013, 99)

On the other hand, many Waorani also accepted the westerner's gifts, as the Waorani needed to subsist. In the words of a Quichwa ally:

> Oil companies threw crackers and tuna from helicopters and took away the Waoranis' autonomy. To live in cities they [the Waorani] need money and they like computers and iPhone 6s . . . The rapid contact [with outsiders] disarticulated them and confused them . . . (Eisenstadt & West 2019, 131)

The Waorani have long debated whether they should cooperate with Western oil drillers for some crumbs from those proceeds or eschew Western offers for their traditional and more sustainable lifestyle. Two outside scholars witnessed the extent of this conflict in a public disagreement between the national Waorani president and vice president over whether the tribe supported oil extraction (Alicia Cahauigia and Moi Enomenga interviews in Eisenstadt & West 2019). While some 60 percent of the Waorani have depended on oil-related jobs for their livelihoods over the last several decades, many Waorani and other indigenous groups complain that they get manual labor jobs clearing the forest, for example, whereas outsiders come in as managers and engineers (Eisenstadt & West 2019, 127–129). Some argue for a conversion of the rainforest economy, and in the following section, we discuss how this might look.

The Amazon's Riches: Environmental Services and Protecting Ecosystems

The Amazon is the lushest forest on earth and by far the world's most important wilderness carbon sink, absorbing about five percent of the world's greenhouse gas emissions (Kaiser 2019). The forest cover removes important CO_2 from the air through photosynthesis and emits oxygen, acting analogously to the human lungs. Burning and deforestation emit the CO_2 that the rainforest has absorbed. Each year, approximately one-sixth of the entire CO_2 pool cycles though the world's forests (Schlesinger & Bernhardt 2013). Intact forests are vital to a functioning carbon cycle.

Spanning 6.7 million km^2 (approximately equivalent to the continental United States), the Amazon Basin contains about 10 percent of the world's biodiversity, including the largest number of freshwater fish in the world, and accounts for about 15 to 16 percent of the world's river-water discharge into oceans. The Amazon contains 90 to 140 billion metric tons of carbon, but deforestation and land conversion release up to 0.5 billion metric tons yearly (World Wildlife Fund 2020). Ecuador contains only about two percent of the Amazon forest, but that constitutes 80 percent of the country's forest (Yale School of Forestry and Environmental Studies 2020).[2] The country is only the size of Arizona but has more biodiversity than the entire United States, and it is the smallest nation in the world that is also one of the 10 most biodiverse (Butler 2016). See Figure 3.1 for more details.

In addition to its unique flora and fauna, as well as an unusually diverse number of ethnic groups, the Ecuadorian Amazon holds oil reserves that are still responsible for over a third of the country's export revenue and much of the national budget's discretionary spending on social programs, although oil revenue declined about six percent with the price drop in 2014. The 2020 COVID-19 pandemic's impact on worldwide oil prices will no doubt have an even greater negative effect on Ecuador's oil revenues. Still, in the early 20th century, President Rafael Correa used oil royalties, which could reach up to 20 percent of proceeds from sale of oil extracted from the Ecuadorian rainforest, to build roads and infrastructure, schools, and clinics. Indeed, poverty decreased from 36.7 percent of the population in 2007 to 21.5 percent in 2017. (World Bank 2018, 38).

While oil production and proven reserves are not within the Top 10 worldwide, the juxtaposition of oil, rainforest diversity, and indigenous communities with autonomous land titles has created tensions. In 2014, oil drilling commenced in Yasuní National Park, a UNESCO Biosphere Reserve and home of nomadic groups of Waorani and other indigenous groups, after a seven-year program to raise billions of dollars to "keep the oil in the soil" failed. In 2016, a community of Shuar people, in the southern Amazon region, were forcedly expelled from their village, Nankintz, by military agents making way for a Chinese-owned oil drilling camp (Eisenstadt & West 2019, 84–85).

In Ecuador, between 1990 and 2018 nearly 2 million hectares of native forests were deforested, representing a 13 percent decrease in the landmass covered by native forests. (Redacción 2019a). Most deforested land is converted into farming land, while some is set aside for mining and oil extraction. Ecuador's share of the Amazon Basin rainforest is about six percent, although it possesses perhaps the greatest biodiversity of all the countries in the Amazon Basin, as it slopes down from the high Andean peaks through forested foothills to the swampy river basins.

Paradoxically, agricultural lands, as a percentage of total landmass, have also decreased dramatically in Ecuador, especially since 2013. It is difficult

Figure 3.1 This map conveys the low number of settlements in the eastern half of Ecuador, where rivers run into Peru (and then Brazil). The majority of the country's 16 million people live on the coast (where Guayaquil has nearly three million people) and in the high Andean valleys anchored by Quito, the nation's capital, with a population of nearly two million. The indigenous population consists of some 14 ethnic groups, although 60 percent of this population are the Quichwa people of the Andean Central Valley. Some 24 percent of the nation's indigenous language speakers live in the Amazon, in 10 ethnic nationalities, and the rest live in coastal areas (International Work Group for Indigenous Affairs 2020).

to disentangle the effects of policies aimed at ecological protection and the effects of transitioning from an agricultural economy to industrialization. Countries have sought to move away from strong regulatory frameworks to protect forests from deforestation and instead moved toward neoliberal, market-based solutions. In the early 1990s, and in part propelled by the Kyoto Protocol on climate change, many countries sought to create an emissions trading systems (see Chapter 7), where companies paid for any emissions over their corporate limit and could offset emissions by purchasing CO_2 sinks, like rainforest preserves. While emissions trading systems found some success in Europe during the 1990s as established in the Kyoto agreement, many other countries, like the United States, never set these up. Policymakers in Ecuador and the other Amazon nations, however, believed that if polluting companies invested in their rainforests, more such lands could be preserved.

Another relevant example of a market-based solution is the Payment for Environmental Services (PES). PES aims to correct the market failures produced by the positive side effects that forests and other natural ecosystems provide not being included in most balance sheets (World Wildlife Fund n.d.). These "positive externalities" (i.e., benefits from saving forests, which are not presently calculated in economic models but should be) include natural resources produced by a lake, oxygen and carbon capture produced by a forest, aesthetic beauty and leisure activities (like camping and hiking) provided by a forest, and so on. Many of these resources do not have an established market price, thus evaluating their value is complicated.

PES is an important methodology for providing market-based solutions to environmental degradation. In general, environmental services can be divided into four broad categories:

1. *Support of services:* These include services necessary to sustain the natural conditions needed to support the provision of environmental services.
2. *Provision of services:* The natural resources extracted from the environmental services. These are commonly referred as extraction resources and include the wood removed from forests, petroleum extracted from wells, water flows used for dams, and so on.
3. *Regulating services:* The capacity of an ecosystem to support the earth's natural process. This includes an ecosystem's role in climate change, desertification, and the like.
4. Cultural services: These include benefits related to aesthetics, human appreciation for a forest, and so forth.

PES seeks to value the social benefits of the natural resources provided. Effective pricing mechanisms allow for accurate valuation of resources and to assess the impacts of, for instance, intense deforestation on otherwise invaluable social benefits. It also provides proper incentive mechanisms for countries to avoid free-riding (i.e., getting the benefits without paying) on climate change mitigation efforts by paying for the provision of the public good.[3] These market-based solutions provide a Coase-like approach[4] to mitigating the externalities produced by intense deforestation.

The Role of REDD+

All that said, the UN, under its Framework Convention on Climate Change (UNFCCC), has developed the Reducing Emissions from Deforestation and Forest Degradation (REDD+) program to build on the experience from PES. Following the "fee for services" logic, REDD+ seeks to provide countries with useful mechanisms to abate deforestation. Specifically, REDD+ seeks to provide clear economic incentives and opportunities to communities that are directly tied to the forests and are invested in protecting these ecosystems. In other words, the program wants to make the preservation of forests an economic priority (i.e., pay heavily forested nations not to deforest).

For more than 10 years, REDD+ has provided technical assistance to governments and communities to maintain and protect ecological systems (Frechette et al. 2016). Specifically, its goal is to strengthen the technical and institutional capacities of developing countries to fight deforestation and mitigate carbon emissions produced by the reduction in native forest lands. Additionally, REDD+ lays the foundation for PES schemes in countries where these have not yet been implemented. REDD+ has also connected developing countries with investments from developed countries; as of 2014, its largest contributor to an international trust fund established to incentivize forest preservation was the government of Norway. The program's governing body is complex, with actors from country governments, civil society, and indigenous people (Frechette et al. 2016).

REDD+ has developed three forms of financing. These funding mechanisms depend in large part on donations by developed country governments. According to Hall (2012), these are:

1. *REDD+ and the National Appropriate Mitigation Actions model:* This provides direct donations from developed country partners within an established abatement roadmap by the National Appropriate Mitigation Actions. This mechanism is appropriate for providing REDD+ startup funds. However, it may not serve as appropriate, independent, unconditional funding for REDD+ programs, since much of the control over funding lies with the donor.
2. *UN COP-mandated funds:* These are centralized funds under control of the UN, which channels funds to REDD-receiving country institutions. This funding mechanism extends existing agreements between countries to

develop an environmental protection framework and give more control to countries implementing REDD+ strategies over how the funds are distributed.

3. *REDD+ Market model:* This highly complex model of transferring financial resources to REDD+ programs requires methods to evaluate the monetary value of the carbon reductions stemming from REDD+ programs. Measuring carbon emissions is complex; thus, this funding mechanisms requires a sophisticated methodology to account for both abatement and the dollar value of the carbon emissions abated. This funding mechanism requires strong national government and local government coordination and support resources that, if implemented, lay the foundations for strong institutional infrastructure to support future REDD+ programs within the country.

REDD+ funding mechanisms have been criticized on two fronts. According to Seiwald (2012), one fault in the funding mechanisms is that they require offsets to stem from net emissions and do not necessarily promote offsets from pollution sources. In other words, REDD+ aims at reducing carbon emissions from deforestation and funds protection efforts, but it does not fund efforts to transition the structures of the economy into greener spaces. This means that REDD+ countries could increase their carbon emissions from dirty energy production, for instance, and increase their REDD+ investment to "offset" their carbon emissions. Accounting for carbon reductions from reduced deforestation relative to increased emission from dirty energy sources is complex and resource intensive. Measuring carbon emissions requires investments in measurement equipment. It also requires setting up accounting mechanisms that would allow regulators to assign carbon impact to emitters.

Another critique of the REDD+ funding structures is that they are not sustainable. A 2014 external evaluation (UN REDD 2014) identified REDD+'s funding through bilateral/multilateral nation investments as being unsustainable in the long run and a threat to the program's sustainability. Loss of funding to countries managing REDD+ programs could lead to the dismantling of the program, as local governments have no capacity to fund payments for ecological services and forest protection. In other words, nations with international REDD+ funding that pays forest dwellers not to deforest might, in the future, lose funding. And when the forest dwellers chop down their trees, the accumulated benefits from preserving trees through that moment are mostly nullified.

REDD+ has had strong support from national governments. The program is considered to be strategically relevant and an important tool in aligning the REDD+ country's environmental protection goals with the overall goals of REDD+ (Frechette et al. 2016). However, there are concerns that as mechanisms to fund REDD+ get more complex, and as oversight becomes more involved, countries may not report carbon offsets from reduced deforestation stemming from REDD+ (Seiwald 2012). Again, these concerns are linked to the market-based financing structures that integrate the complexity of financial markets in trading carbon offsets in conjunction with relying on own-country inventory and reporting of these offsets.

REDD+ in Ecuador: The Socio Bosque Program

A key Ecuadorian REDD+ program, the Socio Bosque program, was launched by the Ecuadorian Ministry of the Environment (Ministerio del Medioambiente Ecuatoriano, or MAE) in 2008 resulting in the protection of nearly 900,000 hectares of native forest by 2011 (FONAFIFO et al. 2012). Many of the benefits of Socio Bosque accrued to indigenous communities, which were able to earn cash income by preserving rainforest areas of their communal lands. Socio Bosque is a nationwide program providing financial incentives to landowners who promise to conserve native forests and avoid extraction and deforestation (Raes 2014).

Through conservation agreements, Socio Bosque paid landowners some 60 dollars annually for each hectare of forest cover protected between 1 and 20 hectares and 30 dollars per hectare of forest cover protected for between 21 and 50 hectares. The sum may not seem large, but given that the minimum wage in Ecuador was about 400 dollars per month in 2020, a 20-hectare holder could make nearly three months in minimum wage merely for not cutting his forest areas. The program also asked landowners to conceive of alternative sustainable management and monitored compliance through satellite imagery. The implementation of Socio Bosque was followed by a large drop in agricultural land-use intensity. The program's objectives included not only forest conservation, but also an increase in wealth transfers to alleviate poverty among indigenous communities in rural Ecuador (Raes 2014).

The institutional governing body of Socio Bosque is the MAE. The MAE collects applications for the Socio Bosque program and processes the documentation requirements. Following an evaluation based on the environmental and poverty impact of each application, the MAE conducts field verification of the information provided, and if the application is successful, an agreement is drafted and monetary incentives are transferred from the Ecuadorian government (funds donated by European donors) to indigenous and other communities.

Socio Bosque has different requirements for participating in the program, depending on the ownership structure of the property. According to the MAE, individuals who own properties must provide, among other identifying documents, a title of ownership for the property, a georeferenced map of the areas to be entered into conservation, and an investment plan for the incentives.

For communal properties and those managed by indigenous communities, legal representatives must present a title of ownership, a georeferenced map of the areas to be entered into conservation, statutes of the community, and the Acts of the Assembly where entry into Socio Bosque was decided by the community.

An application is evaluated based on a three-tier points system. The first tier evaluates the deforestation pressures a community suffers as a function of how difficult it is to access the territory. Generally speaking, the harder it is to reach the territory, the lower the deforestation threat, relative to the capacity of indigenous communities to stave off illegal deforestation and formal deforestation as connectivity increases throughout the country. This last point is important, as communities use Socio Bosque funds to develop communities adjacent to the protected property. There may be incentives to exploit the protected forest once the Socio Bosque funds have been disbursed. The rationale for this threat is the relationship between building up infrastructure and deforestation (Seiwald 2012).

The second tier evaluates the potential ecological services provided by the property in the Socio Bosque application. This evaluation is based on the biodiversity represented in the territory relative to existing national protected areas. The more under-represented the biological systems are in existing national protected areas, the more points the application gets. Another item evaluated is the carbon-sink capacity and the hydraulic regulation of the territory. Assessing the value of carbon captured by a forest is less complex than defining the value of the services provided by the forest as a whole. It may problematic to place prices on different elements of an ecosystem that indigenous communities view as a sacred system, which in turn may result in a negative perception by indigenous communities of the Socio Bosque program (Seiwald 2012).

Finally, a third evaluative tier is the poverty level. This element seeks to satisfy the aim of providing financial transfers to poverty-stricken indigenous communities that may deforest to sustain their economy. This element, however, provides the fewest points within the evaluation (only 3 of the 22 total possible points from each tier).

Critiques of Socio Bosque

Three important critiques of the Socio Bosque program are its financial sustainability in the long term, its capacity for true community engagement in the decision-making process (the FPIC question), and the complexity and length of the contracts that it provides. One example of flawed community engagement is the oil extraction negotiations conducted by the Sápara community of Ecuador and Peru. In 2012, the Sápara community had rejected offers by Chinese petroleum corporations to explore and exploit large swaths of their territories. Despite this generalized rejection of these offers, Sápara leaders supporting extraction bypassed the community's wishes and, in conjunction with the national government, tried to move forward with extraction (Eisenstadt & West 2019, 140).

Socio Bosque has not been successful in fostering a sustainable finance structure. Former program director Max Lascano indicated (Eisenstadt and West 2019) that part of the issue was securing funding from external funders but also, and more importantly, assuring permanent funding from the Ecuadorian government. As budgets reflect importance, providing Socio Bosque with a permanent line item within the national budget would cement the project into a staple of Ecuadorian environmental policy. Socio Bosque has not secured long-range funding, however and eventually was unable to make payments during the 2016–2018 period (Eisenstadt & West 2019, 174). The program has started again, but a lack of sustained payments has reduced incentives for Socio Bosque partners to continue in the program, hurting efforts to reduce deforestation in Ecuador.

Initially, the Socio Bosque program was financed by the Ecuadorian government through the National Environment Fund (Fondo Ambiental Nacional, or FAN). The FAN was supported by donors like the German and Norwegian governments starting in about 2010 (Martin 2015, 135). However, external funding grew tight. A 2016 earthquake and the fall in oil prices generated a shortfall in government financing, and discretionary programs like Socio Bosque suffered. Nearly 45,000 families reported late payments of their incentives as the program shut down for two years (Zambrano 2017).

Socio Bosque, to remain sustainable, requires consistent funding, provided by, for example, green taxes and extraction licenses (Fehse 2012). Although there are economic effects of imposing additional taxes on resource extraction, there is also a general consensus that extraction produces important negative externalities that should be abated. Extractive costs could be diminished through taxes paid by extractors as a function of the intensity of extraction, increasing the marginal cost of extraction and thus making it less attractive to extract and, more importantly, increasing the resources available for deforestation abatement as extraction increases.

This pattern of sidelining communities is prevalent in Ecuador, not only in the extraction of petroleum but also in mining and silviculture. The Ecuadorian government has used promises of development to lure communities into accepting pernicious extraction agreements, at times with the aid of tribal leaders. There is not only the problem with bending the arm of anti-extraction advocates from tribal communities through populist promises, but also the moral hazard of allowing individual representatives to negotiate the terms of extraction, at times to their own benefit. There is some evidence that Socio Bosque has favored more pro-extractivist community leaders, as a means of guaranteeing support for the national government's greatest discretionary revenue source (oil royalties), as well as accusations of kickbacks on behalf of extracting companies to community leaders (Eisenstadt & West 2019, 142).

As the Waorani tribe has shown, it may be costly for extracting companies to bypass communities, as courts continue to set precedents supporting indigenous communities. Perhaps because of the need to price payments for environmental services schemes that underpin their capacity to produce economic benefits, the cultural value of native forests may be quite difficult to quantify. The voices of communities expressing the sociocultural importance of these forests have not been heard in the past, and their political pull may not be as powerful as the national government or investors with access to the government's ear. This is not to say that tribal communities should be ignored, or that they are even unified in what they want.

Recall that the Waorani were divided about whether the benefits of oil drilling in their communities outweighed the costs, at least in the short term. The Waorani experience indicates that the value of native forests as human habitats may often be underestimated, and that courts are ready to recognize this with rulings. The Waorani experience also demonstrates the need for FPIC, that the communities seeking to preserve their forests rather than submit them to extraction or subsistence farming deserve an authentic say These groups, like the Waorani and the Sápara, should not rely on proxy representatives or on co-opted leaders willing to serve as representatives of the revenue-starved state seeking oil royalties.

Contracts are complex and require both complete information and incentives for enforcement that may leave tribal communities in a deterred position relative to the MAE, which issues the incentive contracts for Socio Bosque. What is key for PES systems overall are simplified contracts complemented by systematic review of requirements for participating in the program (FONAFIFO et al. 2012). Socio Bosque provides a simple contracting system. However, the contracts themselves place a heavy burden on landowners and tie up their lands for extensive time periods (Seiwald 2012). Additionally, communities may be exploited by outsider "consultants" who better understand MAE contracts and by the institution issuing the contract, which may exploit this position rather than benefit the community (Fehse 2012). Because of the sometimes-precarious economies of communities entering into the Socio Bosque program, contracts the government proposes to these desperate landowners may in fact be too burdensome for some.

Furthermore, Socio Bosque can unilaterally terminate the contract without sanctions (Seiwald 2012). Considering the small per hectare incentives that landowners receive, it may become unattractive for existing Socio Bosque participants to continue in the program or for new communities and landowners to negotiate new contracts. And Socio Bosque contracts may deter larger landowners and communities altogether (Seiwald 2012).

What Has Socio Bosque Done Right?

The Socio Bosque program has been able to foster strong support from the Ecuadorian government and the MAE. This political momentum has been crucial in sustaining government financing and in providing for a straightforward implementation (Fehse 2012). MAE has set up a separate Socio Bosque corporation that manages the activities of the program, giving it a level of autonomy needed to maintain flexibility.

Socio Bosque not only addresses environmental degradation and climate change, but also directly addresses poverty alleviation. The program explicitly targets low-income landowners and indigenous tribes and makes direct money transfers to these communities for their development. The program also requires that investment decisions for communities follow FPIC consultation procedures. If executed correctly, these procedures help communities decide what to do with Socio Bosque resources (Fehse 2012).

There is also evidence that Socio Bosque has reduced land conflicts in Ecuador. However, such conflicts had been prevalent in communal contracts rather than private or semiprivate land contracts (Jones et al. 2020). This may be tied to the program's requirements to conduct community consultation processes.It is not clear whether the reduction in land conflicts also translated into fewer inter- and intratribal conflicts or conflicts between tribes and oil drillers and the government.

Conclusion

Market-based solutions for mitigating the pernicious effects of deforestation are viable. However, as with most market-based solutions, important issues arise when determining pricing of the environmental services provided. In the context of REDD+ and, more specifically, Socio Bosque, setting up payments for environmental services is complex and nuanced.

Key elements of a successful PES system are the inclusion of communities, transparency in how contracts are negotiated, and accountability mechanisms to avoid the moral hazard prevalent throughout the implementation process. Communities must have their voices heard so as not to be marginalized in the decision-making process. Tribal communities are especially vulnerable to being steamrolled, as their way of life and community structures may provide for slower, more deliberate decision-making. Their cultural and social valuation of the protected lands must also be taken into consideration, as they convey the sensitivity that local and national governments have with hard-to-access, remote communities.

Contracts should be simple. Socio Bosque has partially gotten this solved; however, simple contracts can still be burdensome. To that point, national and local governments should provide ample support in developing and implementing accountability metrics required by the Socio Bosque contracts in order to proceed with incentives.

High penalties for noncompliance may work in traditional contracting but given the characteristics of rural communities and tribal communities, the government has an extra duty to provide support through negotiation and implementation of contracts. All parties must be in full understanding of the conditions.

National and local governments should also support communities in creating governing bodies for decision-making and shy away from influencing the decisions of communities by pressuring individual decision-makers or luring them with attractive side-deals. Mitigating moral hazard is also the responsibility of communities, which must strive to hold their leaders accountable through internal mechanisms and the formal local and national courts. Moral hazard can create long-lasting problems in communities that are disconnected from the decisions that their elected leaders make. Thus, increasing community participation and facilitating spaces for community engagement are key. The national and local governments also have a role to play here by providing physical space for communities to congregate and facilitators for discussions when needed.

Overall, PES has the potential to provide agile solutions to climate change. Like any market-based solution, there needs to be strong oversight and community participation. Socio Bosque has shown that these programs do work. However, future evaluations of the program are needed, and the Ecuadorian government must set up information systems to facilitate the evaluation of this program by external researchers.

EXERCISE

SCENARIO FOR CLASSES

After reading the case study, and carefully considering the advantages and disadvantages of REDD+ programs like Socio Bosque in Ecuador, students should divide into teams of three. Each team will have the following representatives: the Ministry of Hydrocarbons, which is interested in maximizing oil revenue; the MAE, which administers Socio Bosque and seeks to maximize forest conservation; and the Ministry of Culture, which possesses the office attending to indigenous patrimony and culture, which previously had its own ministry). The year is 2025, and international donors are trying to decide whether to give funding to:

1. Help improve the lives of indigenous peoples and others who live in areas directly impacted by extraction, funding clinics, housing, and schools based on national government planning and Government of Ecuador execution.

Continued

2. Bolster Socio Bosque by paying current program recipients to survey their lands to better establish the inventory of how many trees are being conserved and, thus, how much CO_2 is being offset.
3. Give the funds directly to indigenous communities so that they can monitor FPIC processes and also undertake environmental conservation efforts in culturally sustainable ways, tied to local customs and interests as determined by tribal leaders directly.

Starting questions for consideration would include the following:

1. What are the pros and cons of REDD+ at the international level?
2. What about with REDD+ as implemented in Ecuador's Socio Bosque program?
3. Should programs like Socio Bosque primarily benefit environmental conservation or sustainable development?
4. Which should be accorded a higher priority by the Government of Ecuador: garnering discretionary income by maximizing oil income to help the poor, conserving the rainforest and hence mitigating climate change, or making sure the cultures of those who "live off the rainforest" are preserved?

CASE STUDY 3.1 NOTES

1. Fernando Hernandez is candidate for MPP at American University. He currently works on policies focusing on child and family welfare in Washington, DC.
2. Brazil contains 64 percent of the Amazon rainforest, Peru 10 percent, and Bolivia, Colombia, and Venezuela 6 percent each. Ecuador has the sixth-largest proportion of the Amazon rainforest (Yale School of Forestry and Environmental Studies 2020).
3. In this case, the public good is clean air, and one country alone may not want to carry all the burden of developing GHG mitigation. In fact, if one country overly reduces their carbon emissions, another country may decide that it need not mitigate and thus would free ride on the first country's efforts. This may lead the abating country to forego additional abatement efforts as they too may decide to free ride. As with all public goods, payment mechanisms and incentives to avoid free riding are key for sustainable provisions of these goods.
4. The Coase Theorem suggests that externalities can be mitigated by defining the ownership of the externality and providing payment transfers between those involved. Given no transaction costs, Coase's theorem provides a framework for payments to communities due to a particular environmental degradation or, conversely, to avoid such degradation.

CASE STUDY 3.1 REFERENCES

Butler, Rhett A. 2016. "The top 10 most biodiverse countries: What are the world's most biodiverse countries?" Accessed June 19, 2020, at https://news.mongabay.com/2016/05/top-10-biodiverse-countries/

Eisenstadt, T. A., and K. J. West. 2019. *Who Speaks for Nature? Indigenous Movements, Public Opinion, and the Petro-state in Ecuador*. Oxford University Press, New York.

Fehse, Jan. 2012. "Private Conservation Agreements Support Climate Action: Ecuador's Socio Bosque Programme." *Climate and Development Knowledge Network*. Accessed June 23, 2020, at cdkn.org/wp-content/uploads/2012/09/Ecuador-InsideStory__WEB2.pdf

FONAFIFO, CONAFOR and Ministry of Environment. 2012. *Lessons Learned for REDD+ from PES and Conservation Incentive Programs. Examples from Costa Rica, Mexico, and Ecuador.* Accessed December 19, 2020, at https://www.forest-trends.org/publications/lessons-learned-for-redd-from-pes-and-conservation-incentive-programs/

Frechette, A., M. de Bresser, and R. Hofstede. 2014. "External evaluation of the United Nations Collaborative Programme on reducing emissions from deforestation and forest degradation in developing countries (the UN-REDD Programme)." Volume I–Final report. Accessed December 19, 2020, at https://www.unenvironment.org/resources/other-evaluation-reports-documents/external-evaluation-united-nations-collaborative

Hall, A. 2012. *Forests and Climate Change: The Social Dimensions of REDD in Latin America.* Northampton, Massachusetts: Edward Elgar Publishing.

International Work Group for Indigenous Affairs (IWGIA). 2020. "Indigenous Peoples in Ecuador." Accessed June 19, 2020, at https://www.iwgia.org/en/ecuador.html

Jones, K. W., N. Etchart, M. Holland, L. Naughton-Treves, and R. Arriagada. 2020. "The impact of paying for forest conservation on perceived tenure security in Ecuador." *Conservation Letters.* Accessed December 19, 2020, at https://conbio.onlinelibrary.wiley.com/doi/abs/10.1111/conl.12710

Kaiser, Anna Jean. 2019. "AP explains: Role of the Amazon in global climate change." *Associated Press*, August 27. Accessed June 19, 2020, at https://apnews.com/384fdb5ee7654667b53ddb49efce8023

Kimerling, Judith. 2013. "Oil, contact, and conservation in the Amazon: Indigenous Huaroani, Chevron, and Yasuní." *Colorado Journal of International Environmental Law and Policy* 24(1):44–115.

Martin, Pamela L. 2015. "Leaving the oil under the Amazon: The Yasuní-ITT Initiative as a postpetroleum model?" In *Ending the Fossil Fuel Era*, edited by Thomas Princen, Jack P. Manno, and Pamela L. Martin. Massachusetts Institute of Technology Press, Cambridge MA.

Raes, L. 2014. "The Socio Bosque Program for Rainforest and páramo Conservation, Ecuador." *Economics of Ecosystems and Biodiversity (TEEB).*" Accessed on December 19, 2020, at https://www.researchgate.net/publication/270508075_TEEB_case_The_Socio_Bosque_Program_for_rainforest_and_paramo_conservation_Ecuador_Available_at_httpdocteebweborgwp-content uploads201409The-Socio-Bosque-Programme-for-Rainforest-and-Paramo-Conservation

Riederer, Rachel. 2019. "An uncommon victory for an indigenous tribe in the Amazon." *The New Yorker*, May 15. Accessed June 23, 2020, at www.newyorker.com/news/news-desk/an-uncommon-victory-for-an-indigenous-tribe-in-the-amazon

Schlesinger, William H., and Emily S. Bernhardt. 2013. *Biogeochemistry: An Analysis of Global Change.* 3rd ed. Elsevier Academic Press, Amsterdam.

Seiwald, M. 2012. "REDD and indigenous peoples: The Socio Bosque programme in Ecuador. Issues on Climate Change and Development." Saarbrücken, Germany: AV Akademikerverlag.

United Nations Development Program, Latin America and the Caribbean, 2019. "Ecuador receives US$ 18.5 million for having reduced its deforestation.". Accessed June 23, 2020, at www.latinamerica.undp.org/content/rblac/en/home/presscenter/pressreleases/2019/ecuador-receives-us--18-5-million-for-having-reduced-its-defores.html

World Bank. 2018. "Ecuador systematic country diagnostic." Accessed June 19, 2020, at http://documents.worldbank.org/curated/en/835601530818848154/pdf/Ecuador-SCD-final-june-25-06292018.pdf

World Wildlife Fund. 2020. "Inside the Amazon." Accessed June 19, 2020, at https://wwf.panda.org/knowledge_hub/where_we_work/amazon/about_the_amazon/

World Wildlife Fund. n.d.. "Payments for ecosystem services." Accessed June 30, 2020, at wwf.panda.org/knowledge_hub/where_we_work/black_sea_basin/danube_carpathian/our_solutions/green_economy/pes/

Yale School of Forestry and Environmental Studies. 2020. "Ecuador—Forestry governance," Accessed June 19, 2020, at https://globalforestatlas.yale.edu/amazon-forest/forest-governance/forest-governance-ecuador

Zambrano, Ricardo. 2017. "Plan Socio Bosque continuará, pese a pagos con retrasos." April 9. *El Universo*, Accessed December 20, 2020, at https://www.eluniverso.com/vida/2017/04/09/nota/6128902/plan-socio-bosque-continuara-pese-pagos-retrasos

4

The Domestic Politics of Climate Change

Environmentalists Versus Energy Interests and the Fight for Public Attention

Nearly two-thirds of all methane and carbon dioxide (CO_2) emissions between 1854 and 2010, from the Industrial Revolution to the present, can be attributed to 90 entities, known as the "Carbon Majors" (Heede 2014, 229)." ExxonMobil was number two on the list, generating some 3.22 percent of the Carbon Majors' total, but Chevron was number one, with 3.52 percent (see a list of the most notorious "Top 10" in Table 4.1). So why is ExxonMobil getting so much more negative attention?

It seems that ExxonMobil may have deceived shareholders and the public and left a footprint (a large carbon footprint, to be sure) of evidence behind. According to New York State Attorney General Barbara Underwood, who filed a lawsuit in 2018:

> Investors put their money and their trust in Exxon—which assured them of the long-term value of their shares, as the company claimed to be factoring the risk of increasing climate change regulation into its business decisions. . . . Instead, Exxon built a façade to deceive investors into believing that the company was managing the risks of climate change regulation to its business when, in fact, it was intentionally and systematically underestimating or ignoring them, contrary to its public representations. (cited in Milman 2018)

The lawsuit is winding its way through the US legal system, which will no doubt take years, and ExxonMobil, on its company website, "unequivocally rejects allegations that ExxonMobil suppressed climate change research" and insists that "ExxonMobil is a responsible participant in the discussion on climate change—we will continue to research the issue, support energy efficiency, work to reduce emissions, pursue new technologies and advocate for effective policy approaches" (ExxonMobil 2018). The company did agree to join a fossil fuel industry Oil and Gas Climate Initiative in 2018, but that group has been accused by critics of being a "greenwash" (Vaughan, 2018).

In addition, ExxonMobil had actively promoted climate denial for decades, despite knowing of the risks since 1981, and only adopted a more accepting position on climate change in 2007.

Following the Big Tobacco Model? ExxonMobil and the "Selling" of Climate Change Doubt

Using some of the approaches mentioned in Chapter 1 to sow doubt about climate change science, ExxonMobil operatives apparently took a page from the campaign in the 1970s to disqualify mounting evidence of the now-obvious conclusion that cigarette smoking is unhealthy: "Doubt is our product, since it is the best means of competing with the 'body of fact' that exists in the minds of the general public. It is also the means of establishing a controversy" (Brown & Williamson 1969). Along these lines, the ExxonMobil-funded American Petroleum Institute developed a plan in 1998 that helped block ratification of the Kyoto Protocol, which was signed by 192 countries and by US President Bill Clinton. The goal of these lobbyists was to criticize climate change advocates as well as climate change science.

According to the memo by the American Petroleum Institute subcontractor lobbying firm (Union of Concerned Scientists 2007, 44), victory would be achieved in the campaign to discredit climate change when several goals were achieved, including the following:

- "Average citizens 'understand' (recognize) uncertainties in climate science; recognition of uncertainties becomes part of the 'conventional wisdom'"
- "Media coverage reflects balance on climate science and recognition of the validity of viewpoints that challenge the current 'conventional wisdom.'"
- "Those promoting the Kyoto treaty on the basis of extant science appears [sic] to be out of touch with reality."

The fossil fuel-hired lobbyists were not climate scientists who submitted work for peer review, but they did have "the trappings of scientific argumentation—graphs, charts, references, and the like" (Kaiser & Wasserman 2016a), and they worked hard to sow doubts about climate science, creating a debate and then excoriating journalists who did not report a "balance" of both sides.

This campaign in the late 1990s was perhaps the most orchestrated effort by ExxonMobil to obfuscate climate science, but the company had remained mum on the subject for over two decades. In 1977, ExxonMobil Senior Scientist James Black said that "there is general scientific agreement that the most likely manner in which mankind is influencing the global climate is through CO_2 release from the burning of fossil fuels" (Hall 2015). Furthermore, in 1978, Black told the corporation's management that "present thinking holds that man has a time window for 5 to 10 years before the need for hard decisions regarding changes in energy strategies might become critical" (Hall 2015). Several internal company documents in the 1970s and 1980s attested to the existence of climate change, including a 1982 paper by Roger Cohen that, given company researchers had confirmed the scientific consensus on climate change, "our ethical responsibility is to permit the publication of our research in the scientific literature; indeed to do otherwise would be a breach of Exxon's public position and ethical credo on honesty and integrity (Kaiser & Wasserman 2016a)." The company knew better, even while leading the climate denial campaign, and was internally exploring "whether climate change might offer new opportunities for oil exploration and profit (Kaiser & Wasserman 2016b)."

Rex Tillerson, President Trump's first secretary of state in 2017–2018, became ExxonMobil's CEO in 2006 and started shifting the company's position on climate change, possibly fearing that the company's denials might make it susceptible to lawsuits (Kaiser & Wasserman 2016b). Shortly before the 2009 inauguration of President Obama, who championed the fight against climate change from 2009 to 2017, Tillerson announced his company's support for a carbon tax. A carbon tax, which implied paying for expensive measures to mitigate climate change (i.e., reduce emissions) and adapt to inevitable changes, would have taxed consumers of gasoline at the pump, consumers of home electricity, or other energy-requiring utilities. This form of paying for climate change relief contrasts with the creation of an emissions trading system (ETS), whereby companies are given limits on the amount of greenhouse gases (GHGs) they can emit. When they exceed this quota, companies can still emit additional GHGs, but only if they "purchase" additional emissions rights from companies that have used less than their quota or from other "sellers" of emissions rights. (See Chapter 7 for further discussion.)

The point is that very few in the United States supported a carbon tax, in contrast to, say, the United Kingdom, where one was imposed in 2008 as part of that nation's pioneering Climate Change Act calling for important emissions reductions by 2020.[1] In the United States, momentum was gaining for an ETS system (also known as "cap and trade," so called for capping the amount of GHG emissions allowed and then trading to allocate permissions to emit), and critics argue that ExxonMobil advocated for a carbon tax, which they knew would never be implemented, to avoid passage of an ETS, which was possible (Kaiser & Wasserman 2016b).

The US Senate Committee on Environment and Public Works did approve the Lieberman-Warner Climate Security Act late in 2007, but the legislation was not passed by the full Senate. That bill sought to require the majority of US emissions be subject to cap-and-trade provisions similar to those of Europe's ETS. Another measure, introduced by Senators Bingaman and Specter, would have placed a "ceiling" on the price that could be charged for carbon emissions at 12 dollars per metric ton of CO_2. The election of President Obama in 2008 led to several further cap-and-trade initiatives in the US Congress achieving committee consideration. Several proposals to tax CO_2 emissions, as implemented in the United Kingdom in 2013 and a few other nations (e.g., Chile, Ireland, and Sweden), were proposed but not executed. During the 2016 Democratic presidential primaries, progressive candidate Bernie Sanders issued a carbon tax proposal in the US Senate, but it has not received traction (see https://www.carbontax.org/bills/). The panorama in the United States did change in 2020, as all candidates contending in the Democratic primaries to compete against President Trump did feature climate change as a vital part of their platforms. President-elect Biden named climate change as one of the four "great battles of our time," alongside recovering from the COVID-19 pandemic, achieving racial justice, and fostering economic recovery (Doniger and Longstreth, 2020). He named high-level special climate change advisors, including former Obama Secretary of State John Kerry, for the first time ever and mainstreamed climate policy into the core functions of most cabinet positions.

The judicial path to climate action has on some occasions been more effective in the United States than the executive and legislative branches. In 2007, the US Supreme Court ruled 5-4 that the US Environmental Protection Agency (EPA), the strong executive branch agency regulating pollution, must, under the landmark 1970 Clean Air Act, monitor GHG emissions. US President George W. Bush managed to defer implementation, but President Obama was able to implement this court decision, after being further empowered by a later ruling of the US Court of Appeals for the District of Columbia. After these rulings, the US government, under the leadership of President Obama's EPA and the Department of Transportation, implemented gasoline standards as well as standards to reduce carbon emissions from power plants, called the Clean Power Plan. The Clean Power Plan, which would vary emissions reductions by state and introduce statewide ETS, is still being litigated in courts. The Trump administration repealed administrative measures that were part of the plan, which had expected to yield a 32 percent reduction (over 2005 levels) in US GHG emissions by 2030. President Trump declared that the Clean Power Plan sought to kill jobs in coal mining states and that he was going to end the "war on coal" (YaleEnvironment360 2018). While his administration's climate agenda is in its early stages of development, President Biden during the 2020 campaign pledged to return to aggressive emissions-reducing measures, including the introduction of stringent national standards for the generation of clean electricity.

Climate Change, the Tragedy of the Commons, and the Problem of Interest Group Pluralism

Nobel laureate Elinor Ostrom (and predecessors such as Russell Hardin) envisioned solutions to the "Tragedy of the Commons" mentioned in Chapter 1 by changing peoples' incentives. The story she tells is based on the medieval village commons in the United Kingdom, where residents grazed their sheep. The problem Ostrom identified was that these commons (the pasture at the center of the village where everyone's sheep grazed) could readily be overgrazed if people did not limit their sheep's consumption there. But the collective incentive villagers had of keeping the pasture maintained by not overgrazing was offset, and often superseded, by the incentive individual sheep owners had of letting the sheep ruin the commons by overgrazing to the point that no grass was left. The optimistic solution posed by Ostrom, and applied to a range of "public goods" problems, including—famously—fishing in international waters, is that people could be trained to graze but not overgraze by putting rules in place that encouraged people to graze responsibly and to punish them if they overgrazed.

The problem is that around the world, Big Oil and Big Coal seem to "own" the good grazing lands, in the sheep grazing analogy, and even if they don't formally own the lands, they seem to bribe the village mayor—who manages the commons—with legal but corrupt offers to help finance the mayor's next election if s/he "looks the other way" while these private entities degrade the commons and obtain the best resources. Interest group pluralism, the idea that interests should compete in the "marketplace of ideas" (even if large corporations can project their views with ads and social media, while individuals have much diminished projections), favors highly financed "loud" voices over poorly financed "quiet" ones.

Back to Trump's claims that the Obama administration was trying to "kill" coal, Trump was correct that reducing emissions from coal was a high priority of Obama era climate and environmental policy. Indeed, the alleged "revitalization" of the coal industry and Trump's false claim to have created thousands of new coal mining jobs (see Lewis 2018) ran counter to Obama administration policies. For one thing, the renewed emphasis on coal could endanger tens of thousands of people with respiratory problems (YaleEnvironment360 2018). More to the point, the coal industry has obstructed national energy policy in the United States and in other coal-heavy economies like Australia, China, India, and Germany, the last of which is otherwise a world leader in climate change mitigation. Coal does generate about twice as much GHGs as low-grade oil, which generates more GHGs than higher quality crude. On the hierarchy of GHG emissions by fossil fuels, natural gas is the "cleanest" of the fossil fuels, but remember that renewable energy sources (solar, wind, geothermal, and hydro power) emit zero GHGs.

The health liabilities of coal and its climate-adverse effects prompted many in the United States to think we were done with the kind of soot air pollution that during the 19th-century Industrial Revolution, had darkened the US Northwest and Midwest and the United Kingdom, where white moths evolved black wings after the soot had darkened the trees in their habitat for several generations. Not surprisingly, Trump staff had important ties to the coal industry, and his administration received support from the industry, centered in Ohio and Pennsylvania, two of the most important "swing states" among the half dozen that gave Trump his 2016 victory, and in West Virginia, which is less important electorally but is the symbolic heart (or "black lung") of the coal industry. In 2020 Biden won Pennsylvania (losing Ohio and West Virginia, as democrats still tend to do), and some critics argue this may have diminished President Biden's early-in-the-campaign pledge to end fracking, which is also an important industry there.

While the US case perhaps most exemplifies the obstruction by fossil fuel interest groups of climate change action by governments, it is clearly not the only country to have its energy and environmental policies steered by such special interests. A study of some 40 countries demonstrated that their level of ambition to reduce GHGs was driven, to a statistically significant degree, by a few issues: whether the fossil fuel industry receives subsidies (a long-time effect of cozy relations between fossil fuel companies and governments in nations like the United States), a nation's level of emissions, and a country's level of vulnerability to climate change (Eisenstadt and Stevens 2017). The problem for climate change mitigation, of course, is that the climate-vulnerable nations tend to be in the Global South on the "front lines" of climate change, have little industrial development, and thus have low emissions to begin with. This chapter introduces these issues and then profiles several emitting countries (China, the United States, and India), which together account for well over half of the world's emissions (with China and the United States alone accounting for just under half). Next, we survey the components of stronger climate policies and briefly look at how climate vulnerability impacts some of the most vulnerable nations. We conclude with a broader conversation about why some countries have better policies to "fix" climate change than others from a sectoral perspective (considering, e.g., how to reduce emissions in the industry sector worldwide, rather than only considering nation-states as our unit of analysis).

What Causes Some Nations to Have Better Climate Policies?

FOSSIL FUEL SUBSIDIES

The complicity of the US government with ExxonMobil in failing to recognize climate change as a problem requiring a gradual reduction in fossil fuel use over time is not the worst of the sometimes excessively close relationships between Big Oil and government. Indeed, not just in the United States but also in much of the rest of the world, governments actually subsidize coal, oil, and natural gas companies for extracting fossil fuels, as a way of "greasing" economies and creating jobs. But the level of these subsidies is enormous, and perhaps even breathtaking. Estimates exist that energy subsidies worldwide constitute 5.3 trillion dollars per year (6.5 percent of the world's gross domestic product [GDP]), and even more modest projections calculate the amount at 452 billion dollars per year in fossil fuels alone. The International Institute for Sustainable Development has estimated that just eliminating subsidies to fossil fuels would reduce GHG emissions between 11 and 18 percent (Merrill, Brindle, Bassi, and Toft Christensen 2015). Environmentalists, scientists, and academics alike oppose these widespread subsidies of fossil fuels,[2] both because of the pernicious effects of GHG emissions into the environment and because they tend to be socially regressive (Global Subsidies Initiative 2012, Victor 2009), and propose subsidy reforms tailored to nations' specific economic structures (Beaton & Lontoh 2010, Laan & Beaton 2010, Lin & Oyuang 2014).

Several studies find that removing fossil fuel subsidies would bring an important environmental benefit. In a 2011 analysis of non–Organization for Economic Co-operation and Development (OECD) countries, Burniaux & Cheatau (2011) found that removing subsidies would aggregate to a seventh of the effort needed to maintain global warming below 2°C. A 2013 study of OECD countries identified 1200 mechanisms in the world's largest economies that encouraged the production or consumption of fossil fuels (Elgouacem 2020).[3] Even though the G20 (the group of the world's 20 largest economies) have committed to phase out fossil fuels, there has been little progress in this regard (Bast et al. 2015). Why is this? It seems that, as per the ExxonMobil example given above, specialized groups have the resources, organization, and interest to keep these policies in place. Correspondingly, policymakers also obtain political or electoral advantages, acquiring resources for their districts or campaign donations with the goal of staying in power (Baumgartner & Leech 1998, Dur & Bievre 2007, Olson 1984, Victor 2009).

LOW EMISSIONS

As per the Chapter 1 discussion of equity regarding fossil fuel–driven development in the developed and developing countries, the North-South debate, so prominent during the Cold War with regard to geopolitical proclivities toward capitalism or communism, has reemerged with a climate change frame. As stated by Najam (2015):

> Developing countries have consistently contextualized environmental issues as part of the larger complex of North-South concerns, particularly concerns about a ubiquitous international order and their desire to bring about structure change in that order. This has become more poignant in recent years as environmental negotiations on issues such as climate change have become increasingly focused on trade and economic aspects (220–221).

Scholars such as Althor et al. (2016) have pointed out the inequity in levels of emission and "the burden of climate change," and others such as Mbeva and Pauw (2016) have argued that countries need to better differentiate responsibilities. Still others like Gignac and Matthews (2015) argue that a centrally determined carbon budget, specifying emissions allowances to each nation, are really the only fair means of addressing historical inequities.

We addressed this question in Chapter 1 of the book, as one of justice. And here, consider the disconnect between the greatest emitters of CO_2 (Table 4.1) and the world's most vulnerable nations (see Table 4.3). There is almost no overlap, meaning that the creators of the problem are not the same as those who suffer most.

HIGH VULNERABILITY

As discussed extensively in Chapter 2, climate change has dramatically increased the propensity of extreme weather events (EWEs) in some parts of the world. Even more to our point, a 2015 special report on event attribution published in the *Bulletin of the American Meteorological Society* (BAMS) presents a list of events that are attributable to climate change. The findings include that all heat-related events studied by meteorologists that year, without exception, were made more intense or more likely to occur due to human-induced climate change.[4] Equally important is that this association of weather forecasters demonstrated that "attribution science" is now capable of separating the anthropogenic effects from natural drivers (BAMS 2015).

Long held as the purview of meteorology and ecology, not much has been written in the social sciences about vulnerability to climate change as an object of study. Vulnerability assessments and sectoral exposure assessments abound in the natural and social sciences, but little attention has been paid to vulnerability to EWEs as a factor eliciting policy action (rather than just "emergency responses" in times of urgent need). While studies do exist considering the role of governments in responding to such emergencies, theoretical positions linking vulnerability to EWEs and longer-term climate change policy actions are scarce. However, empirical studies that link EWEs to economic losses abound.

For example, Surendran Nair et. al. (2020) find a trend of dramatically increasing economic losses from extreme weather events worldwide based on local climate hazards, socioeconomic exposure and the social vulnerability of communities. Some authors have identified the oil and gas sectors as particularly threatened by EWEs, especially given that much of the sector's infrastructure is located in low-lying areas or offshore (Cruz & Krausmann 2013, 41–42). Others have analyzed the impacts of climate change on the energy sector (Mabey & Mitchell 2010) and in transportation (Savonis et al. 2008). With a focus on the United Kingdom's construction sector, Alshebani

Table 4.1 **Characteristics of the World's Biggest Carbon Emitter Countries**

	Percentage of World Emissions (in 2016)	*GDP per Capita (US$ in 2017)*	*Population Density (people per km² of land area in 2017)*	*Population Disaster Displaced (in 2017)*	*Fragile State Index (in 2019)*	*Fossil Fuel Consumption (in 2014)*	*Renewable Consumption (in 2014)*
China	29.16	8,826.99	147.67	4,473,000	71.1	87.67	12.22
United States	15.26	59,531.66	35.61	1,686,000	38.0	83.09	8.75
European Union	3.26	33,723.27	120.90			71.17	16.22
India	6.98	1,942.10	450.42	1,346,000	74.4	73.58	36.65
Russia	4.70	10,743.10	8.82	5,900	74.7	92.14	3.42
Japan	3.47	38,428.10	347.78	21,000	34.3	94.40	5.63
Brazil	1.4	9,821.41	25.04	71,000	72.0	59.11	41.84
Indonesia	1.44	3,846.86	145.73	365,000	70.4	66.09	37.45
Canada	1.62	45,032.12	4.04	85,000	20.0	73.19	22.02
Mexico	1.34	8,910.33	66.44	195,000	69.7	90.28	9.76
Iran	1.88	5,593.85	49.83	225,000	83.0	99.02	0.94
South Korea	1.71	29,742.84	527.92	4300	33.7	82.40	2.84
Australia	1.14	53,799.94	3.20	31,000	19.7	93.39	9.28
Saudi Arabia	1.82	20,849.29	15.32	100	70.4	99.93	0.01
South Africa	1.34	6.151.08	46.75	15,000	71.1	86.79	16.59

and Wedawatta (2014) find that small and medium-sized businesses lack the strategies to deal with EWE and are therefore less productive and have fewer capabilities to sustain the businesses in time.

In their Nationally Determined Contributions (NDCs), most countries explicitly consider vulnerability as a liability. With data from the NDC tracker (European Commission and European Environment Agency 2020), Table 4.2 shows the percentage of countries, categorized per level of income, that identified different areas of vulnerability in 2018. For example, 87 percent of the low-income nations explicitly point out in their NDC that the agriculture sector is threatened by the effects of climate change. A link exists between a nation's vulnerability and its environmental policies (Eisenstadt, Fiorino and Stevens 2019). Perhaps it is the climate vulnerable nations, rife with EWEs that undermine economic stability and peoples' day-to-day lives, that best prioritize fortifying environmental policies, such as those relating to climate.

What different nations consider to be vulnerable may vary, however, so it is necessary to use a single cross-cutting measure of vulnerability. Unfortunately, the NDCs themselves are not a source that indicates the economic or human costs paid due to recent EWEs. Over 77 percent of the low-income countries and 91 percent of the high-income countries do not report a specific figure that calculates the costs resulting from EWEs. Regarding the projected costs of future climate-related hazards, only 16 percent of the low-income countries include such provisions, while no high-income countries consider it in their NDC (European Commission European Environment Agency 2020 and Pauw et al. 2018).

Better data on the precise relationship between climate change and EWEs is needed, as is data to allow us to fully consider the effects of climate change on peoples' standards of living and regional economies. International insurance claims by EWE-impacted people have generated data used by one think tank to estimate the most vulnerable nations over the last 20 years, using number of fatalities as well as material damage. Those results are presented in Table 4.3.

Table 4.2 **NDC Mentions of Vulnerability by Sector**

Type of Nation	*Agriculture*	*Water*	*Ecosystems*	*Health*	*Coastal Zones*
Low income	87.08%	70.95%	48.41%	64.54%	38.68%
Lower middle income	76.45%	72.58%	52.92%	52.92%	37.27%
Upper middle income	62.29%	56.58%	45.31%	39.60%	28.33%
High income	25.53%	41.15%	29.38%	20.57%	20.57%

Source: Pauw et al. 2016.

Table 4.3 **Germanwatch Global Climate Risk Index (CRI): 10 Countries Most Affected From 1998 to 2017 (Annual Averages)**

	Percentage Losses per Unit GDP	*Death Toll*	*Percentage of World Emissions (in 2016)*	*GDP per Capita (US$ in 2017)*	*Population Density (in 2017)*	*Population Disaster Displaced*	*Fragile State Index (in 2019)*	*Fossil Fuel Consumption (in 2014)*	*Renewable Consumption (Per Capita)*
Puerto Rico*	4.20	150.05	0	N/A (US)	N/A (US)	86,000	N/A (US)	N/A (US)	1.63
Honduras	1.85	302.45	0.03	2,480.13	82.81	8945	77.8	52.48	54.04
Myanmar	0.66	7,048.85	0.07	1,256.66	81.72	351,000	94.3	44.29	66.13
Haiti	2.64	281.30	0.01	765.68	398.45	15,000	99.3	22.02	78.39
Philippines	0.58	867.40	0.37	2,988.95	351.87	2,529,000	83.1	62.43	28.58
Nicaragua	1.01	163.60	0.01	2,221.81	51.67	20,000	78.1	40.70	51.84
Bangladesh	0.64	635.50	0.24	1,516.51	1,265.04	946,000	87.7	73.77	37.63
Pakistan	0.57	512.40	0.54	1,547.85	255.57	1,800	94.2	61.59	46.60
Vietnam	0.51	296.40	0.54	2,342.24	308.13	633,000	66.1	N/A	37.04
Dominica	21.20	3.35	0.00	6,719.34	98.57	35,000	N/A	N/A	8.61

Sources: Death toll and GDP loss data from https://germanwatch.org/sites/germanwatch.org/files/Global%20Climate%20Risk%20Index%202019_2.pdf, 7; for all other data, see Table 3.1.

*While Germanwatch rated Puerto Rico as the most vulnerable country, the organization acknowledged that it is a US protectorate rather than an independent nation.

LEVELS OF DEMOCRACY

If a strong fossil fuel lobby is related to weakening a nation's environmental policies and climate vulnerability is related to helping nations strengthen them, does a nation's level of democracy affect its environmental policies? The theoretical case for linking democracy to better environmental performance generally (including in the climate change space) is based on such issues as the free flow of ideas, the ability of voters to hold leaders accountable for problems affecting health and well-being, and regimes' need to deliver public goods to constituents. These visions claim that authoritarian governments, in contrast, worry about delivering private goods to elites, with less need to satisfy the broader public; lack the flow of and access to information that supports demands for change; involve less accountability; and are weaker in their capacity for technological innovation.

Several studies have considered the effects of democracy on environmental policies, or outputs, but only a few have focused on environmental performance, or outcomes. For example, Payne (1995, 45–46) claims that the free flow of information and the responsiveness and accountability of democratic governments to public opinion promote the formation of environmental advocacy groups, which in turn encourage environmental legislation. Likewise, Bättig and Bernauer (2009) argue that democracies adopt more climate change mitigation measures than nondemocracies, since the median voter demands them as public goods. Using cases from Eastern Europe and Latin America, Hochstetler (2012) argues that democracies do have stronger environmental policies, although she acknowledges that no quantitative studies have established a firm pattern. Esty and Porter (2005) argue that democracies show more effective governance in terms of a professional civil service, stronger legal system, more accountability, and less corruption, which in turn lead to more stringent environmental policies.

Li and Reuveny (2006, 941–943) used a variable consisting of two measures of regime type, one dichotomous (democracy and nondemocracy) and one continuous (to distinguish among degrees of democracy). Contrary to the already-mentioned works, their study did focus on outcomes—that is, on environmental performance measured in such terms as CO_2 and nitrogen oxides (NO_x) emissions. They concluded that a rise in democracy levels reduces environmental degradation, although the results vary across pollutants. Similarly, Gallagher and Thacker (2008) used a "democracy stock" measure for the "accumulation" of democratic institutions over time and gauged its effect on two pollutants' emissions, sulfur dioxide (SO_2) and CO_2. They found that a higher democracy stock can help reduce emissions, but only in the long run. In contrast, Midlarsky (1998) found that democratic countries tend to have higher CO_2 emissions and deforestation rates. Another recent study (Eisenstadt, Fiorino and Stevens 2019) found no relation between democracy and environmental performance. That study found that while quality of governance matters, democracy, per se, does not. It seems that nondemocrats (authoritarians but also, increasingly, "hybrid" rulers who have extensive discretion but rely on popular support to renew their mandates) also want clean environments, as dirty ones might reflect poorly or have adverse consequences on their regimes.

THE CLIMATE LEADERSHIP VACUUM

Climate leadership has come from Europe since the 1990s. Having considered the US case and the reversal of leadership from Obama to Trump (and a new switch from Trump to Biden), and at least mentioned the historic leadership of the European Union, let us address in a bit more detail the positions of the two other largest emitters, China and India. While the European Union launched a common carbon market in the 1990s as part of the world's most comprehensive effort to stem climate change, some individual member nations have been even more progressive. Giddens (2012, 79–82) singles out Sweden as perhaps the vanguard environmental nation that became active in the decades after the 1973 energy crisis in expanding the role of renewable energy sources, adopting carbon taxes, and trying to construct more efficient utility grids and infrastructures. While other European nations, like Denmark and Germany, also led this movement, we should consider the nations that matter most for global emissions such as China, India and the United States.

A strong caveat is in order—namely, that nations' climate policies are strongly correlated, as already mentioned, with their energy production, and that this may be even more true of nations dependent on coal, such as Australia, China, and India (and the United States, where coal is a lower overall component of the energy economy but coal-producing and "fracking" states, including the politically "purple" states of Ohio and Pennsylvania are extremely important). Germany's climate policies are also hindered by the prominence of coal production in its economy. Nuclear energy, considered a "clean" alternative energy source by some but not by others (due to the radioactive waste it creates, which must be stored and deactivated over time), also plays an outsized role in the energy mix of several countries, including France and Japan (even after the deadly 2015 Fukushima reactor explosion), which also seems to constrict the choices available to those nations. The US oil lobby may be the biggest energy interest outside the Middle East and a few other oil-corrupted nations, such as Nigeria, Russia, and Venezuela. In some of these nations, like the United Arab Emirates, Qatar, and Saudi Arabia, monarchs derive their very political legitimacy from the distribution of oil largesse to citizens. Oil-rich nations like Canada, Mexico, and Norway tend to have stronger regulatory systems and/or state oil companies, which must respond directly to government pressures. China and India both suffer the misfortunes of having what little natural resource endowments do exist mostly as coal, which generates by far the worst CO_2 emissions.

China as Unlikely Emerging International Leader in Emissions and in the Green Economy

As the world's largest CO_2 emitter, accounting for 23.43 percent of all emissions in 2017,[5] China has experienced health-choking air pollution in cities like Beijing, Hefei, and Urumqi. The World Health Organization (2016) reports that 40 percent of the seven million people killed by air pollution globally in 2012 lived in China, while Cohen et. al. (2017) ranked ambient particulate matter pollution as a leading contributor to death in that country (and worldwide). Recently, however, China has also emerged as a leader in the Green Economy and in the establishment of a subnational "cap-and-trade" system, even as the country's environmental crises continue, exacerbated by large-scale migrations from the countryside to the cities. This after a period during which China had to reconcile differences between its increasing role on the world stage, where China has lagged, and domestic responsibilities, where the country has undertaken extensive reforms (Conrad 2012). Energy security concerns had grown in the 1990s as the country's economy started to take off, prompting the nation in 2002 to restructure government institutions for energy governance (Held et al. 2013, 38–39). Initially, these policies sought, as in Europe, to improve energy efficiency and increase the share of renewables, but over time, leadership realized it could be on the right side of the international climate discussion. Thus, China adopted tighter restrictions on emissions, even as its negotiators continued to blame the West for GHG generated climate change.

While China was viewed in the past as a laggard in reducing CO_2 emissions, and one of the main culprits (along with the United States and India) in failing for over a decade at the United Nations to negotiate a successor to the Kyoto Protocol (from which China was immune to CO_2 reductions anyway), the nation's leadership has recently assumed a more proactive position regarding the reduction of CO_2 emissions, in part by establishing seven subnational pilot carbon markets and a national ETS, which was launched during the13th Five Year Plan (2016–2020). China also pledged in a bilateral deal with the United States to impose an upper bound to its emissions by 2030 at the latest.[6] Furthermore, it pledged to increase its share of non–fossil fuels to 20 percent of the country's energy mix by 2030. The subnational carbon markets are viewed as a lynchpin in reaching these goals, although the nation has done little to transition from coal-burning plants, which produce roughly twice the CO_2 emissions as oil (Climate Home News 2015).

Given China's unique position as the largest national emitter of GHGs globally, its continued rates of rapid economic growth (as evident in an eight-to-nine percent annual GDP growth rate between 2010 and 2014), and its heavy reliance on fossil fuels (especially coal) to sustain this growth, the political and economic incentives to minimize the stringency of a national emissions trading scheme, comparatively speaking, are immense. This is compounded by factors specific to the Chinese context, including a lack of transparency in political decision-making such that the process of determining pilot scheme emissions caps, the setting of benchmarks, and the development of rules and methodologies remain, in many cases, unclear and uncertain. This context has in turn impeded the development of clear market signals for participants, hampering market liquidity and increasing levels of noncompliance, for which penalties are inconsistent and/or weak. As the world's largest GHG emitter, second-largest economy, and traditionally, a leader of the developing world's advocacy for "development (with pollution) over mitigation," China's decision to implement seven subnational ETS in 2016 demonstrated the country's willingness to accept leadership on the issue. China has also become a pioneer of "green" economic development, viewing renewable technologies like solar energy as an important emerging market where the government can push its socialized economy to be an early mover.

Climate Footprint "Foot-Dragging" in India, the "Conscience" of the Developing World

Throughout much of the 1990s, India's rate of emissions grew dramatically, and negotiators made strong claims that this was a by-product of desperately needed economic development and that climate equity needed to be a foundational pillar of international climate accords such as the Kyoto Protocol. Indeed, during the decades of Jawaharlal Nehru's reign as prime minister from the 1940s to the 1960s, India assumed leadership of the Group of 77, which were nations caught in the middle during the Cold War between Western democratic capitalism, led by the United States and Western Europe, and communist centralized economies, led by the Soviet Union. The legacy of those decades of independence led India to champion the developing countries again when climate change emerged as an international issue in the 1990s.

India was credited with helping craft the language assigning "common but differentiated responsibilities" to nations based on their levels of development, and on insisting that industrial nations possessed "historic responsibilities" for emissions (Atteridge 2013, 56–57). As aptly characterized by Atteridge (2013, 57–58), India's climate diplomacy was centered on a call for equity in emissions cuts between developed and developing nations; a lack of trust in international institutions, based on the notion that industrial nations sought to have nations like India bear the brunt of emissions cuts; concerns about an increased demand for more energy as the middle class in India increased; and the fact that small and medium-sized enterprises dominated energy production, and these firms sought to avoid emissions cuts as they were inefficient and not readily able to change.

In the early 2000s, India tightened energy standards to promote economic development even given India's lack of natural resource endowments. If the nation was to prosper without oil deposits (and with coal), the Indian power sector would have to stretch more out of less. The government continued to press for climate equity and differentiated responsibilities for developing countries, and as part of the demonstration of equity, the prime minister announced in 2007 that India's per capita emissions would never exceed those of the industrialized countries. Some took this public statement as an assertion of a new climate policy in India, which the nation did start to pursue at the international level, as climate policy becoming "a tool for pursuing objectives such as greater international status, enhancement of regional security and protection of national economic interests" (Atteridge 2013, 61). Indeed, India's subtle but steady change was something of a bellwether of successful negotiations. Other member-states at the United Nations reasoned that if India, the world's foremost nationalist climate laggard, could take a forward-looking position, then perhaps negotiations could go forward after all.

The election in 2014 of Narendra Modi, whose ambitions for stronger regional leadership in Asia and a larger worldwide role for India, furthered the country's ambitions of climate leadership. And yet India's Paris Agreement NDC was considered unambitious, and in 2016, India released a national energy policy counting more heavily on expanding coal-sourced energy, even as it also continued ratcheting up the role of renewables. As noted by analysts (Mohan & Wehnert 2019, 280), a "significant mismatch" has formed between India's international climate pledges and its domestic energy plans, "mirroring the broader problem with inconsistency of talk and action in global climate policy" (280). While off-grid renewable energy systems, often powered by photovoltaic solar cells, are increasingly disrupting traditional energy grids powered by coal and oil, coal miners are concentrated in regions—as they are in the United States, Australia, and other coal-producing countries—with traditional political clout and do not want to sacrifice what have been profitable industries.

India—and China—also contends with the world's largest population increases. Rural-to-urban migration seems to increase energy consumption as a new generation of middle-class Indian and Chinese workers want air conditioners and cars. And why shouldn't they? Both of these nations, along with the US behemoth (which still consumes over 20 percent of the world's energy while having 5 percent of the world's population), are the world's largest CO_2 emissions. But while the United States and Europe have already industrialized (sending the original CO_2 emissions into the atmosphere starting about 170 years ago to initiate the mess we are all in), China and India are still in that process. And they want to raise the standards of living of their people, even as we are all called upon to sacrifice.

Reconsidering the Sources of Emissions: Evaluating Emissions by Sector (Rather than Country)

The Top 20 polluter nations emit some 93 percent of GHGs, meaning that it might be fair—depending on how we define "fairness"—to target them for the lion's share of emissions reductions. But what would they need to do in order to reduce emissions? At this point, it may be beneficial to consider emissions reductions through another optic, that of economic sectors. According to Henson (2019, 49), the responsibility—by economic sector—for GHGs, is as follows:

1. Industry, 29 percent
2. Agriculture, forestry, and land use, 20 percent
3. Buildings, 18 percent
4. Transportation, 15 percent
5. Other energy/power, 15 percent
6. Waste and wastewater, 3 percent

Harvey et al. (2018, 14), dividing the emissions "pie" somewhat differently, argued that the energy sector (overall) constitutes 74 percent of emissions, with industrial processes constituting 20 percent. To Harvey et al., the way forward to reducing emissions is straightforward: "Implement policies that reduce emissions in the electricity [energy], industry, transportation, and building sectors in the top 20 emitting countries" (15).

Having explored leading emitter countries, in this section we consider briefly industry, agriculture/land use, energy/power, buildings, and transportation.

INDUSTRY

This category, the largest contributor to GHG emissions, includes emissions from fossil fuel combustion for industrial production and emissions from industrial processes. Harvey et al. (2019, 215) estimate the share of emissions from the industrial sector will be about 36 percent by 2050. Efficiency improvements in industrial production will seemingly require a comprehensive set of policies, including, as stated by Harvey et al. (2019). "education and technical assistance, financing, financial incentives, mandatory equipment standards, energy management system promotion, energy audit and energy manager requirements, and politics to support research and development" (217). More specifically, industrial plants need to be overhauled, especially in developing nations where plants are likely older and less efficient. This is especially GHG threatening in nations where coal-powered plants are prominent, such as middle-income and developing nations like Colombia, India, Indonesia, and Kazakhstan, but also in more developed nations, such as Australia, Russia, Ukraine, South Africa, and the worst coal polluters: China and the United States. Waste heat recovery, the installation of properly sized industrial motors, and the use of high-efficiency compressed air systems, for example, are among the needed upgrades. Harvey et al. (2019, 232) cite China's "Top-10,000 Program" as the best example of such a system, integrated at the national level, as it reduced energy use per unit of GDP in recent years even as China's overall energy use increased dramatically (as did economic growth).

The other way GHGs are generated in industry is through "by-product," non–energy producing means of leaking gases into the atmosphere, such as through livestock manure release, natural gas leaks and petroleum processing discharge, cement production, crops and fertilizer, refrigerants, propellants and electoral insulators, and so on. Many of these processes leak methane, although refrigerants are being phased out through the 2016 Kigali Amendment to the Montreal Protocol, discussed in Chapter 2. Improved monitoring of industrial processes, performance standards, and financing for antiquated, high-polluting processes would seem, as in the case of industrial production of energy, to be the best means of diminishing the carbon footprint of industrial processes. In addition to the Montreal Protocol example, Zimbabwe has initiated a program to create biogas (by decomposing organic material, especially methane) to be used for cooking, lighting, and powering appliances (Harvey et al. 2019, 249).

AGRICULTURE, FORESTRY, AND LAND USE

As discussed in Chapter 2, the conversion of land from one use to another disturbs carbon in soil. Carbon loss from soils occurs when forests are cleared and again when soil is tilled or overturned. Tilling exposes deep organic matter to ambient oxygen and therefore allows it to oxidize and produce CO_2. Additionally, farming itself causes the soil matrix (both mineral and organic components) to wash away at between 10 and 20 times (no till) to 100 times (conventional till) the rate of soil formation. We have to bear in mind this "mechanical" loss will appear as reduced carbon in soils, but that the carbon is not lost as CO_2. Rather, it is washed into streams or groundwater. This is why measuring actual CO_2 flux, not just looking at carbon loss alone, is necessary.

The Intergovernmental Panel on Climate Change (2019) estimates that 23 percent of total anthropogenic GHGs (methane, nitrous oxide, and CO_2), as measured by CO_2 equivalents, is derived from agriculture, forest management, and other land-use activity (AFOLU). In terms of actual gases, 13 percent of all CO_2, 44 percent of methane, and most of the nitrous oxide (82 percent) is from AFOLU. Although methane and nitrous oxide have a much lower concentration in the atmosphere than CO_2, they are both much more potent GHGs.

BUILDINGS

The building sector will likely account for about 26 percent of emissions in 2050, according to Harvey et al. (2019, 199). Means of improving thermal and energy efficiency of buildings are emerging, and building codes and appliance standards must be tightened so that carbon emissions can be diminished.

TRANSPORTATION

This sector is changing quickly with the market introduction and mainstreaming, over the last five years, of electric cars and other technological innovations. According to Harvey et al. (2019, 121–122), the increasing need for freight transportation and increases in auto ownership, however, mean that without additional policies, the transportation sector will continue to be responsible for about 14 percent of cumulative emissions by 2050, but could account for some 7 percent of the reductions needed to meet the target of restricting the global temperature increase to 2°C by the end of this century. This would be achieved through vehicle performance standards, fuel fees and "feebates" (i.e., more sophisticated variants on carbon taxes on fossil fuels), the promotion of electric vehicles, and encouraging investment in low-carbon cities that diminish demand for fossil fuel–based transportation options.

Vehicle performance standards, both to improve energy efficiency and to reduce emissions, have proven successful in China, Japan, and the United States. The best such programs seem to rachet up fuel efficiency requirements over time. Fuel fees are taxes on vehicle fuels, which could also play an important role in reducing emissions, as could complementary feebates or rebates to purchasers of efficient vehicles. Fuel fees have proven to be unpopular in many countries, as they amount to a sales tax on gasoline, although a small federal tax – and some state taxes - does exist in the United States (where pressure exists to raise it considerably, even though it was last raised—and only slightly—in 1993). France initiated a popular feebate auto sale subsidy in 2008, which handily rewards purchasers of fuel-efficient cars. Policies promoting use of electric vehicles include the institution of an infrastructure of charging stations (starting to appear in major urban areas), sales incentives, and public education on how these vehicles save money on fuel and can be powered carbon free.

Another, more nebulous—but critical—policy strategy, referred to as "urban mobility," considers better design of more compact urban cores, improvement of public transportation, and encouragement through planning of more walkable, and bikeable urban areas. Large new electric vehicle public transit, such as the Trans Milenio system in Bogotá, Columbia, and the rapid transit system in Guangzhou, China, are diminishing traffic and emissions, and Copenhagen, Denmark has encouraged mass bicycle traffic through the construction of special bike paths, where some 45 percent of residents commute to work and school (Harvey et al. 2019, 196). In Chapter 8, we address national and subnational carbon pricing policies, research and development to achieve technological solutions, and other policies for decarbonization.

ENERGY/POWER

According to Harvey et al. (2019), the power sector will grow to some 30 percent of GHG emissions by 2050. Those authors recommend increasing renewable use by requiring renewable portfolio standards or feed-in tariffs, along with encouraging utilities to pursue cleaner options. Renewable portfolio standards would require load-serving entities—like utilities—to procure a percentage of their power generation from renewable sources by a certain date. Compliance with the standards would be assessed by requiring the load-serving entities to hold credits for units of renewable energy generated or supplied to them through the grid. Feed-in tariffs would "fix" prices for each unit of renewable energy, often in a manner that makes these more competitive with still-cheaper fossil fuel technologies. Harvey et al. estimate that such energy/power sector policies could contribute 21 percent of emissions reductions to meet the United Nations Framework Convention on Climate Change (UNFCCC) target of restricting the rise in average temperature to 2°C by the end of the century.

In addition to subsidizing renewable energy "sourcing" and mandating renewable supply as a portion of total energy output by utilities, Harvey et al. (2019) propose a set of emissions-reducing policies that governments can undertake. These include promoting grid flexibility to diminish costs of "ramping up" and "ramping down" as usage levels change dramatically in the course of a day, and integrating clean energy (renewables and nuclear energy) prominently into the mix. Additionally, utilities and other power generators could be organized into performance-based regulation (rather than the "cost of service" model, which allows utilities to earn returns on capital investments and pass operational expenses on to customers) that could grant premiums to utilities for reducing emissions to regulator-established goals. Adding competition to the old "command and control" monopoly utility providers would also encourage utilities to explore increasingly cheap renewable options, such as solar energy and wind, especially if renewable portfolio standards and/or feed-in tariffs exist to make renewables relatively cheaper. Governments could also serve a role by "retiring" coal power plants, which have paid the sunk costs of construction but are expensive in terms of marginal cost when their GHG emissions are also considered. If companies were allowed to sell old assets they had not yet depreciated on their record books to bond-holders, they could invest the capital in cleaner power-generation techniques.

Conclusions: How Can Domestic Energy Interests Be Tamed?

The term "polycentrism" arose in the literature on environmental policy and politics and is attributed to Elinor Ostrom (2010) and others. To Brown and Sovacool (2011), "Polycentrism posits that when multiple actors at a variety of scales [from local to global] must compete in overlapping areas, they can often promote innovation as well as cooperation and citizen involvement" (238). To these analysts, policies need global, central, homogenizing elements but also local, decentralized, and heterogeneous components. Political scientists might simply view polycentrism as pluralist interest articulation in a porous policy environment, where pluralism, in the Dahl (1961) tradition, is the form of interest articulation where every group is free to state its position and contest policies in the intellectual marketplace.

APPENDIX A:

Explanatory Note on Sources for Tables 4.1 and 4.2

Percent of world emissions: "Each country's share of global carbon dioxide (CO_2) has been calculated by Our World in Data based on annual national emissions data published by the Global Carbon Project (GCP). This is calculated as each country's share of the sum of all country emissions. This is calculated by dividing each country's emissions divided by the sum of all countries' emissions in a given year. This does not include international aviation and shipping ('bunkers') and 'statistical differences'." Source is https://ourworldindata.org/co2-and-other-greenhouse-gas-emissions

GDP per Capita: GDP per capita is gross domestic product divided by midyear population. GDP is the sum of gross value added by all resident producers in the economy plus any product taxes and minus any subsidies not included in the value of the products. It is calculated without making deductions for depreciation of fabricated assets or for depletion and degradation of natural resources. Data are in current US dollars. Source is https://data.worldbank.org/indicator/ny.gdp.pcap.cd

Continued

Population Density (people per skm^2 of land area): Population density is midyear population divided by land area in square kilometers. Population is based on the de facto definition of population, which counts all residents regardless of legal status or citizenship except for refugees not permanently settled in the country of asylum, who are generally considered part of the population of their country of origin. Land area is a country's total area, excluding area under inland water bodies, national claims to continental shelf, and exclusive economic zones. In most cases, the definition of inland water bodies includes major rivers and lakes. Source is https://data.worldbank.org/indicator/en.pop.dnst

Population Disaster Displaced 2017: This is the Internal Displacement Monitoring Centre estimate of the number of persons newly displaced from their homes by the threat or impact of a disaster. The figure for Honduras is the average of 2016 (890) and 2018 (17,000), since there is no data for Honduras for 2017. Source is http://www.internal-displacement.org/database/displacement-data

Fragile State Index: The twelve indicators, upon which the Fragile States Index is based, cover a wide range of state failure risk elements, such as extensive corruption and criminal behavior, inability to collect taxes or otherwise draw on citizen support, large-scale involuntary dislocation of the population, sharp economic decline, group-based inequality, institutionalized persecution or discrimination, severe demographic pressures, brain drain, and environmental decay. States can fail at varying rates through explosion, implosion, erosion, or invasion over different time periods. Sources are https://fragilestatesindex.org/indicators/ and https://fragilestatesindex.org/data/

Fossil Fuel Consumption 2014: Fossil fuel comprises coal, oil, petroleum, and natural gas products. There is no fossil fuel per capita. What the World Bank reports is overall/total energy use per capita. For fossil fuel, what exists is the percentage of the total, and this is the same for renewables consumption as well. Data here are fossil fuel energy consumption (percentage of total). Source is https://data.worldbank.org/indicator/eg.use.comm.fo.zs

Renewable Consumption 2014: Renewable energy consumption (percentage of total final energy consumption). Renewable energy consumption is the share of energy from renewables energy in total final energy consumption. Source is https://data.worldbank.org/indicator/eg.fec.rnew.zs

NOTES

1. The United Kingdom's exit from the European Union and the election of conservative Prime Minister Boris Johnson in 2019 have cast doubts on these policies, although in 2020, Johnson did give a speech favorable to policies addressing climate change.
2. A subsidy is the "difference between the price facing consumers or producers and a specified 'optimal' benchmark price" (Piotrowski et al. 2010, 4). In general, government subsidies lower the cost of producing or consuming a good or service. A fossil fuel subsidy works by lowering the cost of fossil fuel energy production, raising the price received by producers, or lowering the price paid by consumers.
3. These policies include, for example, transfers of funds to producers or certain tax exemptions, among others.
4. This effect was clearly discernible for events that were strongly influenced by the El Niño phenomenon in 2015.
5. By comparison, the second-largest CO_2 producer, the United States, produced 14.69 percent of the world's emissions in 2017, according to the World Resources Institute data accessed January 5, 2021 at: https://www.climatewatchdata.org/ghg-emissions?chartType=percentage&regions=WORLD%2CUSA&sectors=614
6. The US target is to reduce emissions by 26 to 28 percent below 2005 levels by 2025. See the White House Brief "U.S.-China Joint Announcement on Climate Change and Clean Energy Cooperation." Accessed January 5, 2021, at https://obamawhitehouse.archives.gov/the-press-office/2014/11/11/fact-sheet-us-china-joint-announcement-climate-change-and-clean-energy-c

REFERENCES

Althor, G., Watson, J. & Fuller, R. 2016. Global mismatch between greenhouse gas emissions and the burden of climate change. *Scientific Reports* 6, 20281. https://doi.org/10.1038/srep202816.

Alshebani, Mohammed N. and Gayan Wedawatta, 2014. "Making the Construction Industry Resilient to Extreme Weather: Lessons from Construction in Hot Weather Conditions," in *Procedia Economics and Finance* 18: 635-642. https://doi.org/10.1016/S2212-5671(14)00985-X

Atteridge, Aaron. 2013. "The Evolution of climate policy in India: Poverty and global ambition in tension." In *Climate Governance in the Developing World,* edited by Held, David, Charles Roger, and Eva-Maria Nag. Polity Press, New York.

Bas Elizabeth, Alex Doukas, Sam Pickard, Laurie, van der Burg and Shelagh Whitley. 2015. "Empty promises: G20 subsidies to oil, gas and coal production." Overseas Development Institute Working Paper. Accessed December 20, 2020 at https://www.odi.org/publications/10058-empty-promises-g20-subsidies-oil-gas-and-coal-production

Bättig, Michèle B., and Thomas Bernauer. 2009. "National Institutions and Global Public Goods: Are Democracies More Cooperative in Climate Change Policy?" *International Organization* 63, 2: 281-308. Accessed December 23, 2020 at http://www.jstor.org/stable/40345935.

Baumgartner, Frank R. and Beth L. Leech. 1998 *Basic Interests: The Importance of Groups in Politics and in Political Science.* Princeton: Princeton University Press.

Beaton, Christopher and Lucky Lontoh. 2010. "Lessons Learned from Indonesia's Attempts to Reform Fossil-Fuel Subsidies," Report by International Institute for Sustainable Development. Accessed December 20, 2020 at https://www.iisd.org/publications/lessons-learned-indonesias-attempts-reform-fossil-fuel-subsidies

Brown, Marilyn A., and Benjamin K. Sovacool. 2011. *Climate Change and Global Energy Security: Technology and Policy Options.* Massachusetts Institute of Technology Press, Cambridge, MA.

Brown and Williamson. 1969. "Smoking and health proposal." Accessed December 20, 2020, at https://www.sourcewatch.org/index.php/Smoking_and_Health_Proposal

Burniaux, Jean-Marc and Jean Chateau. 2011. "Mitigation potential of removing fossil fuel subsidies, a general equilibrium assessment," The Organisation for Economic Co-operation and Development (OECD) Working Paper. Accessed December 20, 2020 at www.oecd-ilibrary.

org/economics/mitigation-potential-of-removing-fossil-fuel-subsidies_5kgdx1jr2plp-en;jsessionid=1blm941lppm8n.delta

Climate Home News. 2015. "Burnt data: China coal use since 2000 far higher than estimated," accessed December 23, 2020 at https://www.climatechangenews.com/2015/09/17/burnt-data-china-coal-use-since-2000-far-higher-than-estimated/

Cohen, Aaron J, Michael Brauer, Richard Burnett, H. Ross Anderson, Joseph Frostad, Kara Estep, Kalpana Balakrishnan, Bert Brunekreef, Lalit Dandona,Rakhi Dandona, Valery Feigin, Greg Freedman, Bryan Hubbell, Amelia Jobling, Haidong Kan, Luke Knibbs, Yang Liu, Randall Martin, Lidia Morawska, C Arden Pope III, Hwashin Shin, Kurt Straif, Gavin Shaddick, Matthew Thomas, Rita van Dingenen, Aaron van Donkelaar, Theo Vos, Christopher J L Murray and Mohammad H Forouzanfar. 2017. "Estimates and 25-year trends of the global burden of disease attributable to ambient air pollution: an analysis of data from the Global Burden of Diseases Study 2015," in *The Lancet* 389, 10082: 1907–1918 (May 13). DOI:https://doi.org/10.1016/S0140-6736(17)30505-6

Conrad, Bjorn. 2012. "China in Copenhagen: Reconcling the "Beijing Climate Revolution" and the "Copenhagen Climate Obstinancy." *China Quarterly* 210:435–455. DOI: 10.1017/S0305741012000458

Cruz, Ana & Krausmann, Elisabeth. 2013. "Vulnerability of the oil and gas sector to climate change and extreme weather events," *Climatic Change.* 121. 10.1007/s10584-013-0891-4.

Dahl, Robert A. 2005. 2nd ed. *Who Governs? Democracy and Power in an American City*. New Haven: Yale University Press. .

Dür Andreas and Dirk De Bièvre. 2007. "The Question of Interest Group Influence," in *Journal of Public Policy* 27:1; 1-12.

Eisenstadt, Todd and Daniela Stevens. 2017. "Assessing the Stringency of Nationally Determined Contributions after Paris 2015: Building Blocks or Stumbling Blocks in Global Efforts to Combat Climate Change?" presented at the International Studies Association Annual Meeting, Baltimore, February 23, 2017.

Eisenstadt, Todd A., Daniel Fiorino and Daniela Stevens. 2019. "National Environmental Policies as Shelter from the Storm: Specifying the Relationship Between Extreme Weather Vulnerability and National Environmental Performance" (with Dan Fiorino and Daniela Stevens), in *Journal of Environmental Studies and Sciences* 9:1 (March: 96–107). DOI: 10.1007/s13412-018-0523-4

Elgouacem, A. 2020. "Designing fossil fuel subsidy reforms in OECD and G20 countries: A robust sequential approach methodology," *OECD Environment Working Papers*, No. 168, OECD Publishing, Paris, https://doi.org/10.1787/d888f461-en.

Esty, Daniel., and Michael Porter. 2005. "National environmental performance: An empirical analysis of policy results and determinants." *Environment and Development Economics, 10*(4), 381–389. doi:10.1017/S1355770X05002263

European Commission and European Environment Agency. 2020. "NDC Explorer" accessed December 23, 2020 at https://klimalog.die-gdi.de/ndc/#NDCExplorer/worldMap?NDC??income???catIncome

ExxonMobil. 2018. "ExxonMobil's perspectives on climate change: Our climate science history." Accessed October 29, 2018, at https://corporate.exxonmobil.com/en/current-issues/climate-policy/climate-perspectives/our-climate-science-history

Gallagher, Kevin P. and Strom C. Thacker. 2008. "Democracy, Income, and Environmental Quality," Political Economy Research Institute Working Paper, University of Massachusetts, Amherst. Accessed December 23, 2020 at https://scholarworks.umass.edu/cgi/viewcontent.cgi?article=1135&context=peri_workingpapers

Giddens, Anthony. 2012. *The Politics of Climate Change*. 2nd ed. Polity Press, New York.

Gignac, Renaud and H Damon Matthews. 2015 *Environmental Research Letters* 10 075004. Accessed December 21, 2020 at https://iopscience.iop.org/article/10.1088/1748-9326/10/7/075004

Global Subsidies Initiative of International Institute for Development Studies. 2012. Reforming-Fossil-Fuel Subsidies to Reduce Waste and Limit CO2 Emissions while Protecting the Poor." Report by International Institute for Sustainable Development. Accessed December 20, 2020 at https://www.iisd.org/gsi/sites/default/files/ffs_apec.pdf

Hall, Shannon. 2015. "Exxon knew about climate change almost 40 years ago." *Scientific American* October 26. Accessed October 29, 2018, at https://www.scientificamerican.com/article/exxon-knew-about-climate-change-almost-40-years-ago/

Harvey, Hal, Robbie Orvis, and Jeffrey Rissman. 2019. *Designing Climate Solutions: A Policy Guide for Low-Carbon Energy*. Washington, DC: Island Press.

Heede, Richard. 2014. "Tracing anthropogenic carbon dioxide and methane emissions to fossil fuel and cement producers, 1854–2010." *Climatic Change* 122:229–241.

Held, David, Charles Roger, and Eva-Maria Nag. 2013. "A Green Revolution: China's Governance of Energy and Climate Change," in Held, David, Charles Roger, and Eva-Maria Nag, eds.. *Climate Governance in the Developing World*. Malden, MA: Polity Press. 29–52.

Henson, Robert. 2019. *The Thinking Person's Guide to Climate Change*. Boston: American Meteorological Society.

Herring, Stephanie C., Martin P. Hoerling, James P. Kossin, Thomas C. Peterson, and Peter A. Stott. 2015. "Explaining Extreme Events of 2014 from a Climate Perspective." *Bulletin of the American Meteorological Society* 96(12): S1–S72. DOI: https://doi.org/10.1175/BAMS-ExplainingExtremeEvents2014.1

Hochstetler, Kathryn. 2012. "Democracy and the Environment in Latin America and Eastern Europe," in Steinberg, Paul F. and Stacy D. Vandeveer, eds. *Comparative Environmental Politics: Theory, Practice, and Prospects*. Cambridge, MA: Massachusetts Institute of Technology Press. 199–230.

Intergovernmental Panel on Climate Change. 2019. "Summary for policymakers." In *Climate Change and Land: an IPCC Special Report on Climate Change, Desertification, Land Degradation, Sustainable Land Management, Food Security, and Greenhouse Gas Fluxes in Terrestrial Ecosystems,* edited by P. R. Shukla, J. Skea, E. Calvo Buendia, V. Masson-Delmotte, H.-O. Pörtner, D. C. Roberts, P. Zhai, R. Slade, S. Connors, R. van Diemen, M. Ferrat, E. Haughey, S. Luz, S. Neogi, M. Pathak, J. Petzold, J. Portugal Pereira, P. Vyas, E. Huntley, K. Kissick, M. Belkacemi, and J. Malley. Accessed December 24, 2020 at https://www.ipcc.ch/srccl/

Kaiser, David, and Lee Wasserman. 2016a. "The Rockefeller Family Fund vs. Exxon." *New York Review of Books*, December 8. Accessed October 29, 2018, at https://www.scientificamerican.com/article/exxon-knew-about-climate-change-almost-40-years-ago/

Kaiser, David, and Lee Wasserman. 2016b. "The Rockefeller Family Fund takes on ExxonMobil." *New York Review of Books*, December 22. Accessed October 29, 2018, at https://www.nybooks.com/articles/2016/12/22/rockefeller-family-fund-takes-on-exxon-mobil/

Laan, Tara, Christopher Beaton, and Bewrtille Presta. 2010. "Strategies for Reforming Fossil-Fuel Subsidies: Practical lessons from Ghana, France and Senegal," Report by International Institute for Sustainable Development. Accessed December 20, 2020 at https://www.iisd.org/publications/strategies-reforming-fossil-fuel-subsidies-practical-lessons-ghana-france-and-senegal

Lewis, Nicole. 2018. "Fact-checking the Trump administration's claims on 'saving' coal country." *The Washington Post*, January 18. Accessed November 1, 2018, at https://www.washingtonpost.com/news/fact-checker/wp/2018/01/16/fact-checking-the-trump-administrations-claims-on-saving-coal-country/?utm_term=.d4ad30b656e5

Li, Quan and Rafael Reuveny. 2006. "Democracy and Environmental Degradation," *International Studies Quarterly*. 50, 4: 935–956, https://doi.org/10.1111/j.1468-2478.2006.00432.x

Lin, Boqiang and Xiaoling Ouyang. 2014. "A revisit of fossil-fuel subsidies in China: Challenges and opportunities for energy price reform," in *Energy Conversion and Management* 82:124–134. DOI: 10.1016/j.enconman.2014.03.030

Mabey, Nick and John Mitchell. 2010. "Investing for an Uncertain Future: Priorities for UK Energy and Climate Security," Chatam House Briefing Paper 2010/4. Accessed December 23, 2020 at https://www.chathamhouse.org/sites/default/files/public/Research/Europe/bp0710_mitchell mabey.pdf

Mbeva, Kennedy Liti and Pieter Pauw. 2016. "Self-Differentiation of Countries' Responsibilities: Addressing Climate Change through Intended Nationally Determined Contributions." German Development Institute Discussion Paper 4/16. Accessed December 21, 2020 at https://www.die-gdi.de/uploads/media/DP_4.2016.pdf

Merrill, Laura, Richard Bridle, Andrea M. Bassi, Lasse Toft Christensen. 2015. "Tackling Fossil Fuel Subsidies and Climate Change: Levelling the energy playing field," Report by International Institute for Sustainable Development. Accessed December 20, 2020 at https://www.iisd.org/publications/tackling-fossil-fuel-subsidies-and-climate-change-levelling-energy-playing-field

Midlarsky, Manus I. 1998. "Democracy and the Environment: An Empirical Assessment." *Journal of Peace Research*, vol. 35, no. 3: 341–361. *JSTOR*, www.jstor.org/stable/424940.

Milman, Oliver. 2018. "A new climate change lawsuit takes aim at Exxonmobil." *Wired*, October 24. Accessed October 29, 2018, at https://www.wired.com/story/a-new-climate-change-lawsuit-takes-aim-at-exxonmobil/

Mohna, Aniruddh, and Timon Wehnert. 2019. "Is India pulling its weight? India's nationally determined contribution and future energy plans in global climate policy." *Climate Policy* 19(3):275–282.

Najam, Adil. 2015. "The view from the South: developing countries in global environmental politics." In *The Global Environment: Institutions, Law, and Policy,* edited by Regina Axelrod and Stacy VanDeveer. CQ Press, Los Angeles.

Olson, Mancur. 1984. *The Rise and Decline of Nations: Economic Growth, Stagflation, and Social Rigidities*. New Haven: Yale University Press.

Ostrom, Elinor. 2010. "Polycentric systems for coping with collective action and global environmental change." *Global Environmental Change* 20(4):550–557.

Pauw, W.P., Klein, Richard J. T. Klein, Kennedy Mbeva, Adis Dzebo, Davide Cassanmagnago and Anna Rudloff 2018. Beyond headline mitigation numbers: we need more transparent and comparable NDCs to achieve the Paris Agreement on climate change. *Climatic Change* 147, 23–29. https://doi.org/10.1007/s10584-017-2122-x

Payne, Rodger A., 1995. "Freedom and the Environment," *Journal of Democracy* 6(3): 41–55.

Savonis Michael J., Virginia R. Burkett, Joanne R. Potter. 2008. "Gulf coast study: Phase I—Impacts of climate change and variability on transportation systems and infrastructure." United States Climate Change Science Program Working Paper. Accessed December 23, 2020 at http://www.iooc.us/wp-content/uploads/2010/09/Impacts-of-Climate-Change-and-Variability-on-Transportation-Systems-and-Infrastructure-Gulf-Coast-Study-Phase-I.pdf

Union of Concerned Scientists. 2007. "Smoke, mirrors & hot air: How ExxonMobil uses Big Tobacco's tactics to manufacture uncertainty on climate science." Accessed October 29, 2018, at https://www.ucsusa.org/sites/default/files/legacy/assets/documents/global_warming/exxon_report.pdf

Vaughan, Adam. 2018. "ExxonMobil agrees to join oil and gas climate change alliance," in *The Guardian*. Accessed December 20, 2020 at https://www.theguardian.com/business/2018/sep/20/exxonmobil-joins-oil-gas-climate-change-alliance-global-warming

Victor, David. 2009. "The Politics of Fossil Fuel Subsidies." Report by International Institute for Sustainable Development. Accessed December 20, 2020 at https://www.iisd.org/gsi/sites/default/files/politics_ffs.pdf

World Health Organization. 2016. *Ambient Air Pollution: A Global Assessment of Exposure and Burden of Disease*. Geneva: World Health Organization. Accessed December 23, 2020 at https://apps.who.int/iris/bitstream/handle/10665/250141/9789241511353-eng.%20pdf?sequence=1

YaleEnvironment360. 2018. "Trump administration rolls back the Clean Power Plan." Accessed November 1, 2018, at https://e360.yale.edu/digest/the-trump-administration-rolls-back-the-clean-power-plan

Case Study 4.1

CAN STATE ACTION OVERCOME NATIONAL INACTION ON THE EMISSIONS-REDUCING CLEAN POWER PLAN?

LAUREL WOLF[1]

The debate was sharp over a hypothetical lunch at the American Governors' Association as they grappled with a huge shift in climate policies. Even friends at this made-up affair (to provide you with a role-playing opportunity, as elaborated later) like Democratic Governors Andrew Cuomo (of New York) and Steve Bullock (of Montana) had bickered about two major developments in the first half of 2017: the US plans to leave the Paris Climate Agreement and the Trump administration's repeal of the Clean Power Plan (CPP). Although they agreed that climate change was a pressing issue, each had their own ideas about exactly how much action should be taken. Others, like Governors Mary Fallin of Oklahoma and Kay Ivey of Alabama, had always opposed climate action and were excited to not have to worry about the climate commitments anymore. You will be asked to debate the impacts of national climate policy shifts on your state, but bear with us for a few pages as we explain the national shift in climate policy from Obama to Trump and then summarize your state's position.

Passed in 2015 by President Obama, the CPP sought to regulate greenhouse gas (GHG emissions from power plants, with each state designing its own plan for meeting emissions-reduction targets). With the CPP being challenged in the courts, a politically divided Congress, and a president whose stated goal was to "end the war on coal" (Volcovici et al. 2017), there seemed to be no policy effort to reduce US carbon emissions. However, the CPP did require states to submit their own plans for their part in achieving emissions reductions, which many had already begun or almost finished creating when the order was scuttled. These state-determined plans and the politics behind them can help us understand what paths states may take to effect large-scale change both locally and nationally.

Could Governors Cuomo and Bullock agree on policies to replace the CPP in their respective states, or cooperate to achieve even better results? Could either of them convince Governors Fallin and Ivey to take any action at all? To understand the differences between states on climate action, particularly under the CPP, this case study examines why the Obama administration enacted this rule, what the rule entailed, and what outcomes were expected after the Trump administration replaced the CPP rule with a weaker one called the Affordable Clean Energy (ACE) rule. The Biden administration will bring

International Context—What Reductions Need to Be Made?

Climate change is an international problem, which is why many of the early solutions proposed were focused at the international level, in the hopes of high-level government support for bold action. Even now, much of the climate information that we consume comes from places far across the globe that are experiencing calamity as a result of climate change.

The United States is one of the largest annual and historical emitters of GHGs. It is responsible for the second-largest annual total carbon emissions and has one of the highest per capita carbon footprints in the world. Specifically, the United States was responsible for 16 percent of CO_2 emissions in 2016 and was among the highest per capita emitters (Union of Concerned Scientists 2019).

KYOTO PROTOCOL

The United Nations Framework Convention on Climate Change (UNFCCC) has brought nations together since 1992 to attempt to limit the amount of CO_2 and other GHGs that are emitted into the atmosphere. One of the first and most important agreements made under the UNFCCC was the Kyoto Protocol, an international agreement of 192 parties that was adopted in 1997. In this agreement, developed countries would set emissions targets, which would then be international law. It was an international effort, and only the United States sat it out when President George W. Bush pulled out of the negotiations (Arroyo 2019, 434).[2] This refusal to commit to climate action set an unfortunate precedent in the United States, while most of the rest of the developed countries of the world worked together to try and curb emissions.

PARIS AGREEMENT

As the final commitments under the first and second periods of the Kyoto Protocol were set to expire by 2020, world leaders scrambled to renew the agreement—and strengthen it. After several international UNFCCC

Conferences of the Parties (COPs) failed, there was finally an agreement reached in Paris in 2015, which came to be known as the Paris Agreement, setting bold standards for action that involved developed, developing, and least developed countries. The objective was limiting global warming to less than 2.0°C, and to pursue additional efforts to limit the rise to 1.5°C.

Under the Obama administration, the United States entered the agreement, which it considered an executive agreement (Lee & Madulin 2016), rather than a treaty, and thus would not have to be ratified by Congress. President Obama signed an executive order entering the United States into the agreement in April 2016 and committing billions of dollars to international climate action. In 2018, the United States was almost halfway to the original US Paris Agreement emissions reduction pledge of 26 to 28 percent below 2005 levels by 2025, but the laws that currently exist at the federal and economy level are likely to only get the United States to 60 percent of its original goal (Bloomberg Philanthropies 2018, 9), an effort that was further hampered by President Trump's active attack on the Obama administration's signature strategy to reduce emissions, the CPP.

National Context—What Have We Done (or Not Done)?

Under Trump, the United States withdrew from the United Nations Paris Agreement to regulate emissions, meaning there was little the international community could do to push the federal government to take serious action to curb carbon emissions. This is beginning to change under the Biden administration, which has pledged to immediately rejoin, and "double down" on efforts to mitigate climate change. During Trump's presidency (2016-2020) the United States took no significant climate action at the federal level. The primary environmental laws that still regulate much of the pollution in the United States are relics of the first wave of these laws, passed in the 1970s. These laws were (to some extent) effective in their historical context, greatly cleaning up air pollution and to a lesser extent water pollution, as well as introducing the environment and conservation as concepts into US policy. These policies were passed by bipartisan coalitions and by majorities that would be unheard of today, and they have been the building block for many other federal environmental actions as presidents and legislators have struggled to rebuild these coalitions.

AIR POLLUTION LAWS AND REGULATIONS

The 1970s were a time of significant policy action in the United States as political and policy actors sought to address the results of an industrialized and urbanizing society. This decade was one of the "rare occasions in which the problem, policy, and politics streams converged" (Kraft 2018, 119), and with the pressure of public opinion to do something about pesticides and toxic chemicals, policymakers were able to pass several landmark environmental protection laws. There had been some environmental laws at the national level in the 1960s, including the Clean Air Act of 1963, but most of those laws were modest in size and scope and did little to help states deal with problems that were increasingly crossing local and state boundaries. President Richard Nixon (1968–1974) passed these laws through Congress after environmental concerns came to the fore following the publication of Rachel Carson's dramatic book on pesticides, *Silent Spring*; the increase in acid rain and other air pollution, especially in the US Northeast; and the largest oil spill to date off the coast of California in 1969 (Hamilton 2019).

In the case of the Clean Air Act of 1970, lawmakers saw the need for a more sweeping and forceful law because of the severity of the problem of air pollution and the failure of existing laws to properly address it (Kraft 2018, 154). The public wanted to know that their health was being protected, and the piecemeal efforts of the states and national governments simply weren't enough. The nationalization of environmental protection policies in the 1970s was a result of strong public and bipartisan political support, a visible concentration of problems, the ineffectiveness of prior state or national laws, and the increasingly transboundary nature of environmental problems.

Development of new air-quality legislation on the federal level was stalled until November 15, 1990, when Congress finally passed the Clean Air Act Amendments of 1990 (Environmental Resources Management [ERM] 2018, 3). With these amendments, the Clean Air Act did more than any previous law to reduce air pollution, and it has been one of the most successful US environmental policies. It introduced important concepts such as point or nonpoint sources for emissions,[3] and it helped establish use of certain standards to protect human health and permitting systems for polluting facilities (ERM 2018, 5). The law allowed the US Environmental Protection Agency (EPA) to figure out exactly which pollutants were doing the most harm, and then set standards for polluters to follow to reduce the health and environmental impacts that had been so visible in the 1960s and 1970s. However, it has not been until very recently that the US government recognized the danger to human health and public welfare from GHG emissions in the 2007 Supreme Court case *Massachusetts v. EPA*, which ultimately resulted in the EPA's CPP.

PARTISAN AND BIPARTISAN EFFORTS TO TACKLE THE ISSUE

One of the first major US political figures to acknowledge the problem of CO_2 emissions from fossil fuels was President Lyndon Johnson in 1965, when he mentioned the issue in an address urging the American people to take greater pride and interest in their country's natural heritage: "This generation has altered the composition of the atmosphere on a global scale through radioactive materials and a steady increase in CO_2 from the burning of fossil fuels." Action to curb air pollution was also taken by Republican President George H. W. Bush when the 1990 Clean Air Act Amendments were passed to tackle acid rain from coal-fired power plants (Kraft 2018, 154). Democratic President Bill Clinton proposed a tax on dirty energy, which passed the House but was rejected by the Senate after vehement opposition from the fossil fuel industry (Rosenbaum 1993).

Under President George W. Bush, the United States announced that it would not implement the Kyoto Protocol. The Bush administration also failed to regulate GHGs under the Clean Air Act, which led a group of nongovernmental organizations, states, and municipalities to sue the EPA after it denied a petition brought by several groups to regulate GHGs as air pollutants (68 Fed. Reg. 52,922). In 2007, the Supreme Court decided in their favor in *Massachusetts v. EPA* (549 U.S. 497, 528-529) (Meltz 2014).

During this time, climate action became increasingly politicized and divisive, with climate action seen as a value only held by Democrats and many Republicans taking a decidedly anti-climate stance. There were, however, important exceptions to this rule, including the positions of the late Senator John McCain of Arizona (Shogren 2017) and Senator Lindsey Graham of South Carolina (Dumain 2019). Despite famous climate deniers in the Republican Party, there are officials in each party who care about the issue, and some who even agree on how to fix the problem, such as the Climate Solutions Caucus in the House of Representatives and in the Senate.

This back and forth between the federal government, states, environmental groups, and the fossil fuel industry gave conflicting signals to state and local governments, businesses, and the American people. There was a great need for climate legislation that would achieve real results, but there was much uncertainty over how emissions reductions could happen without a clear national climate policy.

WAXMAN-MARKEY (2009)

This obvious and pressing need is what brought the American Clean Energy and Security Act to Congress in 2009. Better known as "Waxman-Markey" for its sponsors, Representatives Henry Waxman and Ed Markey, the bill was a sweeping climate policy that included a cap-and-trade mechanism to reduce GHGs 80 percent by 2050. The bill was complex, accounting for cost mitigation measures, a minimum price floor, allowances for offset credits, protection from leakage for fuel-intensive industries, minimum renewable energy standards for utilities, and money for research and development into renewable energy (Robinson 2010, 128). The bill passed the House by a slim margin in 2009 (Robinson 2010, 127) but failed in the Senate.

The failure of the bill was a blow to many who hoped for a national climate policy, but it had faced fierce opposition not only from Republicans and the fossil fuel industry, but also from environmental groups who saw it as too watered down (Walsh 2010). In the end, the bill had many critics. And one explanation for its failure was the outsized representation of fossil fuel interests in the Senate, where fossil fuel–producing states are overrepresented (Rabe 2015, 58)[4]. Another argument was that the public was not supportive of the bill, and that climate skepticism was high at the time (Hoffman 2011, 79; Rabe 2015, 57; Robinson 2010, 130). Finally, as mentioned, many critics thought that the bill did not go far enough, and that a much more stringent law was required (Robinson 2010, 22).

However, on the heels of the Waxman-Markey failure came an EPA finding that would pave the way for much broader government regulation of GHG emissions. The same year as the Waxman-Markey defeat, the EPA found the GHGs constituted a threat to human health and welfare, meaning that under the Clean Air Act they should be regulated (Doniger & Lynch 2017). In 2009, the EPA found that CO_2 accounted for 82 percent of US GHG emissions, and that fossil fuel–fired power plants in particular made up 31 percent of US GHG emissions (EPA 2015). The Clean Air Act requires the EPA to create a list of stationary sources that "cause, or contribute significantly to, air pollution which may reasonably be anticipated to endanger public health or welfare," for which the agency must then develop standards (Harvard Law Review 2014). This "endangerment finding" represented an important development for climate policy, as it allowed federal action on the issue with the potential to bypass Congress through executive action.

The Clean Power Plan

With the new endangerment finding in force, the Obama administration developed a federal rule under the powers of the Clean Air Act, which includes broad language requiring the EPA Administrator to regulate "air pollution which may reasonably be anticipated to endanger public health or welfare" (Clean Air Act §202(a)(1), 42 U.S.C. §7521). The plan was first proposed by the EPA in 2014, and President Obama unveiled the final plan in August 2015, just a few months before the historic Paris Agreement was negotiated under the UNFCCC. Anticipating climate commitments for the United States under the agreement, Obama was keen to get a federal policy in place to regulate GHGs, knowing that some progress had already been made as in 2015 the US GHG levels were 11 percent below 2005 levels, while the country's GDP and population had increased during that time (EPA 2015). The plan was framed not only as a climate policy, but also as a policy that would spur innovation and reduce health impacts from pollution (Lavelle 2015).

The plan itself was highly detailed, establishing state GHG emissions standards under a national plan to reduce these, and states were afforded many options in designing their own state-specific plan (EPA 2015). The CPP encouraged states to use a system that suited their own needs, including making existing coal-fired power plants more efficient, increasing their reliance on natural gas, and increasing the use of renewable energy. If states did not wish to make their own plan, they could simply follow the EPA's federal standard and require that all power plants meet national requirements. In an effort to boost collaboration and cost-efficiency, the EPA also allowed states to submit plans together or by trading allowances (EPA 2015). These plans were due to the EPA by September 2016, or September 2018 with an extension, giving states ample time to create them. Overall, the rule would have significant achievements, including emissions reductions that ranged 6 to 20 percent below 2012 levels (24–35 percent below 2005 levels) by 2032 (Wright & Kanudia 2016, 450), negligible increases in utility bills (increases less than 1 percent in 2020 and 2030 compared to baseline) as energy efficiency overall decreases bills (EPA 2015), and major benefits to public health.

Despite the rule coming from a federal agency that had pursued the policy because of a legal requirement in the Clean Air Act, the reaction from many states and the fossil fuel industry was swift. Challenges began before final rule was even published in the *Federal Register*, but the Supreme Court said it would not decide until the final rule was issued (CRS 2017, 9). Despite this negative reaction, many praised the rule for its flexibility, including economists who were encouraged by the potential for state collaboration, the option of emissions trading schemes, and each state's ability to find the most cost-effective option (Fowlie et al. 2014, 816). Yet there were also important concerns raised by the rule, especially with the outsized role that natural gas was assumed to play in the fast transition away from coal (Peters 2017). Additionally, each state's design for their plan might have led to "leakage" across state lines, meaning that business would just chase the state where it was absolutely cheapest to continue polluting (Wright & Kanudia 2016, 450).

When the final rule was published on October 23, 2015, it was again assailed by interests seeking to protect fossil fuel industries (CRS 2017, 10), with some states saying that the rule violated states' rights and took away their traditionally held power to control their energy supply (CRS 2017, 23; Glicksman 2017, 295). Over 20 states, municipalities, and organizations then sued the EPA over the rule, and because of the great number of individual cases, they were mostly all combined into the case of *West Virginia v. EPA*.[5] In response, 18 states joined EPA Administrator Gina McCarthy in defending the rule.

Fossil fuel interests and state party control played an important role in the roll-out of the CPP. The State of New York was an early defender of the CPP and also happened to be under Democratic control, as well as a relatively low

GHG emission producer. Under the CPP, New York would only have to achieve a 19 percent reduction in GHGs to be compliant. Montana, on the other hand, had competing incentives when the CPP came into effect, as it was one of the states that would need to make the largest percentage reduction in emissions and had mixed-party state leadership (EPA 2015, 20). States were divided, and much of the CPP's success hinged on the Supreme Court's interpretations of the legality from state challenges (Harvard Law Review 2014).

This is where the governors differed at the hypothetical Governors Association Lunch, which is where you come into the story. Some wanted to abandon all of the planning that they had started, saying it was a waste of resources. Others saw it as a "make lemonade" moment, where they essentially now had some state options for tackling climate change already at their disposal without having to pass legislation or take action on their own. And some states vowed to go a step further, planning to take their climate plans and run with them over the objections of the federal government.

MONTANA

The Climate Stakes

Despite the unknown fate of the CPP, in 2017 states were still expecting to have to comply with the rule and submit state plans. Although Montana's governor was a Democrat, the state was politically mixed, and many opposed the plan. To understand the future of state, and maybe even federal, action on climate change, we must see what states are doing, where the cuts will hurt the most, and where action has only tepid support.

Montana is known for its natural beauty, but it has also been a historically important contributor to climate change as it mines and burns a huge amount of coal. From 1990 to 2005, a 14 percent increase in GHG emissions moved the state from net carbon sink (because of its vast wild lands) to net carbon emitter. The state now averages net emissions of approximately 12 million metric tons of CO_2 equivalents per year and has a high rate of GHG emissions per capita—nearly double the national average—because of their fossil fuel (coal) production industry, substantial agricultural industry, large distances for transportation, cooler climate, and low population base (Governor's Climate Change Advisory Committee 2007). Overall, Montana in 2017 obtained 49 percent of its energy from coal, 39 percent from hydroelectric, 8 percent from wind, 4 percent from petroleum and natural gas, and it exports a large amount of coal-fired energy to neighboring states (EIA 2018).[6] In Fiorino's (2016, 20) ACE ranking comparing state carbon emissions, energy use, and other air pollutants, Montana ranked almost last, in 48th place, among the states.

Despite the important economic benefits of this industry in the state, Montana also has a lot to lose in the face of climate change. Glacier National

Park is located in Montana, as are many other large national parks that fuel the state's booming tourism industry (EIA 2018). It is home to the last remaining viable and self-sustaining megafauna (e.g., American bison, grizzly bear, and gray wolf) in the lower 48 United States (Pederson et al. 2009, 135). With glaciers diminishing, fishing opportunities contracting due to rising stream temperatures (Pederson et al. 2009, 149), and wildfires ravaging the state (Thackeray 2012, Whitlock et al. 2017), Montana's temperatures are expected to continue to rise dramatically, perhaps even 1.8 times faster than the global average (Pederson et al. 2009, 146). All of these effects will cost Montanans dearly, with the suppression of 2012 fires alone costing taxpayers about 113 million dollars, not including the private property damaged and destroyed (Thackeray 2012).

Montana's geography, however, lends itself well to renewable energy. The state's vast plains provide it with some of the best wind resources in the nation, which are already being exploited, with plans for a closed-loop, pumped hydro storage facility planned just northwest of Billings (EIA 2018). State laws are also slowly urging utilities to switch to renewables, as their current renewable resource standard requires retail electricity suppliers to generate at least 15 percent of the electricity they sell in-state from renewable energy resources (Long 2015, 3). Furthermore, coal is facing a decline, as neighboring states Oregon and Washington seek to go 100 percent renewable, and their energy sourced from units at the Colstrip plant in Montana will be decommissioned soon (Cates-Carney 2016b, Whitney 2016b).

Reaction to CPP

Montana in many ways resembles the country as a whole when it comes to climate change: a big fossil fuel extractor, refiner, and user with important natural and human heritage to protect and a very promising clean energy future if the politics can align properly. To date, however, Montana's politicians have achieved only piecemeal steps toward curbing GHG emissions, and the CPP would have been an important driver of change.

When the CPP was first issued, Governor Bullock called targets for his state in the rule unfair (MTPR News 2016), and his Republican attorney general joined the lawsuits that became *West Virginia v. EPA* (Cates-Carney 2016a). However, the governor also wanted Montana to write its own plan to comply with the rule to avoid potentially harsh compliance requirements in case the EPA instead determined the scope of his state's cuts (Rinfret & Pautz 2017). He also acknowledged that, despite his opposition to the rule, his state needed to begin to tackle climate change (Cates-Carney 2016a). As a presidential candidate in the Democratic primaries in 2019, Bullock staked out a strong position on working to diminish climate change, claiming to be a Democratic centrist in a Republican state who could get things done (Astor 2019, Bullock 2019).

In assessing what the CPP would mean for the state, the Montana Bureau of Business and Economic Research provided a gloomy outlook, forecasting the death of the coal industry in the state (Whitney 2015). Republicans in Montana blasted the plan, saying it went too far and was the wrong way to go about things. "Utilities across the country have made resource planning decisions based on this regulation and in some cases a carbon-cost component is already baked into the rates customers pay for electricity," Public Service Commission Chairman Brad Johnson of East Helena said. "Those decisions can't be easily undone, and ratepayers will continue to suffer higher electricity prices as a result" (Puckett 2017).

Nevertheless, Governor Bullock established a bipartisan council of stakeholders to begin the CPP planning process, despite criticism from environmentalists that the majority of that council had direct ties to or were supportive of the coal industry (MTPR News 2016). Not only did the Governor want to deal with the CPP on his terms, he was also fending off a Republican challenger in a state where climate action, and anything that might affect the coal industry, was a very sensitive matter (Whitney 2016a).

NEW YORK

The Climate Stakes

In contrast to Montana, New York has a lot going for it on the climate front: state leadership supportive of climate legislation, public support for climate action, and relatively few power-plant sources of GHG emissions. In the ACE Index, New York is ranked first (Fiorino 2016, 19). These factors positioned the state to develop a strong plan for the CPP and be a leader of other states across the country. New York is also experiencing the effects of climate change (Department of Environmental Conservation 2019), with increasing temperatures (particularly in the winter), increasing seasonal variability of precipitation, rising sea levels, and increasing instances of Lyme disease (EPA 2016b). However, New York has been a leader on climate action and has continued to do so through the CPP and its own independent state initiatives.

Part of why New York was so well suited to the CPP is its energy landscape. For one thing, the majority of its energy is supplied from out of state, meaning that few coal-fired power plants within the state needed an overhaul (EIA 2019b). Another advantage is that, because of the density of New York City, the state is one of the most energy efficient in the nation, consuming less total energy per capita than any other state besides Rhode Island (EIA 2019b; Rabe 2015, 67). The state's energy mix was also already suited to the CPP, with a majority of the electricity consumed in the state produced by natural gas and a large portion also coming from renewable sources (EIA 2019b).

The policy landscape in New York was just as well set up for the CPP, with ambitious renewable energy policies and more on the way. New York adopted its first renewable portfolio standard in 2004, requiring 29 percent of the state's electricity needs to be met with renewable energy. When the policy expired in 2016, a more stringent standard replaced it, requiring utilities and other retail electricity suppliers to acquire 50 percent of the electricity sold in the state from clean energy resources by 2030 (EIA 2019b). Importantly, New York is a leader in the Regional Greenhouse Gas Initiative (RGGI) emissions trading system, which the state credits with helping drastically reduce emissions from power plants while the economy has continued to grow (New York State 2015).[7]

Governor's Reaction

In New York, the reaction to the announcement of the CPP was very warm, with Governor Andrew Cuomo having released just earlier that year a comprehensive clean energy plan. The Governor's response to the CPP was positive, in stark contrast to his fellow Democratic Governor in Montana. In addition to this plan, the state was already part of the RGGI, a mandatory market-based program to reduce carbon emissions.

OKLAHOMA AND ALABAMA—THE CLIMATE DENIERS

The Climate Stakes

The southeastern United States. has experienced mixed climate effects in recent years. The area has not yet experienced the rise in temperatures that other states have (EPA 2016a). However, the states have been pummeled by weather disasters, including torrential rains, flooding, drying soils, sea level rise, tornadoes in the non-Gulf states, and severe hurricanes in those on the Gulf of Mexico (Burkhalter 2019). The changing climate is likely to increase damages from tropical storms, reduce crop yields, harm livestock, increase the number of unpleasantly hot days, and increase the risk of heat stroke and other heat-related illnesses (EPA 2016a).

Both Oklahoma and Alabama are also large exploiters of fossil fuels but have important potential for renewable energy. Alabama supplies a huge amount of hydroelectric power (almost eight percent of the state's net electricity generation), while having ports that also serve as huge import and export hubs for coal. Importantly, Alabama is not only a big exporter of energy but also a big user, ranking among the top quarter of states in energy consumption per person, primarily because of the high energy use in its active industrial sector (EIA 2019a). While coal was once an important source of power, nuclear and natural gas now supply most of the state's electricity, with many coal power plant closings planned for the next decade.

Oklahoma is also a major energy producer, the country's sixth-largest crude oil–producing state in 2017 and, at the same time, the fourth-largest natural gas producing state. In the same year, Oklahoma also ranked second in net electricity from wind generation, which is rapidly surpassing its share of electricity produced from coal, at least in terms of electricity generation (if not production for export to other states and internationally). And wind is not the only viable clean energy option for Oklahoma, as the state also has important solar energy potential and generates more than a third of its electricity from renewables (EIA 2019c).

Reaction to the CPP

Both Alabama and Oklahoma opposed the CPP (E&E News 2016b, Pillion 2019) and joined the lawsuit against it. Alabama was slated to reduce its emissions rate by 33 percent from 2012 to 2030 (Swartz 2016). Although Alabama gets a majority of its electricity from coal, the largest utility, Alabama Power Co., buys wind power from other states, and its parent company, Southern Co., operates across several US states, making a regional linkage program a very promising option under the CPP.

The state's elected officials, however, strongly opposed the plan. State Attorney General Luther Strange of Alabama said that the plan would "force millions of Americans to pay higher electric bills in order to satisfy the political agenda of the Obama administration's war on coal (Pillion 2015)." A Republican Congressman urged governors and attorneys general across the United States to delay submitting state implementation plans until the legal challenges to the rule had been exhausted.

In Oklahoma, State Attorney General Scott Pruitt (before he became EPA Administrator and then resigned the position in disgrace during the Trump administration)[8] fought to overturn the law with the support of Governor Mary Fallin (E&E News 2016a). Under the CPP, Oklahoma must cut its emissions rate about 32 percent below 2012 levels by 2030. In April 2016, Fallin signed an executive order to prevent Oklahoma from submitting a state plan to implement the CPP. The Governor and Attorney General pushed a state bill as a way to help Oklahoma navigate future legal battles over the proposed CPP. (Energy Wire 2016).

"Officials in Alabama canceled two CPP meetings and froze discussions until the Supreme Court weighs in on the regulation." State Air Division Chief Ron Gore said, "We plan to do minimal work on it until it is reactivated by the Supreme Court." Michael Teague, the state's Secretary of Energy and Environment, said in a statement that Oklahoma continues to engage with the Southwest Power Pool and stakeholders on effects of the plan. "Oklahoma no longer faces a September [2016] compliance date and can focus on assisting the attorney general on overturning this rule" (E&E News 2016a).

Current Status—Trump and Beyond

With the election of Donald Trump in 2016, the future of many federal policies was thrown into doubt, as many of his campaign slogans had focused on repealing any laws seen to be "impeding" certain sectors' economic interests. The fossil fuel industry was no different, with Trump generally opposed to climate action and particularly opposed to any Obama administration regulations (Glicksman 2017, 292). On the campaign trail, Trump promised to end the "war on coal" (Tollefson 2017, 311), revive the coal industry, and restore mining jobs, which was particularly popular in coal-producing and -burning states of the Midwest, where he scored some of his most important electoral victories (Dlouhy 2019, Volvovici et al. 2017). Can the Biden administration reverse course at the federal level? And even if Biden's administration succeeds, what will this mean for recalcitrant governors? Is federalism conducive to implementing strong, national policies such as those needed to stem climate change?

REPLACING THE CPP

In March 2017, President Trump directed then-Secretary of the EPA Scott Pruitt to review any agency actions that "potentially burden the development of domestically produced energy sources . . . especially oil, gas, coal, and nuclear" (EPA 2015, 40)." The same year, Trump began the formal process for withdrawing from the Paris Agreement, which he had promised in his campaign, making the United States the only Top 20 economy on earth not formally committed to the international agreement (Schear 2017). These two actions dealt a huge blow to US climate efforts.

In the review of the CPP rule, the EPA under President Trump decided not to repeal the rule altogether, but to roll it back and replace it with a much weaker standard called the ACE rule that would essentially allow states to determine what (if any) action they would take to regulate GHG emissions from power plants (Rott 2018). Part of the reasoning for this was that the EPA is still required to regulate GHG emissions because of its 2009 endangerment finding. If it failed to do so, then the EPA would open itself and the states to legal challenges under the Clean Air Act (EPA 2015, 32).

MONTANA REACTION

While Montana was not friendly to the CPP, many Montanans are committed to climate action, including the CPP-reluctant Governor Bullock. The governor announced the creation of a Climate Solutions Council in July 2019 and that the state would be joining the US Climate Alliance (Perry 2020), and also began to make an economic argument against coal, saying that the jobs were disappearing more due to cheap natural gas than any regulation (Yamanaka 2016). Some Montanans took an economic outlook on the repeal of the CPP. As Ben Reed, a Billings energy developer and member of the Northern Plains Resource Council, said, "Clean energy is one of the fastest growing sectors of the American economy, boosting manufacturing and employment with numbers unmatched by nearly every other economic sector" (Puckett 2017).

In 2019 Trump managed to replace the CPP with his much weaker Affordable Clean Energy (ACE) rule, which sought to reduce power sector emissions by a fraction of the Obama administration's objectives. But President Biden has pledged much more stringent emissions reductions and regulation than Obama. The question may be whether he can implement his agenda through executive orders, as his two predecessors have, or whether he will enjoy legislative branch cooperation. And how will stringent Democratic Party federal reforms play in GOP-governed states? After Trump's swift attack on the CPP, many states decided that they no longer had to plan for a transition to clean energy, and Montana's CPP Council, for example, met only twice before being disbanded (Caina Calvan 2016). In terms of planning for Montana's energy future, this may be a big step backward, as the signals given to the energy sector have been wildly uneven in recent years (Lutey 2017).

Despite many signs of interest toward climate action in Montana, the state remains woefully behind when it comes to planning for dealing with climate change. Coal plants in the state are beginning to close due to unrelated pressures, and the jobs that were once central to Montana's economy are drying up due to a global shift away from coal that will not be reversing its trend anytime soon (Lutey 2017).

NEW YORK REACTION

In New York, reaction to the Trump administration's repeal of the plan was in stark contrast to the reaction in Montana. One of the primary responses is that the New York State Attorney General Letitia James said that she planned to sue the Trump administration over the ACE rule, with other states and municipalities joining the fight (Irfan 2019). On top of this aggressive response to the federal changes, New York has been consistently adopting rigorous and groundbreaking policies to tackle climate change.

Governor Andrew Cuomo denounced the Trump administration's decision (Georgetown Climate Center 2017). Cuomo has made climate action a central piece of his platform, and he recently signed a bill that targets net-zero emissions by 2050, with an interim target of 40 percent emissions reductions from the 1990 level by 2030 (Roberts 2019). The bill was a big win for environmental groups, who lobbied hard to ensure that climate action was done by the legislature and not through an executive action that could be undone with a new administration. While the bill was not exactly what they had wanted when first proposed, with new offset allowances included for industries where a zero carbon alternative has not been identified (e.g., cement mixing), it is still a huge victory that delivers on Cuomo's climate promises (EIA 2019b, Roberts 2019).

OKLAHOMA AND ALABAMA REACTIONS

Based on these states' reaction to the announcement of the CPP, and on its subsequent suspension in the courts, it is easy to see what the reaction might be from Governors Ivey and Fallin. In Alabama, the new Trump rule was warmly received (Klump & Northey 2019). In Oklahoma, the repeal of the rule may not have much effect, as state power suppliers are still working to clean up their coal-fired power plants.

Alabama is one of a group of states that was projected to not meet its climate commitments with the repeal of the CPP (Plumer & Popovich 2017). Despite the unraveling of the CPP, however, Oklahoma might still hit its targets prescribed by the plan because of coal plant retirements (Harball 2015), and will likely beat its targets even without the CPP. Nonetheless, Oklahoma Senator James Lankford said, "I agreed with President Trump when he eliminated the overreach of the previous administration, and I continue to support the idea that the government should not attempt to regulate power generation through executive action like the CPP without authorization from Congress" (Lankford 2019). Overall, Oklahoma and Alabama were receptive to Trump's new rules and happy to be rid of the CPP, even if regulators had already taken at least some steps to comply.

CONCLUSION

The Trump administration did not seem particularly concerned with international commitments to address the climate crisis, and replacing the CPP moved the United States significantly further away from the goal reducing its GHG emissions. Despite the rewrite of the CPP, there is little the Trump administration can do to reverse an international shift away from coal and toward cheaper, renewable sources of electricity such as wind and solar, as well as natural gas (Dlouhy 2019). Part of this shift has been the price of natural gas, which has fallen precipitously since the introduction of the "fracking" process of extraction, and significant increases in energy efficiency, mandated in large part by earlier administrations

An increasing number of states also are taking important actions pursuant to the goals of the CPP on their own, and pledging independently to continue to address GHG emissions (Bloomberg Philanthropies 2018). Particularly at the state level, governors have participated in a multitude of alliances and other collective actions. For instance, the US Climate Alliance now accounts for over 51 percent of the US population, while the "Under2" coalition and Power Past Coal Alliance have also enjoyed success in attracting members in the Trump era (Arroyo 2019, 434). More states are also looking to cooperate on GHG emissions, (Arroyo 2019, 439). Independently, states like New York, California, Nevada, Hawaii, Washington, New Mexico, Washington, DC, and Puerto Rico all now aim for 100 percent clean power (Roberts 2019), and by 2025, most of the US population will live in states with a carbon price (Lempert et al. 2019, 9). After President Trump's announcement of his plan to leave the Paris Agreement, thousands of municipalities, tribes, mayors, cities, states, companies, and other organizations signed the "we are still in" pledge to hold themselves to the reductions outlined in the agreement.

Can States Take It All the Way?

Not all states, however, have the leadership or political support for bold climate policies, and for the United States to fill the national gap left by the CPP, there must be other, less politically charged options for these states to pursue. Two examples of these types of policies are energy efficiency standards and Renewable Portfolio Standards, which have been adopted in a broad majority of states across political lines (Database of State Incentives for Renewables and Efficiency 2019).[9] Energy retrofit programs for buildings and low-income communities are also popular, as well as national programs such as AgStar, which provides support for biogas recovery from farms (Bloomberg Philanthropies 2018, 57, 71).

States could achieve serious reductions by committing to strong climate action, or even by undertaking politically feasible measures of their own while international trends and pressures continue to make coal and oil seem like increasingly worse investments. A 2017 analysis determined that at least 25 states were likely on track to beat their CPP emissions goals even after the Trump administration's replacement of the policy, and 10 more states might be close to their goals (Plumer & Popovich 2017). "I'm not going to build new coal plants in today's environment," said Ben Fowke, CEO of Xcel Energy, which operates in eight states and uses coal for about 36 percent of its electricity production. "And if I'm not going to build new ones, eventually there won't be any." (Volcovici et al. 2017) Even if we simply implemented the policies that are already on the books more intensely, the United States could achieve significant reductions in GHG emissions (Rabe 2015 59).

Unfortunately, piecemeal state efforts are a far cry from the national standard advocated by the CPP, and a patchwork of policies is simply not a cost-effective way of achieving these reductions (Glicksman 2017, 300; Rabe 2015, 59). Montana's entwined relationship with fossil fuels makes it difficult for the state to take the bold action needed to protect employment as coal jobs die off; to protect its natural resources as climate change ravages its forests, mountains, and rivers; and to prepare for a new, clean energy economy. New York has positioned itself exceptionally well for this transition despite a lack of federal leadership, but it has done so primarily because of large, climate-friendly Democratic Party majorities in state offices. There is no real roadmap for states like Montana that now lack a plan and political will.

Final Thoughts

The many good intentions of local and state governments are an important step in tackling climate change nationally, but they do not provide the broad policy directives that the CPP was designed to harness. Despite this, two-thirds of voters now say they are concerned that climate change will affect them personally (Quinnipiac 2017), and in a poll of self-described Democrats, it is now the most salient issue (Meyer 2020). While technology can facilitate some of this transition, it can't all be done without adequate policy drivers (Lempert et al. 2019, 1). A true decarbonization of the US economy requires a broad suite of policies that drive investment and action by setting goals, targeting resources, providing incentives, and ensuring a level playing field (Bloomberg Philanthropies 2018, 9; Lempert et al. 2019, 1).

EXERCISE

SCENARIO FOR CLASSES

You are a state official at the National Governor's Association meeting on climate action after the end of the CPP, where governors from a majority of the states have decided that they want to coordinate plans and negotiate actions in the absence of a strong national policy on climate change. Choose a state, briefly research the state's energy background at eia.gov/state/, and decide what the energy priorities of your state are. Decide if your state is a *bold climate actor, tepid climate actor,* or *climate refuser*. Try and find consensus through debate on *at least two* state policies that can be adopted by all states and *one additional policy* that could be adopted by the tepid climate actor states (like Montana).

Preparation questions include the following:

- Where does your state get the majority of its energy?
- Does your state produce or mine its own energy, or does your state import it from elsewhere?

Continued

- What are some of the most carbon-intensive industries in your state (i.e., what organizations will you make angry with bold climate action)?
- What are some organizations or businesses in your state that would benefit from strong climate action?
- What policies have already been adopted?
- What was the state's reaction to the CPP and the general mood on climate action?

CASE STUDY 4.1 NOTES

1. Laurel Wolf is a graduate student at American University's School of Public Affairs, focused on mitigating and adapting to climate change. She has worked as a grassroots organizer on climate issues in the United States, on agroforestry projects in Senegal as a Peace Corps Volunteer, and is a project assistant with Environmental Incentives.
2. The US Senate failed to ratify the Kyoto Protocol, even though the administration of President Bill Clinton (1992–2000) actively participated in negotiations and Clinton signed the agreement.
3. These two categories separate the recognized sources of air pollution. Point source means that the pollution emissions can be traced back to a single area of emissions, such as a stack, building ventilation, or vent on a storage tank. A nonpoint source might include volatile organic compounds released from wastewater treatment plants or leaks from valves, flanges, or pumps. The protection of health and human welfare are tied to the National Ambient Air Quality Standards that represent a maximum concentration of a pollutant in the air past which humans and the environment may suffer adverse effects. Permitting requirements include a complex set of regulations regarding construction of new pollutant emissions sources or the modification or expansion of existing sources, referred to as "New Source Review" (ERM 2010).
4. The biggest producers of fossil fuels (with important exceptions like Texas, Pennsylvania, and Illinois) tend to be rural states (like Kentucky, West Virginia, Montana, Wyoming, North Dakota, Alaska, and Utah), where a small population is represented by two Senators the same as much larger states. This means that states with relatively few people and economies dependent on the fossil fuel industry have the same amount of clout in the Senate as areas with huge populations (MacNeil 2017, 59).
5. *West Virginia et al.* v. *EPA*, No. 15-1363 (D.C. Cir. docketed October 23, 2015). Petitioners joining West Virginia in the suit included Texas, Alabama, Arizona, Arkansas, Colorado, Florida, Georgia, Indiana, Kansas, Kentucky, Louisiana (Department of Environmental Quality), Michigan (Attorney General), Mississippi, Missouri, Montana, Nebraska, New Jersey, North Carolina (Department of Environmental Quality), North Dakota, Ohio, Oklahoma, South Carolina, South Dakota, Utah, Wisconsin, and Wyoming, as well as Dixon Bros., Inc.; Nelson Bros., Inc.; Wesco International, Inc.; Norfolk So. Corp.; Joy Global, Inc.; Gulf Coast Lignite Coal; and Peabody Energy Corp. in Support of Petitioners (filed February 23, 2016). The court heard oral arguments in 2016 but put the case on hold after the election of President Trump. The U.S. Court of Appeals dismissed the various cases as moot in September 2019 because of the Trump administration's replacement of the CPP with the ACE rule (Gilmer 2019).
6. Montana is the sixth-largest coal producing state by metric tonnes (34,792 in 2016) (MacNeil 2017, 58)
7. RGGI is the first mandatory, market-based effort to reduce GHG emissions in 10 Northeastern and Mid-Atlantic States, including New York. It is implemented in New York by Department of Environmental Conservation and New York State Energy Research and

Development Authority (New York State 2015). States sell nearly all emission allowances through auctions and invest proceeds in energy efficiency, renewable energy, and other consumer benefit programs. These programs are spurring innovation in the clean energy economy and creating green jobs in the RGGI states (Arroyo 2019, 438; www.rggi.org).

8. Scott Pruitt's time at EPA was marked by ethical issues, including flying first class on official business at taxpayer expense, a $50-per-night condo in a home co-owned by the wife of an oil lobbyist, and a 24-hour, 20-person security detail. (Friedman 2018). Pruitt also used his time to strike a hard-line, anti-regulatory stance at EPA, which earned him support at home in Oklahoma (Fausett 2018)
9. Energy Efficiency Resource Standards require electricity and natural gas distributors to achieve a percentage of energy savings relative to a baseline. These policies are adopted at the state level, and in many cases, the utility develops demand-side management programs, which might include financial incentives to consumers to install energy efficient appliances (Database of State Incentives for Renewables and Efficiency, 2019). A Renewable Portfolio Standard (RPS) simply mandates that energy providers source a certain percentage of their energy from renewable sources, and even fossil fuel–supporting states like Texas have such a standard (Rabe 2015, 71).

CASE STUDY 4.1 REFERENCES

Arroyo, Vicki. 2019. "From Paris to Pittsburgh: US state and local leadership in an era of Trump." *Georgetown Environmental Law Review*, 31(3), 433.

Astor, Maggie. 2019. "Where Steve Bullock stands on the issues." *New York Times,* May 14. Accessed June 23, 2020, at www.nytimes.com/2019/05/14/us/politics/steve-bullock-on-the-issues.html.

Bloomberg Philanthropies. 2018. *Fulfilling America's Pledge: How States, Cities, and Businesses Are Leading the United States to a Low-Carbon Future*. Accessed June 23, 2020, at www.bbhub.io/dotorg/sites/28/2018/09/Fulfilling-Americas-Pledge-2018.pdf

Bullock, Steve. 2019. "Opinion: A crazy idea to get things done in Washington." *CNN.com*, October 3. Accessed June 23, 2020, at www.cnn.com/2019/10/03/opinions/steve-bullock-end-permanent-campaign-opinion/index.html

Burkhalter, Eddie. 2019. "Introduction: The impact of climate change on Alabama." *Alabama Political Reporter.* July 9. Accessed June 23, 2020, at www.alreporter.com/2019/07/09/introduction-impact-of-climate-change-on-alabama/

Caina Calvan, Bobby. 2016. "Montana legislative panel suspends work on Clean Power Plan." *Montana Public Radio,* February 17. Accessed June 23, 2020, at www.mtpr.org/post/montana-legislative-panel-suspends-work-clean-power-plan

Cates-Carney, Corin. 2016a. "Clean Power Plan decision gets mixed reaction in Montana." *Montana Public Radio*, February 10. Accessed June 23, 2020, at www.mtpr.org/post/clean-power-plan-decision-gets-mixed-reactions-montana

Cates-Carney, Corin. 2016b. "Some Montana Politicians Not Ready to Leave Coal Behind." *Montana Public Radio*, January 13. Accessed June 23, 2020, at www.mtpr.org/post/some-montana-politicians-not-ready-leave-coal-behind

Database of State Incentives for Renewables and Efficiency. 2019. "Renewable portfolio standards and clean energy standards updated June 2019." Accessed June 23, 2020, at ncsolarcen-prod.s3.amazonaws.com/wp-content/uploads/2019/07/RPS-CES-June2019.pdf.

Department of Environmental Conservation. 2019. "Impacts of climate change in New York." Accessed June 23, 2020, at www.dec.ny.gov/energy/94702.html.

Dlouhy, Jennifer A. 2019. "Trump makes his biggest move yet to try to save coal plants." *Bloomberg,* June 18. Accessed June 23, 2020, at www.bloomberg.com/news/articles/2019-06-18/trump-s-biggest-move-to-end-war-on-coal-won-t-rescue-industry.

Doniger, David, and Lissa Lynch. 2017. *EPA's Endangerment Finding: The Legal and Scientific Foundation for Climate Action*. Accessed June 23, 2020, at www.nrdc.org/sites/default/files/epa-endangerment-finding-fs.pdf.

Dumain, Emma. 2019. "Can Lindsey Graham help improve the GOP's image on the environment?" *McClatchy DC*, July 10. Accessed June 23, 2020, at www.mcclatchydc.com/news/politics-government/congress/article232467697.html.

E&E News. 2016a. "E&E's power plan hub: Alabama." Accessed June 23, 2020, at www.eenews.net/interactive/clean_power_plan/states/alabama.

E&E News. 2016b. "E&E's power plan hub: Oklahoma." Accessed June 23, 2020, at www.eenews.net/interactive/clean_power_plan/states/oklahoma.

Energy Information Agency (EIA). 2018. "Montana—State profile and energy analysis." Accessed June 23, 2020, at www.eia.gov/state/analysis.php?sid=MT#5.

Energy Information Agency (EIA). 2019a. "Alabama—State profile and energy analysis." Accessed June 23, 2020, at https://www.eia.gov/state/?sid=AL.

Energy Information Agency (EIA). 2019b. "New York—State profile and energy analysis." Accessed June 23, 2020, at www.eia.gov/state/analysis.php?sid=NY.

Energy Information Agency (EIA). 2019c. "Oklahoma—State profile and energy analysis." *Accessed June 23, 2020, at* https://www.eia.gov/state/?sid=OK.

Energy Wire. 2015. "Oklahoma response to Clean Power Plan gains traction." Accessed June 23, 2020, at www.eenews.net/energywire/stories/1060017423.

Fausset, Richard. 2018. "Scott Pruitt, fallen E.P.A. Chief, may rise again in Oklahoma." *New York Times,* July 8. Accessed June 23, 2020, at www.nytimes.com/2018/07/08/us/pruitt-oklahoma.html.

Fiorino, Dan. 2016. "The Pilot Eco-Efficiency Index: A new state environmental ranking for researchers and government." *Fall Research Conference, Association for Public Policy Analysis and Management (APPAM)*, Washington, DC. Accessed June 23, 2020, at https://www.researchgate.net/publication/315736703_The_Pilot_Eco-Efficiency_Index_A_New_State_Environmental_Ranking_for_Researchers_and_Government.

Fowlie, Meredith, Lawrence Goulder, Matthew Kotchen, Severin Borenstein, James Bushnell, Lucas Davis, Michael Greenstone, Charles Kolstad, Christopher Knittel, Robert Stavins, Michael Wara, Frank Wolak, Catherine Wolfram 2014. "Environmental economics. An economic perspective on the EPA's Clean Power Plan." *Science (New York, N.Y.)* 346(6211):815–816. DOI: 10.1126/science.1261349.

Friedman, Lisa. 2018. "The investigations that led to Scott Pruitt's Resignation." *New York Times*, April 18. Accessed June 23, 2020, at www.nytimes.com/2018/04/18/climate/scott-pruitt-epa-investigations-guide.html.

Georgetown Climate Center. 2017. "State reactions to Trump repealing the Clean Power Plan." Accessed June 23, 2020, at www.georgetownclimate.org/articles/state-reactions-to-trump-repealing-the-clean-power-plan.html.

Gilmer, Ellen M. 2019. "DC circuit scraps Clean Power Plan litigation." *Bloomberg Environment*, September 17. Accessed June 23, 2020, at news.bloombergenvironment.com/environment-and-energy/d-c-circuit-scraps-clean-power-plan-litigation.

Glicksman, Robert. 2017. "The fate of the Clean Power Plan in the Trump era." *Carbon & Climate Law Review* 11(4):292–302. DOI: 10.21552/cclr/2017/4/5.

Governor's Climate Change Advisory Committee. 2007. *Montana Climate Change Action Plan.* Accessed June 23, 2020, at deq.mt.gov/Portals/112/Energy/ClimateChange/Documents/FinalReportChapters.pdf.

Hamilton, 2019. "How California's worst oil spill turned beaches black and the nation green." *NPR*, January 28. Accessed November 28, 2020, at www.npr.org/2019/01/28/688219307/how-californias-worst-oil-spill-turned-beaches-black-and-the-nation-green.

Harball, Elizabeth. 2015. "Some states fighting Clean Power Plan still on track to meet targets—Report." *Climate Wire*, August 14. Accessed June 23, 2020, at www.eenews.net/climatewire/stories/1060023469.

Harvard Law Review. 2014."Utility Air Regulatory Group vs. EPA." Accessed December 23, 2020, at https://harvardlawreview.org/2014/11/utility-air-regulatory-group-v-epa/.

Hegyi, Nate. 2017. "Colstrip's future still uncertain after end of Clean Power Plan." *Montana Public Radio*, October 9. Accessed June 23, 2020, at www.mtpr.org/post/colstrips-future-still-uncertain-after-end-clean-power-plan

Hoffman, Andrew J. 2011. "The culture and discourse of climate skepticism." *Strategic Organization* 9(1):77–84. DOI: 10.1177/1476127010395065

Irfan, Umair. 2019. "Trump's EPA just replaced Obama's signature climate policy with a much weaker rule." *Vox*, June 19. Accessed June 23, 2020, at www.vox.com/2019/6/19/18684054/climate-change-clean-power-plan-repeal-affordable-emissions

Johnson, Lyndon B. 1965. "Special message to Congress on conservation and restoration of natural beauty." Accessed June 23, 2020, at https://www.presidency.ucsb.edu/documents/special-message-the-congress-conservation-and-restoration-natural-beauty

Klump, Edward, and Hannah Northey. 2019. "How officials from 4 states view Trump's new carbon rule." *Energy News Network*, July 1. Accessed June 23, 2020, at energynews.us/2019/07/01/west/how-officials-from-4-states-view-trumps-new-carbon-rule/

Kraft, Michael E, *Environmental Politics and Policy*. 7th ed. Taylor & Francis. New York: Routledge, 2018.

Lankford, James. 2019. "Senator Lankford opposes returning to Obama-era rule." Accessed June 23, 2020, at www.lankford.senate.gov/news/press-releases/senator-lankford-opposes-returning-to-obama-era-rule

Lavelle, Marianne. 2015. "Obama's Clean Power Plan is finally out. But scientists say the work's just begun." *Science*, August 4. DOI: 10.1126/science.aac8965

Lee, Carol E., and William Maudlin. 2016. "US, China agree on implementing Paris Climate-Change Pact." *Wall Street Journal*, September 3. Accessed June 23, 2020, at www.wsj.com/articles/u-s-china-agree-on-implementing-paris-climate-change-pact-1472896645

Lempert, Robert, Benjamin L. Preston, Jae Edmonds, Leon Clarke, Tom Wild, Matthew Binsted, Elliot Diringer, and Brad Townsend. 2019. "*Pathways to 2050 Alternative Scenarios for Decarbonizing the US Economy. Center for Climate and Energy Solutions.* Accessed June 23, 2020, at www.c2es.org/site/assets/uploads/2019/05/pathways-to-2050-scenarios-for-decarbonizing-the-us-economy-final.pdf

Long, Noah. 2015. *Montana's Pathway to Cutting Carbon Pollution*. Accessed December, 23, 2020, at https://www.nrdc.org/sites/default/files/CPP-MT-Compliance-IB.pdf

Lutey, Tom. 2017. "Trump's repeal of Clean Power Plan gets mixed Montana reception." *Billings Gazette*, March 29. Accessed June 23, 2020, at billingsgazette.com/news/trump-s-repeal-of-clean-power-plan-gets-mixed-montana/article_c71f60b9-46a2-5d0b-822b-34833101959f.html

MacNeil, Robert. *Neoliberalism and Climate Policy in the United States: From Market Fetishism to the Developmental State.* New York: Routledge, 2017.

Meltz, Robert. 2014. "Federal Agency Actions Following the Supreme Court's Climate Change Decision in Massachusetts v. EPA: A Chronology." Congressional Research Service Brief. Accessed December 23, 2020, at https://fas.org/sgp/crs/misc/R41103.pdf

Meyer, Robinson. 2020. "Voters really care about climate change." *The Atlantic*, February 20. Accessed June 23, 2020, at www.theatlantic.com/science/archive/2020/02/poll-us-voters-really-do-care-about-climate-change/606907/

MTPR News. 2016. "Environmentalists critical of Bullock's new Clean Power Plan Council." *Montana Public Radio*, January 5. Accessed June 23, 2020, at www.mtpr.org/post/environmentalists-critical-bullocks-new-clean-power-plan-council

Montana Environmental Information Center. 2017. "Montana and the Clean Power Plan." Accessed June 23, 2020, at meic.org/issues/clean-air/montana-clean-power-plan/

MTPR News. 2016. "Environmentalists critical of Bullock's new Clean Power Plan Council." *Montana Public Radio*, January 5. Accessed June 23, 2020, at www.mtpr.org/post/environmentalists-critical-bullocks-new-clean-power-plan-council

New York State. 2015. *The Energy to Lead: 2015 New York State Energy Plan*. Accessed June 23, 2020, at https://energyplan.ny.gov/Plans/2015.aspx

Pederson, Gregory T., L.J. Graumlich, D.B. Fagre, T. Kipfer, and C.C. Muhlfeld 2009. "A century of climate and ecosystem change in western Montana: What do temperature trends portend?" *Climatic Change* 98(1–2)133–154. DOI: 10.1007/s10584-009-9642-y

Perry, Melissa. 2020. "Governor Bullock Announces Montana Climate Solutions Council Requests Feedback on Preliminary Recommendations," February 11 statement by Montana Department of Environmental Quality. Accessed December 23, 2020, at http://deq.mt.gov/Public/PressRelease/governor-bullock-announces-montana-climate-solutions-council-requests-feedback-on-preliminary-recommendations

Peters, Jeffrey C. 2017. "Natural gas and spillover from the US Clean Power Plan into the Paris Agreement." *Energy Policy* 106:41–47. DOI: 10.1016/j.enpol.2017.03.039

Pillion, Dennis. "Alabama among 24 states to challenge EPA's Clean Power Plan." *Alabama.com*, October 23, 2015. Accessed June 23, 2020, at www.al.com/news/2015/10/alabama_joins_multi-state_chal.html

Piotrowski, John M., David Coady, Justin Tyson, Rolando Ossowski, Robert Gillingham, Shamsuddin Tareq. 2010. "Petroleum Product Subsidies: Costly, inequitable, and Rising," IMF Staff Position Note, SPN/10/05 (Washington: International Monetary Fund). Accessed January 5, 2021 at https://www.imf.org/en/Publications/IMF-Staff-Position-Notes/Issues/2016/12/31/Petroleum-Product-Subsidies-Costly-Inequitable-and-On-the-Rise-23584

Plumer, Brad, and Nadja Popovich. 2017. "How will the Clean Power Plan repeal change carbon emissions for your state?" *New York Times*, October 10. Accessed June 23, 2020, at www.nytimes.com/interactive/2017/10/10/climate/clean-power-plan-emissions-your-state.html

Puckett, Karl. 2017. "Montana reaction mixed to proposed repeal of Clean Power Plan." *Great Falls Tribute*, October 10. Accessed June 23, 2020, at www.greatfallstribune.com/story/news/2017/10/10/montana-reaction-mixed-clean-power-plan-repeal/752301001/

Quinnipiac University. 2017. "Two-thirds of US voters take climate personally." Accessed June 23, 2020, at https://poll.qu.edu/images/polling/us/us04052017_Ubgw27pk.pdf/

Rabe, Barry. 2015. "A new era in states' climate policies?" In *Changing Climate Politics: US Policies and Civic Action*, edited by Yael Wolinsky-Nahmias. CQ Press, Washington, DC, pp. 55–80.

Rinfret, Sara, and Michelle Pautz. 2017. "In the eye of the storm: Exploring how Montana and Ohio are framing the debate about the Clean Power Plan rule." *Environmental Practice* 19(1):26–37. DOI: 10.1080/14660466.2017.1275692

Roberts, David. 2019. "New York just passed the most ambitious climate target in the country." *Vox*, July 22. Accessed June 23, 2020, at https://www.vox.com/energy-and-environment/2019/6/20/18691058/new-york-green-new-deal-climate-change-cuomo

Robinson, David. 2010. "US federal energy and climate change legislation: Some lessons to be learned from the Waxman-Markey bill." *Carbon & Climate Law Review* 4(2):127–138. Accessed June 23, 2020, at http://search.proquest.com/docview/734722785/

Rosenbaum, David E. 1993. "Clinton backs off plan for new tax on heat in fuels." *New York Times*, June 9. Accessed November 28, 2020, at https://www.nytimes.com/1993/06/09/us/clinton-backs-off-plan-for-new-tax-on-heat-in-fuels.html

Rott, Nate. 2018. "Trump moves to let states regulate coal plant emissions." *Montana Public Radio*, August 21. Accessed June 23, 2020, at www.mtpr.org/post/trump-moves-let-states-regulate-coal-plant-emissions

Schear, Michael D. 2017. "Trump will withdraw US from Paris Climate Agreement." *New York Times*, June 1. Accessed November 28, 2020, at https://www.nytimes.com/2017/06/01/climate/trump-paris-climate-agreement.html

Selmi, Daniel P. 2015. "Federal Implementation Plans for Controlling Carbon Emissions from Existing Power Plants: A Primer Exploring the Issues." Sabin Center for Climate Change Law, Columbia University Law School. Accessed December 24, 2020, at https://web.law.columbia.edu/sites/default/files/microsites/climate-change/selmi_-_fip_primer_0.pdf

Shogren, Elizabeth. 2017. "McCain resumes his quest for climate change action." *High Country News*, September 14. Accessed June 23, 2020, at www.hcn.org/articles/sen-john-mccain-resumes-his-quest-for-action-on-climate-change

Swartz, Kristi E. 2016. "Defiant Alabama pledges to wait out court ruling." *Energy Wire*, Januray 7. Accessed December 24, 2020, at https://www.eenews.net/energywire/stories/1060030249

Thackeray, Lorna. 2012. "Montana wildfires burn most acreage since 1910; $113M spent to battle blazes." *Billings Gazette*, November 1. Accessed January 30, 2020, at billingsgazette.com/news/state-andregional/montana/montana-wildfires-burn most-acreage-since-m-spent-to-battle/article_88b22157-7b4b-5b0b-9951-9732801e7fe7.html

Tollefson, Jeff. 2017. "US agency moves to revoke emissions limits." *Nature*, October 10. DOI: 10.1038/nature.2017.22813

Union of Concerned Scientists. 2019. "Each country's share of CO_2 emissions." Accessed June 23, 2020, at www.ucsusa.org/resources/each-countrys-share-co2-emissions

US Environmental Protection Agency (EPA). 2015. "Clean Power Plan factsheet." Accessed June 23, 2020, at 19january2017snapshot.epa.gov/cleanpowerplan/fact-sheet-overview-clean-power-plan_.html#print

US Environmental Protection Agency (EPA). 2016a. "What climate change means for Alabama." Accessed June 23, 2020, at https://www.epa.gov/sites/production/files/2016-09/documents/climate-change-al.pdf.

US Environmental Protection Agency (EPA). 2016b. "What climate change means for New York." Accessed June 23, 2020, at 19january2017snapshot.epa.gov/sites/production/files/2016-09/documents/climate-change-ny.pdf

Volcovici, Valerie, Nichola Groom and Scott DiSavino. 2017. "Donald Trump declares end to 'War on Coal,' US energy giants ignore him." *The Independent*, April 5. Accessed June 23, 2020, at www.independent.co.uk/news/world/americas/donald-trump-war-on-coal-end-energy-companies-barack-obama-regulations-scrap-miner-jobs-clean-power-a7668291.html

Walsh, Brian. 2010. "Why the climate bill died." *TIME*, July 26. Accessed June 23, 2020, at time.com/2010/07/26/why-the-climate-bill-died/

Whitlock C., W. Cross, B. Maxwell, N. Silverman, A. A. Wade. 2017. *Montana Climate Assessment*. Montana State University and University of Montana, Montana Institute on Ecosystems, Bozeman and Missoula. 318 p. DOI: 10.15788/m2ww8w

Whitney, Eric. 2015. "BBER defends gloomy Clean Power Plan analysis for Montana." *Montana Public Radio*, November 23. Accessed June 23, 2020, at www.mtpr.org/post/bber-defends-gloomy-clean-power-plan-analysis-montana

Whitney, Eric. 2016a. "Governor Bullock slams coal lease moratorium." *Montana Public Radio*, January 15. Accessed June 23, 2020, at www.mtpr.org/post/gov-bullock-slams-coal-lease-moratorium

Whitney, Eric. 2016b. "Northwestern accepts Governor's Colstrip invitation." *Montana Public Radio*, April 6. Accessed June 23, 2020, at www.mtpr.org/post/northwestern-accepts-governors-colstrip-invitation

World Resources Institute. 2021. "Climate Watch" database of historical emissions. Accessed January 5, 2021 at https://www.climatewatchdata.org/ghg-emissions?chartType=percentage&regions=WORLD%2CUSA&sectors=614

Wright, Evelyn, and Amit Kanudia. "Variation in outcomes and leakage potential across Clean Power Plan compliance designs." *Energy Economics* 60:438–450. DOI: 10.1016/j.eneco.2016.06.016

Yamanaka, Jackie. "Governor Bullock brings energy roundtable to Colstrip." *Montana Public Radio*, May 31. Accessed June 23, 2020, at www.mtpr.org/post/governor-bullock-brings-energy-roundtable-colstrip

5

Short-Term Versus Long-Term Crude Solutions

Peak Oil, Fracking and the "Natural Gas" Bridge, and the Promise of Renewables

Universities have been on the leading edge of social change since at least the Middle Ages, when "Islamic Golden Age" institutions pioneered discoveries in math and medicine, and European universities "stored" and reprocessed the philosophy of Aristotle and the Ancient Greeks, even as the advancement of knowledge (and science in particular) was repressed in some quarters. More recently, university student demonstrations in the United States led the 1960s Peace Movement against the Vietnam War, and in the 1980s, similar demonstrations pushed for divestment by universities and corporations from racist apartheid practices in South Africa. Can university students serve as a similar catalyst for public awareness on climate change by pushing for divesting assets from fossil fuel companies, thus leading the charge toward a decarbonized economy? The early evidence is mixed. The university divestiture movement has intensified over the last decade, seeking to have colleges and other institutions ban investment in fossil fuel companies and thereby "stranding" oil that remains underground (see Case Study 7.1 following Chapter 7). So far, this new student movement seems to have drawn important attention to climate change, but it has not directly helped to reduce emissions.

The 2018 Intergovernmental Panel on Climate Change (IPCC) *Special Report on Global Warming of 1.5°C* was clear in its admonition that carbon emissions must be cut 45 percent by 2030, with investments in fossil fuel extraction and power generation falling by some 850 billion dollars between 2016 and 2050 (and coal extraction falling to zero by 2030).[1] Still, as the IPCC was circulating the report, global demand for oil, gas, and coal was still growing, with fossil fuels accounting for 81 percent of energy use, although the COVID-19 pandemic did diminish fossil fuel consumption in 2020. Still, given the exorbitant costs of transitioning to decarbonized energy sources and the unrivaled and ongoing power of fossil fuel lobbies in industrial countries like the United States, how can we reconcile the need to cut fossil fuel extraction with the need to continue generating economic production using energy? And what role can universities play?

As mentioned, a fossil fuel divestment movement has gained momentum over the last few years, with some 150 universities worldwide committing to selling off corporate investments in fossil fuel companies. The University of Massachusetts voted to divest its endowment of fossil fuel holdings in 2016, and Syracuse University did the same in 2015. Columbia, Georgetown, and Stanford have divested from coal companies, and other large institutional funds, such as the cities of Madison, Wisconsin; Portland, Oregon; and San Francisco, California, have also divested. And in 2018, the sovereign fund of the nation of Ireland voted to divest its fossil fuel company shares within five years. Adding yet more momentum to divestiture, in 2019 Norway's 1.1 trillion dollar sovereign fund—the world's largest—divested from companies dedicated entirely to oil and gas exploration and production, although it did retain investments in companies with renewable energy areas, such as the "carbon majors" British Petroleum and Shell Oil.

But what about the tens of thousands of universities, public institutions, and foundations that have not divested? Many argue that their fiduciary responsibility (of maximizing returns on the endowments they manage in order to give their institutions more options) supersedes the importance of divesting (see Case Study 7.1). And fossil fuel–free "green fund" options are emerging, though it may well be a few more years before such funds offer sufficient choice and diversification to make them true alternatives. As in previous generations of social activism, however, universities are contributing to public awareness that "stranded oil," the notion that some oil deposits need to be left underground, is a reality. Indeed, close assessments reveal that the only means of making needed emissions reductions are to cut subsidies to fossil fuels (discussed in Chapter 7), apply expensive carbon capture and storage technologies (detailed later), and ban extraction of entire stocks of underground fossil fuel assets (i.e., "leave the oil in the soil"). As former US President Barack Obama stated in 2015, "We are going to have to keep some fossil fuels in the ground rather than burn them" (White House 2015).

Having described the political clout of the fossil fuel industry in the United States as part of Chapter 4, we will address policies to fix this in Chapters 8 and 9. But first, let us speak to the technology of "cleaning" fossil fuel emissions, and the prospects for replacing them with renewable energy sources. In this chapter, we review some of the methods for removing carbon dioxide (CO_2) from power plant emissions and from the ambient air. We also review the outlook for various alternative energies as the world tries to decarbonize our energy sources. The prospect for increasing natural gas use as a "bridge fuel," as we transition away from coal for electrical energy production, is also be examined.

Prospects for Cooling the Earth

Prospects for getting ahead of the climate change curve come in two basic forms. The first is geoengineering the atmosphere by adding sulfate (SO_4) aerosol to the stratosphere. This approach will be discussed in Chapter 8. The second involves removing CO_2 from the atmosphere or avoiding its release in the first place. Approaches to removal in turn fall into two broad categories. One is to sequester the CO_2 before it is released to the atmosphere, and the other is to remove the "extra" CO_2 from the atmosphere itself.

CARBON CAPTURE BEFORE RELEASE

Methods for capturing CO_2 before it is released to the atmosphere can involve three basic methods: post-combustion capture and storage, precombustion capture and storage, or oxyfuel combustion. When people talk about "clean coal," these methods are what they are referring to.

Postcombustion involves having fossil fuel power plant emissions passed through chillers or filters that can remove CO_2. If the flue gases (as combustion emissions are referred to) are chilled to the point where CO_2 becomes liquid, it can be collected, pressurized, and injected underground. This "carbon capture and storage" has the potential to reduce future global energy CO_2 emissions by at least 20 percent using current technology (Haszeldine 2009). It is costly, however, and not only in terms of constructing the physical facility. The process itself uses electricity, and it is estimated that 20 to 40 percent of the power plant's electrical output would be used for capturing, pumping, and storing the CO_2 (Haszeldine 2009). The captured CO_2 would also be stored under great pressure more than 800 meters below the surface, and the storage areas would need to be monitored for leaks. Luckily, plenty of storage areas fit the requirements; in fact, the global storage capacity is estimated to be 3.2 trillion tons (sum of storage in 46 natural gas/oil reservoirs, 68 coal bed seams, and 24 saline aquifers). Recall that our current global fossil fuel consumption generates 32.5 billion tons of CO_2 (10.2 billion tons of carbon) per year across all sectors and sources. The US Department of Energy estimates that worldwide emissions will increase to approximately 43 billion tons CO_2 by 2035 (Energy Information Agency [EIA] 2018a. Using those projected emission rates, and assuming we could capture all CO_2, storage capacity could contain 76 years of emissions.

Precombustion involves separating CO_2 from fossil fuels before burning them. The fuel (usually coal, but also natural gas) is reacted with O_2 in a pressurized and carefully monitored chamber. This process does not "burn" or ignite the coal. What is released is CO, hydrogen, and a fuel gas (Wilberforce et al. 2019). The hydrogen/fuel gas is then combusted to generate power. The CO, which was separated from the other gases, is combined with O from steam under pressure, which yields CO_2 and hydrogen. The CO_2 is removed for sequestration, and the hydrogen gas can then be used again as a fuel to generate a second helping of power from the original solid or gaseous fossil fuel. Power generated through precombustion is generally called "Integrated Gasification Combination Cycle" power.

Oxyfuel is a recently developed alternative to the pre- and postcombustion options for removing CO_2. Essentially, it involves combusting the fuel in an almost pure (99 percent) oxygen atmosphere under very high temperatures. Oxyfuel power stations have an air separation unit at the intake that removes almost all gases except oxygen from the inflow to the combustion chamber. This eliminates many of the chemicals usually created by combustion with ambient air, which is 80 percent nitrogen and various trace chemicals. The process produces flue gas[2] that is CO_2, some water, and trace sulfur dioxide (SO_2). The water is easily removed by cooling the flue gas, and the sulfur is removed with limestone scrubbers or other technology that has been used in conventional pollution control for decades. Oxyfuel units are very efficient (90 percent) from a fuel-use perspective, and the capture of CO_2 from the waste gas is much cheaper than with postcombustion units (because of the purity of the flue gas). However, the air separator is expensive, and the technology is mostly in research power plants of 30 megawatts (MW) or less (Wilberforce et al. 2019). Oxyfuels are a promising area of research for alternatives to traditional combustion, but much more work needs to be done to reduce costs.

CARBON CAPTURE AND STORAGE

It may come as a surprise, but the petroleum industry has been sequestering (at least temporarily) CO_2 below ground for decades. Combined with water or brine, CO_2 is injected into the same geological formations where oil is found and, in so doing, increases the amount of extractable oil. Enhanced oil recovery (or EOR, as the process is known) is responsible for most of the underground anthropocentric CO_2 storage today, with individual operations from Saskatchewan, Canada, to Inner Mongolia injecting millions of tons per year. For example, in Tianjin, China, one million tons per year have been injected into oil wells starting in 2016. In Algeria, 1.2 million tons per year of CO_2 from natural gas extraction have been injected into saline water 1,800 m below the surface. The Sleipner Carbon Capture and Storage (CCS) project run by Norway has pumped 17 million tons into saline below the North Sea over the last 20 years (International Energy Agency [IEA] 2017). In the entire United States, a total of 65 million tons of CO_2 are injected into geological formations every year, and 60 million tons of that are for EOR. It is important to note that while some CO_2 is effectively sequestered during EOR, much is recaptured with the extracted petroleum and then reinjected. After all, sequestering the CO_2 is not the petroleum company's objective.

As of 2018, only 30 postcombustion CCS plants were operating worldwide (up from 16 in 2006), so most CCS remains part of petroleum extraction. In an interesting bipartisan twist, however, tax incentive programs for carbon sequestration were passed by the US Congress in 2018. For EOR operations, a credit of up to 35 dollars per ton CO_2 injected is possible. For nonpetroleum companies, up to a 50 dollars/ton tax credit is possible for successful CCS. This may be the beginning of monetizing carbon emissions without cap and trade or imposing taxes.

A few companies have formed specifically to take ambient-air carbon capture out of the R&D laboratory and apply it to larger scale operations, but the efforts are very small at the moment. Perhaps the 2018 tax incentives will continue or increase, but petroleum has tangible value and so companies/governments will invest in EOR to increase yields. Historically, however, they have been less motivated to spur innovation in other sequestration technology. During the last nine years, the US Department of Energy invested 1.7 billion dollars in nine CCS projects, but all except one were deemed too expensive and shut down by 2017 (Mufson & Georges 2019). Until it is widely understood that emissions of CO_2 without any regulation or control merely externalize the costs of the pollution, there will be little incentive for a company, or a country, to invest in capture and storage as the end goal of the business or project.

All of these CCS technologies are expensive and energy intensive (some could, or do, use clean energy for their operations, thereby avoiding the charge of hypocrisy for their claims to be green technologies). For example, adding CCS to an existing power plant, which could capture 82 to 88 percent of CO_2 emissions, would double its size and add between 48 and 123 percent to the cost of producing electricity. The final costs would depend on the method used to sequester the CO_2 and on the cost of fuels (Rubin et al. 2015). Costs come down after the building of the CCS facility and with economies of scale, of course (Haszeldine 2009). The current CCS activity is not keeping pace to meet the 400 million tons per year by 2025 that will be necessary as part of the United Nations effort to keep warming below 2°C (IEA 2017).

In addition to sequestering CO_2 at a concentrated emissions source (where CO_2 can be 300 times higher than the ambient atmosphere), there are ways of capturing it by filtering the air (Keith 2009). It has been known since the 1930s that passing a mix of gasses containing CO_2 through an amine[3] solution or a limestone slurry will remove the CO_2 (the process was developed to remove CO_2 from natural gas) (Rochelle 2009). Unfortunately, the process has been expensive, and unless one was using it to purify or increase yields of natural gas, there has not been a market incentive to turn the technology toward scrubbing CO_2 out of the atmosphere. In recent years, however, venture capitalists, entrepreneurs, scientists, and engineers have started companies that make use of the technology (with massive improvements from the

1930s-era systems, of course). Additionally, methods for removing CO_2 and combining it with H to make new hydrocarbon fuel have been developed. This can almost be thought of as recycling the original fuel. As of this writing, companies like Global Thermostat (United States), Carbon Engineering (Canada), and Climateworks (Switzerland) all scrub CO_2 directly from the atmosphere and sell it to petroleum firms, soda makers, or other industrial companies. If a cap-and-trade scheme or a carbon tax become widespread, this method of capturing CO_2 will become much more mainstream.

ENHANCING NATURAL CARBON SINKS

Interestingly, the efforts to remove CO_2 from the ambient air also include some rather low-tech options that make use of enhancing (one way or another) natural CO_2 fixation. One possible method makes use of enhanced mineral weathering[4] or mineral carbon storage (referred to as "mineral carbonation"). To understand how this works, we need to step back and review a bit of geology.

One of the consequences of having CO_2 in the atmosphere is that it readily combines with water vapor to produce carbonic acid (H_2CO_3). Carbonic acid gives natural rainwater a pH of about 5.6 (so slightly acidic—neutral pH is 7). Carbonic acid is not what contributes to what we call "acid rain," however. Acid rain is caused by the emission of NO_2 (nitrite) and SO_2 into the atmosphere, which form nitric and sulfuric acid, respectively, and originate from internal combustion engines or burning coal. Carbonic acid is "natural" and is actually an important driver of erosion, a process that turns bedrock into soil over geological timeframes. Rocks, such as basalt, that are rich in Ca (calcium)-, Fe (iron)-, and Mg (magnesium)-bearing silicate minerals react with carbonic acid in a moist atmosphere (rain or high humidity). The Ca and Mg are stripped from the silicate minerals and combine with the CO_3 in the carbonic acid, forming what are collectively called "carbonates." This mineral carbonation forms calcite ($CaCO_3$), dolomite ($CaMg(CO_3)_2$), and magnesite ($MgCO_3$) (Rigopoulos et al. 2018). Mineral carbonation reactions contribute to atmospheric CO_2 drawdown over geological time scales (tens of thousands or hundreds of thousands of years). In fact, periods of great mountain building (orogeny) are often associated with low atmospheric CO_2, as the new mineral surface area consumes carbon in the atmosphere. Although the reactions proceed at low rates over vast amounts of time, there has been great interest recently in enhancing mineral carbonation by using waste materials from silicate mining operations (called "quarry fines," these waste materials are similar to "tailings" from metal ore mining operations).

This finely ground material from silicate mining operations has a very large ratio of surface area to volume, which presents much more substrate for reaction than would occur for a natural basaltic hillside. The quarry fines are an unavoidable by-product of mining operations and, unlike many mine tailings, produce zero toxic runoff. To distribute the silicate mineral dust, it would have to be mixed with fertilizers that farmers spread or somehow release over large areas, ideally through some existing practice (e.g., release by crop dusting, liming, passenger aircraft). Enhancing mining activities for the express purpose of generating basaltic/silicate mineral dust would be expensive, however, and it would transform landscapes (as any mining does). Those features of the potential operations should not be overlooked. In addition, the amount of crushed silicate minerals necessary to have an appreciable near-term effect is massive. The amount needed is probably on the order of 100 gigatons/year to offset current CO_2 emissions, which are approximately 34 gigatons/year (a gigaton is equal to a billion tons). For comparison, US production of stone is on the order of 1 gigaton/year, and total world production of coal is about 8 gigatons/year IEA 2020b. Clearly, it would not be possible to offset our emissions through enhanced mineral carbonation alone. It would have to be done in conjunction with other strategies (National Academies of Science 2016).

Another way of using natural mineral carbonation is to pump CO_2 captured postcombustion into deep basalt formations. This process looks very similar to postcombustion capture as described earlier, but in this case, the CO_2 would not be injected into petroleum formations, nor would it be stored under pressure. Instead, the CO_2 would bind with the Ca and Mg in the basalt formation and form new carbonate minerals in situ, which would be stable for the foreseeable future. Basaltic formations make up approximately 10 percent of the continents and almost all the seafloor, so it may be easier to find a suitable location for postcombustion sequestration. Making use of geological formations for carbon storage has been successfully tested on a relatively small scale in Icelandic basalts (Gislason & Oelkers 2014); however, it is estimated to cost between 55 and 112 dollars per ton CO_2. As was the case for the other carbon storage possibilities, until there is a value placed on carbon pollution, it will be difficult to incentivize sequestration (except in the case on EOR, as noted earlier).

CAP AND TRADE AND CARBON TAXES

One method of adding value to CO_2 emissions is through cap and trade (or emissions trading system) arrangements (as mentioned in Chapter 4). Cap-and-trade programs exist where a regulatory body sets an emissions "cap" or maximum emissions target for CO_2 for utilities or other organizations. If a utility's CO_2 emissions exceed the cap, it is either fined or given the option of purchasing carbon credits from another utility company or organization that has emissions lower than the "cap." Cap-and-trade programs recognize that it would be much more difficult for some utilities to reduce their reliance on high-carbon fuels (e.g., many utilities are located away from coastal/

onshore wind or areas where solar photovoltaics are viable). Cap-and-trade programs are designed to meet overall CO_2 emission reduction goals while allowing flexibility for achieving them. If various cap-and-trade structures could put a value on reducing CO_2, then they will likely incentivize investors to do just that, such as by including CCS investments but also increasing efficiency. Pumping CO_2 into geological formations would no longer just be used to enhance petroleum recovery.

Another possibility for valuing carbon is through a CO_2 emissions tax. As elaborated further in Chapter 7, most economists favor a sensibly constructed carbon tax over cap and trade because a tax is predictable. Models can show how much various levels of taxation, be it 10 dollars or 100 dollars per ton, can reduce CO_2 and also impact costs for energy (Rubin et al. 2015). EIA models projected that a 10 dollars per ton CO_2 fee that started in 2014 and rose by 5 percent per year would reduce CO_2 emissions 35 percent by 2040 (relative to 2005 emissions) (EIA 2013). Just for cost perspective, a 100 dollar per ton tax on CO_2 would add about 98 cents per gallon of gasoline. As of 2019, the federal government has an 18 cents per gallon tax, and the average state tax is 24 cents per gallon. So in the United States, you are already paying 42 cents in tax on every gallon you buy (it's a few cents higher for diesel).

Fracking and Natural Gas: The Rise of a Bridge Fuel

In the 1950s, a geologist named M. King Hubbert created a function describing a theoretical resource production rate curve and applied it to petroleum production. Since then, this function has been known as Hubbert's Curve. The curve basically projects the time it takes to deplete a resource at various production rates. Obviously, using any mineral or nonrenewable resource will eventually deplete the resource. And at some point, production rate will reach a maximum and begin to slow; this is peak production and, in the case of petroleum, is referred to as "Peak Oil." There is great interest in figuring out ways of predicting when reserves of minerals, including petroleum, will run out in order to plan for the future.

When Hubbert presented his model in 1956, he projected that peak oil in the United States would occur by 1970 (Hubbert 1956). That did not happen, of course, but as Hubbert himself noted, if either the rate of production or the size of the reserves changes, then the production curve will shift. Reserves are the total amount of a mineral that can be economically extracted given the technology of the day, while a resource of a mineral is the total amount which could theoretically be extracted. Reserves, therefore, fluctuate depending on the market value (if something is more expensive, you can spend more money digging it up), mining or extraction technology, and discovery of new resources. In the case of petroleum, both the technology and the amount of extractable resource have changed in the United States. Up until the early 2000s, production looked a lot like that projected by Hubbert. However, large increases in production starting in 2010 altered Hubbert's curve dramatically (Figure 5.1).

The United States is now the world's largest producer of fossil fuels, outpacing the traditional Organization of Petroleum Exporting Countries (OPEC) nations. The growth in production has been driven by improvements in horizontal drilling (a single drill shaft sunk vertically into the earth that can be connected to many other horizontal drills, like the legs of a spider) and hydraulic fracturing (otherwise known as "fracking"). Although there have been large increases in crude oil production, the real increase has come from fracked natural gas (Figure 5.2).

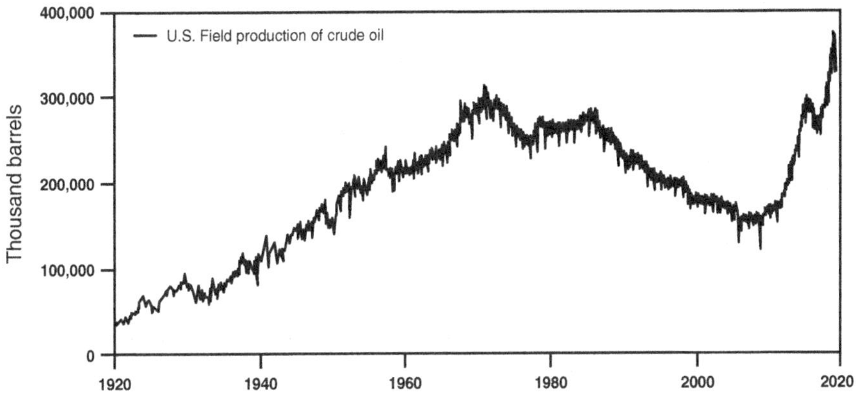

Figure 5.1 US crude oil production in thousands of barrels (a barrel is 42 gallons). The dip in production in the 2000s is the financial crisis. (Figure from the EIA 2020b).

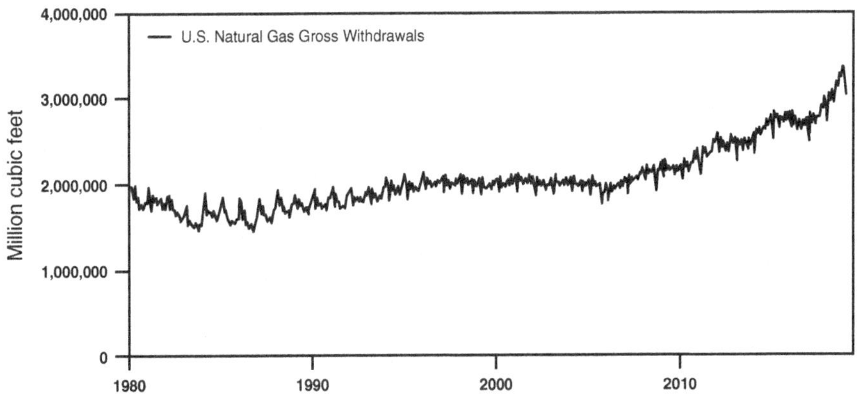

Figure 5.2 Natural gas production in the United States from 1980 to 2019. The rise in production since 2008 is because of hydraulic fracturing coupled to increased demand. (Figure from EIA 2020h https://www.eia.gov/dnav/ng/hist/n9010us2m.htm).

Natural gas produces less CO_2 per unit energy than coal and is cheaper as well (relatively little crude oil is used to generate fuels for electricity production; crude has far more value as refined gasoline and other products). The use of natural gas for generating electricity in the United States has skyrocketed in recent years. In 2000, the United States consumed over 22 trillion cubic feet of natural gas; in 2019, that was up to 30 trillion cubic feet (EIA 2019a). Coal use for electrical power was 964 million short tons in 2001 and 636 million in 2017 (a short ton is 2000 lbs). Between 2010 and 2018, the percent of US electrical production from coal dropped from 45 to 27 percent, while that of natural gas rose from 24 to 35 percent (EIA 2019a). Both of these trends are expected to continue (Figure 5.3).

Since natural gas is cleaner than coal (less unburnt hydrocarbon pollution, fewer cancer-causing combustion by-products, and lower CO_2 emissions), it is sometimes referred to as the bridge fuel toward a low-carbon energy portfolio. As the world pulls back on traditional high-carbon energy sources and invests in ramping up carbon-free alternatives (hydrogen, wind, solar, and hydroelectric), natural gas can be used to bridge the gap while the new energy infrastructure comes online. Policy issues relating to controversies surrounding consideration of a natural gas "bridge" as energy sources diversity are addressed in Case Study 5.1 at the conclusion of this chapter.

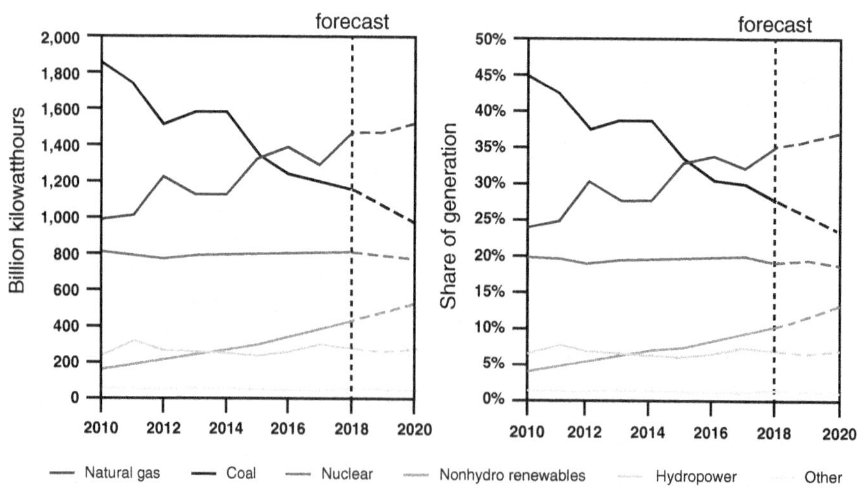

Figure 5.3 US energy mix for generating electricity in billions of kilowatt-hours (left) and percent share of the various sources to the total electrical production. Note the trends with coal and natural gas. (Figure from the EIA 2019c).

Electric Power for All: The Rise of Wind and Solar Plus the Changing Face of Biofuels and Hydropower

Traditionally, robust economies have tended to use a lot of energy. As nations develop their economies, there has always been a positive correlation between energy use and economic growth. Because the two have been perceived as linked, high energy use is sometimes considered a matter of national security (since it grows the economy). However, for developed nations, there has been a decoupling of energy use and economic growth in recent years. In the United States, for example, the split occurred after 1975 and has been increasing since then (Figure 5.4). This shows that technologically advanced nations can experience strong growth without increasing energy use. However, obtaining reliable energy continues to be a paramount national security interest for most nations.

Because the developed world was built using power from fossil fuels, and because fossil fuels are very energy dense (high energy per unit mass), they are the gold standard for power reliability. Obtaining them can be dangerous, however, and the fuels themselves are flammable and explosive in addition to being carcinogenic and toxic. Relying on them can also undermine national security for any country, since obtaining a continuous supply sometimes leads countries to do business with unsavory powers. These powers know they have leverage over their customers and therefore can exert power over policy

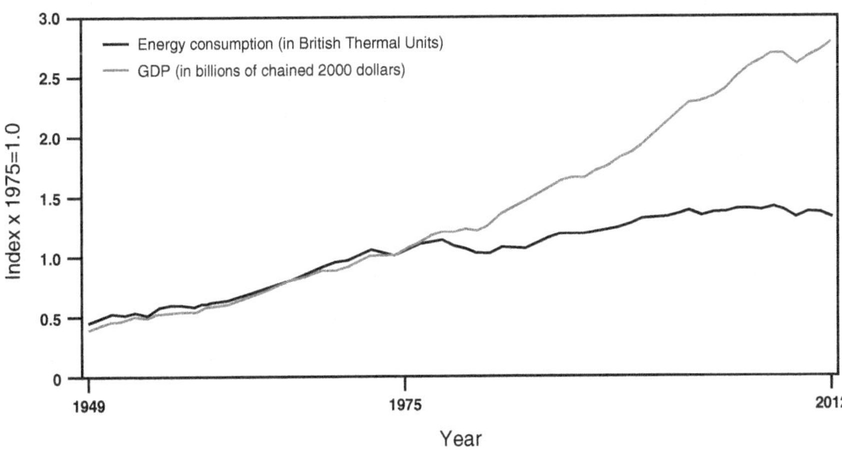

Figure 5.4 Energy use and gross domestic product (GDP) in the United States. Decoupling of energy use and economic growth began in the 1970s, and the divergence is increasing (note the dip in both energy use and GDP associated with the 2008/2009 financial crisis). (Figure from NRDC 2013).

decisions. On top of all of this, the consequences of pollution released from extracting and using fossil fuels are increasingly harmful to the environment. These environmental harms are associated with extracting and transporting the fossil fuels, of course, but the global effects of climate change resulting from the emission of odorless and colorless CO_2 are diverse and widespread. As reviewed in earlier chapters, the atmospheric CO_2 increase has caused the earth to retain more heat energy, which has destabilized most surface ecosystems while making political destabilization more likely, especially in the developing world.

So, it is in this context that alternative, non–carbon based electrical power has become an emerging goal of energy policy initiatives and investment. Indeed, during the seven-year period between 2008 and 2015, global investment in nonhydro renewables rose from 150 billion dollars to approximately 260 billion dollars, while at the same time, fossil fuel power investment held flat at 50 billion dollars. According to Bloomberg New Energy Finance, by 2050 renewable energy will supply 87 percent of electricity in Europe, 55 percent in the United States, 62 percent in China, and 75 percent in India (New Energy Outlook (NEO) 2018). Conversely, by 2050, fossil fuel's share of electricity production will shrink from 63 percent (where it is today) to 29 percent (NEO 2018). The bulk of the renewable power will be from solar and wind.

Sometimes the decline in coal production is seen as economically destructive because of the jobs lost, and this is certainly true for the individual worker. If you quantify jobs in the electrical power generation sector, however, it is clear that many more new jobs are being created in the alternative energy sector than are lost in the coal sector. In 2016, the fossil fuel electricity industry employed 135,800 people, including approximately 86,000 in coal specifically. In nonhydro renewables, approximately 475,000 people were employed that same year, almost four times as many people. The rate of job growth in the electrical generation sector between 2015 and 2016 was 11 percent for fossil fuels and 20.6 percent for nonhydro alternatives (NEO 2018). This is not to say communities that rely on coal jobs are not suffering. They obviously are, and a responsive government should help with retraining options or by diversifying the economy. However, that is beyond the scope of the discussion in this chapter (for more on the economics related to energy, see Chapter 8). We only want to point out that even with real economic losses in coal production, there has been very robust growth in the creation of cleaner electrical power.

Our planet is awash in solar energy. It is the ultimate source of all winds, ocean currents, and biological production (with a few exceptions in the deep ocean and hydrothermal springs). Solar radiation evaporates one meter of ocean water (on average) every year, which drives the global hydrological cycle, not to mention meteorological cycles. Solar irradiance smacks the top of the atmosphere with about 1,361 W/m^2. On average, about 25 percent of that energy (340 W/m^2) reaches the surface, although the actual amounts are variable depending on location and season. How, then, could we capture this energy for our needs? When we talk about "solar power" we usually are referring to photovoltaics (PV),[5] otherwise known as solar panels. However, there is also concentrated solar, which could be important for centralized electric power generation. We will discuss concentrated solar later, and focus on PV for now.

The PV systems are comprised of silica wafers stacked on top of each other like an Oreo cookie. The top of this Oreo is a silica wafer with added impurities that release electrons when exposed to specific wavelengths of light. Those electrons flow through a circuit toward the bottom silica wafer, which has impurities that are "missing" electrons. So, when the PV wafer is struck with light, the electrons spontaneously flow from the wafer with excess free electrons toward the wafer with missing electrons. The electrons move as if on a conveyer belt through the circuit. As this happens, power is generated. The panels generate power with zero moving parts, which means they don't break and have very low maintenance costs. (Some solar systems are mounted on racks that track the sun, so those have moving parts, of course).

Historically, the efficiency of a PV cell was low and the cost was high. In 1979, it cost 79 dollars per watt. Today, the cost is 37 cents per watt. The efficiency of a PV cell in the 1970s was under 10 percent. Today, commercially available PV cells are greater than 22 percent efficient, and efficiencies achieved in research laboratories are as high as 40 percent (King et al. 2007). Over the past 40 years, every time PV capacity doubled, the price dropped by 28.5 percent (NEO 2018). Solar power over that time period has been approaching exponential growth, and in 2018, over 500 gigawatts (GW) were produced globally. In both 2017 and 2018, 100 GW of new capacity came online, largely driven by China. But even with this rapid growth, the solar power's share of the global electrical power capacity is still only 2.6 percent: a drop in the bucket of human energy use (IEA 2019a). The US Energy Information Administration projects that this will rise to at least 11 percent by 2050 (EIA 2018b), and the IEA estimates 16 percent, with the possibility 21 percent if world governments adopt policies that put carbon-free electrical power atop of the list of priorities (IEA 2018b). Barring a major breakthrough in PV technology, however, solar PV will only be a modest component of the low-carbon energy future.

One interesting feature of solar PV is that it can be locally distributed rather than centralized (meaning generated in one location). Combustion and turbine-based power require large power plants, which then disburse the electricity by high voltage lines throughout the grid. Solar power can be centralized (e.g., in a large solar farm), but it doesn't need to be. The PV systems can be attached to a single unit that needs power or can be placed on rooftops for local consumption. In the United States, companies like Sunrun will put solar panels on your roof for very low or no cost and then sell the power generated back to you. Homes and businesses that own solar panels can sell (cash or credit) the excess electricity they generate back to their local utility company (a practice called "net metering"). The distributed nature of solar power makes it a very promising technology for parts of the world with a poorly developed energy infrastructure. Just as the spread of cell phones in sub-Saharan Africa bypassed the need to develop a network of telephone poles and hardwired systems, so distributed solar power might allow communities outside the cities to generated their own local power.

One of the main criticisms of solar has always been related to energy storage—that is, how does one generate power at night or on cloudy days? The answer to that question has always been to store excess production when the sun shines, then release that power later. This has been accomplished in some cases by using the excess power to pump water to a high elevation reservoir, then reclaiming the power by allowing the water to flow back downhill through an electric power generating turbine. There are plenty of places, however, where geography makes this improbable. Therefore, investment dollars have been flowing into building better batteries, and it is paying off. In the United States, cumulative large-scale battery power capacity is approximately 750 MW (EIA 2018b). However, lithium-ion battery capacity is expanding every year, and 1,291 GW of capacity is expected by 2050 (NEO 2018). This massive investment is driving down costs due to increases in efficiency and economies of scale. Prices have dropped dramatically in recent years. In 2012, it would have cost 600 dollars per kilowatt-hour, but that is down to roughly 200 dollars per kilowatt-hour with projections of 70 dollars per kilowatt-hour by 2030 (NEO 2018). That is a 67 percent reduction by 2030 relative to today's costs (NEO 2018).

It must be noted, however, that batteries cannot sustain peak power very long (since they merely release potential energy rather than generating energy). Another critical measurement therefore is cumulative "power capacity," which takes time into account, not just peak power. In the United States, cumulative capacity is approximately 900 MWh. These battery systems are largely used to supply uninterrupted power delivery should there a short-term failure of the main grid for whatever reason. For example, in 2017, Tesla activated the largest lithium-ion battery system on the planet. The system was built in South Australia, where power outages from heat and storms are common. Millions of people had lost power in a statewide blackout during September 2016, and that prompted government officials to find backup systems. Tesla's system, called the Hornsdale Power Reserve, is a 100-MW battery that can power 30,000 homes for eight hours. That power has to be replenished, of course. The system stores energy from an adjacent wind farm, and the batteries are only designed to provide backup power should the generators fail, which they have—twice (Fung 2017). In the future, however, batteries likely will be used not just to ensure uniform delivery of power, but also to store and distribute excess energy generated by alternative sources during periods of optimal wind or sun.

Another type of solar power that fits the centralized power plant model is "concentrated solar" (also called "solar thermal"). Unlike PV, concentrated solar makes use of solar heat energy. Whereas PV systems try to prevent heating (too much heat lowers efficiency for PV), concentrated solar needs to enhance it. This is how it works: Parabolic mirrors are arrayed around a centralized pipe that contains oil or a sodium solution. The mirrors are arrayed so that all the sunlight striking them is reflected directly on the pipe, causing the fluid to become extremely hot (400°C or higher). This hot fluid flows through a heat exchanger, which transfers the heat to water, making steam (all without the fluids mixing). The steam then passes through a turbine the same way steam from a typical large, fossil fuel power plant would.

In 2016, global capacity of concentrated solar electrical power was 4.3 GW (equivalent of about nine large, 500-MW, coal-fired power plants). In 2018, Spain and the United States were the world's largest concentrated solar producers, but they are projected to be overtaken by China by 2023. Examining trends worldwide, growth in concentrated solar is accelerating. Between 2012 and 2017, it experienced 32 percent growth. That should increase to 87 percent between 2017 and 2023, and by 2050, 11 percent of all global electrical power could come from concentrated solar (IEA 2018). Combine this projected growth with that of solar PV, and the IEA (2018) projects that 37 percent of global electrical power will be solar by 2050. This is not a trivial percentage by any measure, yet it reminds us that solar alone will never replace fossil fuels for our electrical power needs.

Wind Energy

The second major player in the advancing capacities of alternative power is wind energy. As with solar power, the world is experiencing rapid change with regards to wind energy cost and deployment. In the United States, wind power rose from 55 million MWh in 2008 to 275 million MWh in 2018 (EIA 2019b). The US Department of Energy projects that by 2030 total wind power capacity will be 224 GW, rising to 404 GW by 2050 (US Department of Energy 2019). Globally, the trends are similarly dramatic. In 2010, global capacity was 194 GW, and that rose to 515 GW by 2017. By 2023, it is projected to be 839 GW (IEA 2018). Growth is 30 percent per year, and every time capacity doubles, the cost goes down by 19 percent (IEA 2018). In 2017, about four percent of global electricity was generated by wind power; by 2050, wind is projected to make up 18 percent of global electrical power (IEA 2018).

Wind turbines may be located on areas of land with consistent wind speeds of greater than 17.1 km/h (onshore wind), or they may be placed in the ocean (offshore wind). The onshore global wind capacity represented 96 percent of wind power capacity in 2017, dwarfing offshore wind. Onshore facilities are earlier to maintain, easier to link to existing grid networks, and cheaper to build than offshore facilities. Onshore can also be combined with other types of land use, since the turbine towers are 80 meters high on average. They are large enough that crops may be grown or livestock grazed under the turbines without harm. Offshore is expensive because marine construction and transmission infrastructure is expensive. Nonetheless, offshore wind is expected to grow from 18 GW in 2017 to 52 GW by 2023 (IEA 2018). Offshore winds are more reliable than onshore winds, and the offshore turbines are operating at peak capacity more often than onshore wind facilities.

Two of the most frequent criticisms of wind power are noise and harm to wildlife, particularly birds. (President Trump stated on April 2, 2019, during the Republican Congressional Committee Spring Dinner, that wind turbine noises cause cancer, but there are zero data supporting his assertion). While the turbines of the 1970s and 1980s were relatively low to the ground and a bit noisy, the onshore turbines of today are far taller, and the turbine blades are much bigger (they don't rotate as fast). So noise has decreased drastically in modern designs. Turbines do kill birds, however, particularly birds of prey who seek out high places to perch and look for lunch. Efforts are being made to minimize bird strikes (i.e., making the top of the turbine prop a less attractive perch or generating a nuisance noise that the birds do not like but humans cannot hear). Even so, it is estimated that between 100,000 and 400,000 birds are killed by turbines worldwide. That is a lot of birds, but let us put the number in perspective. The *Exxon Valdez* oil spill in 1989 killed approximately 500,000 birds. Telecommunication towers kill between 4 and 10 million birds. Cars kill approximately 57 million birds, and glass buildings (windows) kill about 99 million birds. And the largest killer of birds are house cats, which kill between 1.4 and 3.7 billion birds (that is billion with a "b") (Lars et al. 2013). When thinking about environmental harms, risk assessment, or costs and benefits, it is useful to keep things in perspective.

Present and Future of Hydroelectric power

Humans have used hydropower for millennia, initially to mill grain into flour or to do some other type of mechanical work. In 1935, the first hydroelectric plant was built in Scotland, and more were quickly installed on most of the Northern Hemisphere's large rivers. Today, hydropower is by far the largest source of renewable energy. Sixteen percent of global power is from hydroelectric facilities, amounting to 1,200 GW of total capacity. The era of large dam building may be over, however, as most of the appropriate waterways have already been dammed and new large projects in places like China and Brazil are being delayed because of land-use and environmental concerns. Most currently proposed projects are to dam large rivers in low-relief landscapes (as opposed to being in "V"-shaped valleys). This means that large areas of land will flood as the water pools horizontally rather than flooding an existing, steep-sided valley. The tendency of new projects to displace people and flood vast areas of land makes some projects controversial. Therefore, the growth of hydropower is projected to be relatively modest; approximately 123 GW were added by 2023 (IEA 2018). Twenty-percent of that growth is going to be for pumped storage projects, which involve "banking" excess energy produced by pumping water uphill and holding it behind a dam. This high-potential-energy water can then be released when power is needed later (80 percent of the power used to pump the water uphill can be recovered) (Boyle 2012). (This process was described in more detail earlier in the chapter).

Harnessing the power of tides is an attractive area in which to look for alternative energy, since tides are very predictable relative to solar or wind power. However, for tidal electrical power to be economical, you need at least three meters of tidal amplitude, and most coastlines do not experience tides that large. The two largest tidal power stations are the Sihwa Lake Station in South Korea and the La Rance station in France, which produce 254 and 240 MW, respectively. While there are a few other smaller facilities around the world, producing tidal power is not cost-effective now that the price of solar and wind have dropped so dramatically.

Nuclear Power

After hydropower, the next largest source of alternative electrical power is nuclear. Nuclear power is an alternative energy, since it is not fossil fuel–based. However, it is not considered to be renewable, as it does consume uranium fuel (albeit slowly). Worldwide, approximately 11 percent of electrical power is derived from nuclear fission in 450 power plants. The power growth rate is projected to be 1.6 percent per year through 2040 (EIA 2017). Global nuclear capacity from 1955 through 2015 is shown in Figure 5.5.

Most of the world's nuclear power plants use uranium as a fuel. Uranium ores tend to have about 0.7 percent U-235 (meaning that its mass is 235 atomic mass units), which is the fissionable isotope of uranium (the other 93.7 percent is U-238 [238 atomic mass units], which is not usable in reactors). The fuel used in reactors must be "enriched" in U-235 to between three and four percent in order for the self-sustaining chain reactions to take place. When an atom of U-235 is hit with a high-energy neutron, it splits into Kr-92 and Ba-141 and releases heat, radiation, and three neutrons. Those neutrons hit other U-235 atoms, splitting those, and this repeats spontaneously (hence the term "chain reaction").

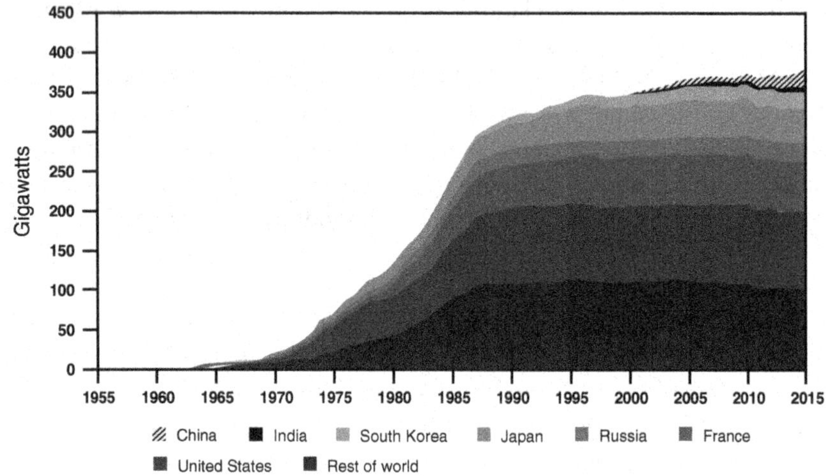

Figure 5.5 Worldwide nuclear power capacity. Approximately 11 percent of global electricity is generated by nuclear power plants. Growth in the future will be dominated by China and India, which are very aggressively investing in non-carbon electrical power for their continued rapid development. (Figure from EIA 2016b).

The enrichment process involves spinning uranium gas in centrifuges to separate the isotopes, which are different masses. Incidentally, this enrichment process is also used when making nuclear bomb material, except bomb-grade uranium needs to be 90 percent U-235. To generate electrical power, the U-235-enriched nuclear fuel is placed into rods (not surprisingly called "fuel rods"), and the rods are grouped alongside "control rods" that absorb neutrons. The control rods can be moved into and out of the reactor vessel where the fuel rods are kept. This allows control of reaction progression, and hence the amount of power the plant produces can be matched with demand. The heat that is released when the U-235 splits is what runs the plant. This heat is used to generate steam, which then turns a conventional turbine, just as you would see in a fossil fuel power plant. If too much heat builds up in the reactor vessel because of a failure of the heat transport system or because of control rod failure, the reactor will melt down.

Meltdowns are what the word implies, the uncontrolled heat melts the fuel rods, and uncontrolled chain reactions start taking place. Many environmental groups and alternative energy promotors traditionally have not supported nuclear energy because of the risks associated with core meltdowns. Indeed, after the Fukushima Daiichi meltdowns in 2011, most of the world pulled back on its nuclear portfolio (China being a large exception). For example, in 2011, Germany produced 25 percent of its electricity from 17 reactors. Today, it produces 12 percent using seven reactors. Coal power plants have replaced the decommissioned nuclear plants. US nuclear capacity from 1960 to 2020 is shown in Figure 5.6.

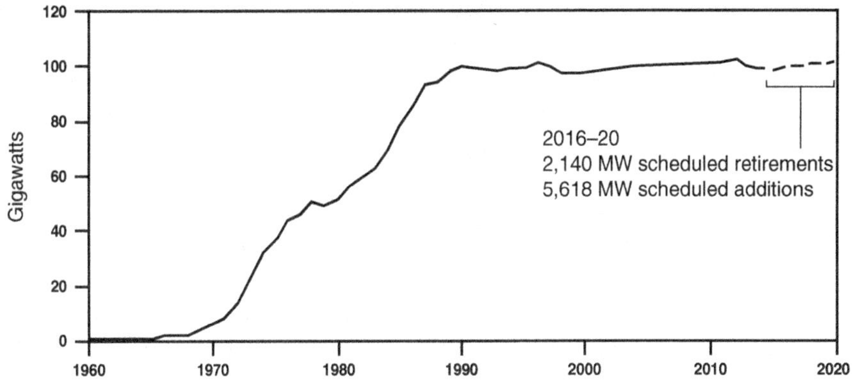

Figure 5.6 US nuclear capacity since 1960. Growth stopped in the late 1980s and early 1990s because of public unease with building new reactors and their expense. Approximately 20 percent of all US electrical power is from nuclear plants. (Figure from the EIA 2015).

Biomass/Biofuel

Biomass was the first source of concentrated and transportable energy for humans. It has never been completely replaced by fossil fuels in many parts of the world. Although wood is mostly made up of carbon-rich lignin (i.e., what makes wood "woody"), it still contains about half as much energy per unit mass relative to coal and therefore is less efficient at producing heat per unit mass. However, it is renewable, draws carbon out of the atmosphere, and is relatively easy to obtain compared to coal in many places. Other nonwoody biomass with less carbon density than wood (mostly cellulose) is also used to generate heat, even in modern power plants. For example, the Netherlands has a 250-MW power plant that burns straw for heat (Boyle 2012). Straw and other cellulose-rich plants can have an advantage over wood, because they grow much faster and can be used in places that are not suited for tree growth. Other types of biomass that are used to generate heat in power plants include algae and municipal waste.

Although biomass (wood/straw/dung) is still used for purposes ranging from cooking to generating heat in power plants, technologies today can also transform biomass into biofuels for other uses. For the purposes of this discussion, a biofuel will refer to a liquid, transportable material, usually used in vehicles. Dominant examples of biomass that can be used to generate biofuels are animal waste digested to biogas (methane) as well as corn, sugarcane, switchgrass, or other fermentable starchy plants (e.g., for ethanol, a gasoline additive or substitute).

Worldwide, biomass energy produced two percent of the electricity (EIA 2019b). Unlike the other forms of alternative energy, biofuel in the developed world is heavily used in the transportation sector. Biomass and biofuels provided about five percent of total primary energy use in the United States in 2017 (EIA 2018c). Of that five percent, almost half (47 percent) was ethanol or biodiesel. The other half was from wood-derived biomass (44 percent), and 10 percent was from biomass in municipal waste (EIA 2018c).

Is a Zero-Carbon Energy Sector Even Possible?

Through the projected growth of various alternative energies discussed so far, only 23 percent of global electrical power infrastructure is projected to rely on fossil fuels by 2050 (EIA 2016a). However, there is no path to complete decarbonization if power demand rises at expected rates and policies are not adopted that speed the transition to renewables or alternative energy. Estimates for electrical power sources in the United States through 2050, based on current trends and policy, show that approximately 43 percent will derive from alternative energy (Figure 5.7). Policy shifts could increase or decrease the projection.

For example, it is possible that policy for enhanced nuclear power could shift to generate 20 percent of global electrical power, but that policy might not be politically possible. Another example of a policy that could dramatically

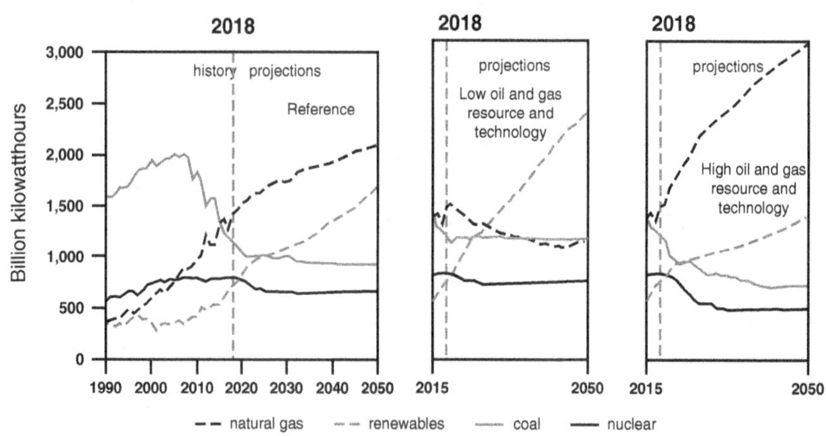

Figure 5.7 Projected sources of US electrical power through 2050. The "Reference" case on the left represents the EIA's best assessment of how US and world energy markets will operate through 2050, based on many key assumptions (EIA 2019c). The middle graph represents a scenario where natural gas becomes more expensive (either reduction in resources or a carbon tax) and substantial investments are made in sustainable technology. The right graph shows projections based on investments enhancing natural gas production and a reduced emphasis on alternative energy technologies (Figure from the EIA 2019a).

alter the projections is the Clean Power Plan put forth by the Obama administration (officially titled "Carbon Pollution Emission Guidelines for Existing Stationary Sources: Electric Utility Generating Units"). This plan would result in reduced power plant carbon emissions 32 percent relative to 2005 by 2030. However, the US Supreme Court blocked its implementation in 2016, and the Trump administration had no interested in pursuing that policy. As of the time of this writing, the regulations in the Clean Power Plan remain "on the books" but in limbo, since a US regulation cannot be discarded without showing a court why it is no longer necessary (just as a new regulation cannot be implemented without showing a court why it is necessary).

Looking globally, an IEA (2017) report declares net-zero CO_2 emissions from the energy sector is possible by 2060, but only with very aggressive and sustained shifts in policy to drive the shift. The proportion of low-carbon primary energy would have to shift from today's 16 percent (global cumulative) to 67 percent to limit global temperature rise to 2°C by the end of the century. Energy efficiency would also have to improve in order to reduce emissions to meet the 2°C target.

On the transportation side of things, decarbonization will take much longer than is projected for electrical infrastructure. Although electric cars are becoming more prevalent (particularly in China), it will take decades to shift transportation services to electrical systems. Policy can speed up or slow down the shift, of course. In the United States, federal tax policy rewarded purchasing electric cars and installing PV or other alternative electrical sources (at least through 2020). If there was a carbon tax, that would add incentives for spending on electrical vehicle infrastructure and increase R&D for more efficient, longer range batteries. In addition to tax policy, regulations regarding fuel efficiency could reduce the consumption of gasoline, saving money for consumers and reducing emissions from the most intractable sector. Under the Obama administration, the United States increased the Corporate Average Fuel Economy (CAFE) standards to between 40 and 41 miles per gallon and limited emissions to 163 grams of CO_2 per mile. With the arrival of the Trump administration in 2017 however, there were systematic attempts to dismantle the CAFE standards, as well as all alternative energy-friendly policies of both the George W. Bush and Obama administrations. It is likely that a Biden administration would push towards greater use of alternative energy and bring back the Obama administration's push for higher CAFE standards.

Transportation energy will continue to be dominated by petroleum fuels as we look toward 2050, although that dominance is expected to decline a bit. The EIA (2016a) projects that the share of petroleum and other liquid fuels will decrease to 88 percent, down from 96 percent today. Although that is an eight percent decline, the total amount of petroleum fuels consumed will rise (based on projections of current consumption) as more of the world develops (Figure 5.8).

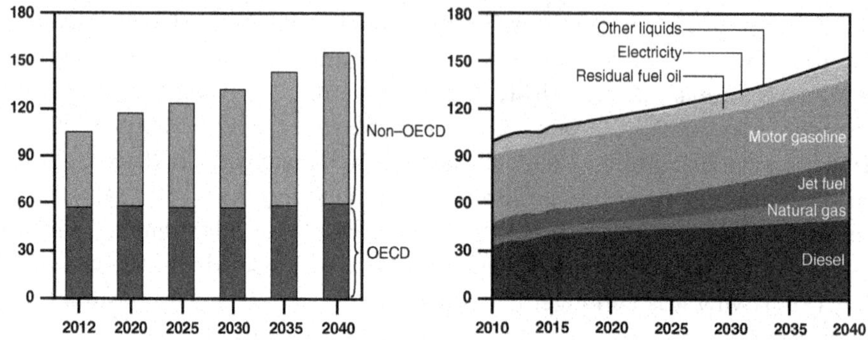

Figure 5.8 The histogram on the left shows historical and projected transportation energy consumed by country grouping. Units are quadrillion British Thermal Units (BTU). Organization for Economic Co-operation and Development (OECD) are the developed countries. (Figure from the EIA 2016a).

The reality is that absent some policy shifts that incentivize a more rapid transition to alternative energy, plus the implementation of carbon sequestration from emissions or ambient air, CO_2 concentration will continue to rise in the atmosphere, and our global experiment will continue. The Paris Agreement was a good first step. Although the agreement was not going to solve the problem, everyone knew that, and the intention was to revisit the goals after examining how effective their various methods to reduce CO_2 emissions were working. In this case, a cliché would illustrate: Every journey begins with a first step. We can decarbonize almost 75 percent of our electricity, and if we accelerate infrastructure development for electric vehicles (as well as incentivize individual investment with tax breaks or similar), we could reduce emissions substantially. Ultimately, reducing the amount of power people use will not be enough. Programs for CO_2 sequestration will be needed in addition to reducing CO_2 production.

COVID-19 Energy and Emissions Impacts

In the first quarter of 2020, 54 percent of the global population was under some sort of stay-at-home order or lockdown. Energy demand and pollution emissions plummeted in an unprecedented manner. Much of the electrical usage during this period resembled weeks of Sundays, with average weekly demand decreasing 25 percent in areas with a full lockdown and 18 percent in areas with a partial lockdown or stay at home order (IEA 2020a). The reduction in energy consumption was not evenly spread across all energy sources, however. Coal was down 8 percent for the first quarter 2020 relative to the same period in 2019. Oil demand was down five percent, since road and air activity declined 50 percent and 60 percent, respectively. Natural gas demand decreased two percent (IEA 2020a). Although obviously driven by the COVID-19 response, some of the declines were also due to a mild winter. Globally, energy demand during the first quarter declined 3.8 percent relative to the same period in 2019.

The outlook for the rest of 2020 at the time of this writing was a global energy demand decline of six percent (IEA 2020c). To put that in perspective, the projected decline is between four and seven times larger than the drop observed in the aftermath of the 2008 financial crisis (Figure 5.9). Interestingly, use of alternatives went up 1.5 percent. This was because new solar and wind projects came online, and alternative energy is not asked to reduce output when demand is low (Figure 5.10) (IEA 2020a). Record high hourly use of these alternative energy sources were observed in both Europe and the eastern United States.

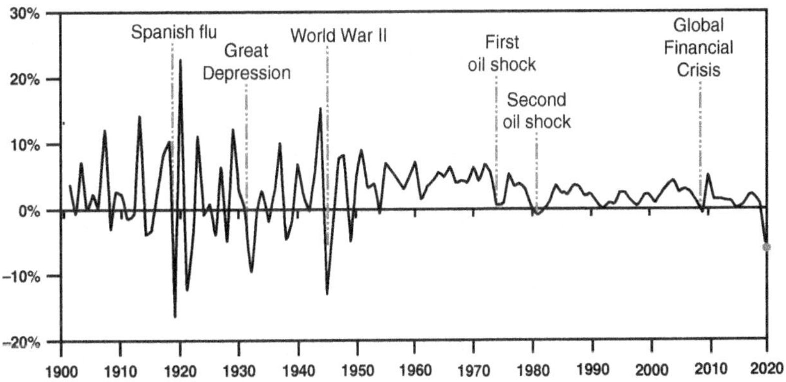

Figure 5.9 Changes in global energy demand during the last 120 years. (Figure from the IEA 2020a).

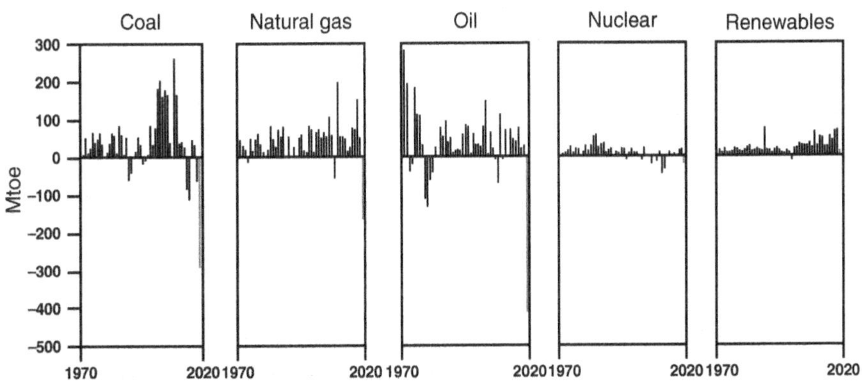

Figure 5.10 Changes in global energy demand by fuel over the last 50 years. Mtoe (y-axis) is million tons oil equivalent (Figure from the IEA 2020a).

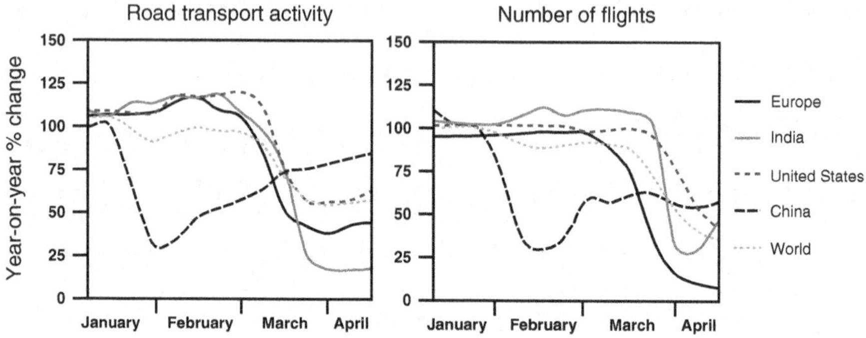

Figure 5.11 Percent change in road and air transportation during the first four months of 2020. (Figure from the IEA 2020a).

Because less fossil fuel was being consumed, daily global CO_2 emissions in April 2020 dropped by 17 percent relative to April 2019, and half of that drop was in the transportation sector (Le Quere et al. 2020). The IEA reports that road and air transport was down in April 2020 approximately 48 and 55 percent, respectively, from what it was in January of that same year (IEA 2020a, Figure 5.11). CO_2 emissions are expected to decline at least eight percent for the year, or 2.6 gigatons less CO_2 than last year. If the projections hold true, 2020 will be the lowest emissions year since 2010. To put the decline in perspective, the 2009 financial crisis lowered CO_2 emissions 0.4 percent (Le Quere et al. 2020). Advocates of alternative energy and policymakers concerned with the looming climate change threat are beginning to see the changes we are being forced to go through as an opportunity to hasten the transition to a much less carbon-intensive global economy.

EXERCISES

Below are two exercises meant to engage readers in thinking about applying real data to solve some problems or better understand trends or perspectives brought up in the chapter. The exercises are simply given as potential food for thought and are a way for you to evaluate data without us (the authors) simply telling you what the data show.

The first exercise looks at how realistic the US obligations under the Paris Agreement obligations would be, and perhaps gives a view to how we could even get ahead of those commitment requirements (as we already know they do not go far enough as goals). Extra investment in alternative energy now may have additional pay back over the long term.

Continued

The second exercise puts the reader in the place of property owners who have been approached with an offer to extract natural gas from their land. In the exercise, the reader will make a value estimate of the natural gas deposit and weigh that estimate with what the mining company is offering. This kind of assessment is being made by tens of thousands of US citizens who live atop shale plays (as gas bearing geological formations are called) made viable by hydraulic fracturing (fracking). The reader may get an appreciation of why so many property owners are keen on granting mineral rights to their land.

Exercise 5.1 US Commitments in the Paris Agreement

In the Paris Agreement, the United States committed to reduce its greenhouse gas emissions by 26 to 28 percent below the 2005 level by 2025. US emissions of CO_2 in 2005 were approximately 6,055 million metric tons (EIA 2011). The CO_2 reduction goals, therefore, are 1,574 million metric tons (from the year's total), making the total CO_2 emissions in 2025 as limited to 4,481 million metric tons.

1. Considering the energy portfolio of the United States and projected changes in energy sources moving toward 2050, how much reduction do you see taking place as natural gas continues to replace coal in electricity generation? (Hint: take a look at information from the EIA).
2. The United States is seeing an increase in renewables (as you have read in this chapter). Given current trends with renewables, how much will carbon emission–free renewable power "save" in terms of carbon emission reductions?
3. Try and find a scenario (again, starting with the EIA is a good idea) where investment in renewable energy becomes a priority of the federal government rather than an afterthought. What would be the projected carbon emission reductions?

Exercise 5.2 Calculating the Value of Natural Gas Extraction for Property Owners.

The advent of fracking and horizontal drilling greatly expanded the production of both natural gas and heavier liquid hydrocarbons in the United States starting in the mid 2000s. As a result of this, the United States became the world's largest petroleum producer in 2012 (Figure 5.12).

The price of natural gas, which fluctuates widely, as do the prices of all hydrocarbon commodities, dropped as vast quantities became available in the marketplace. This was great for consumers (both individuals and utilities looking to get out of coal), but it has been hard on natural gas

Continued

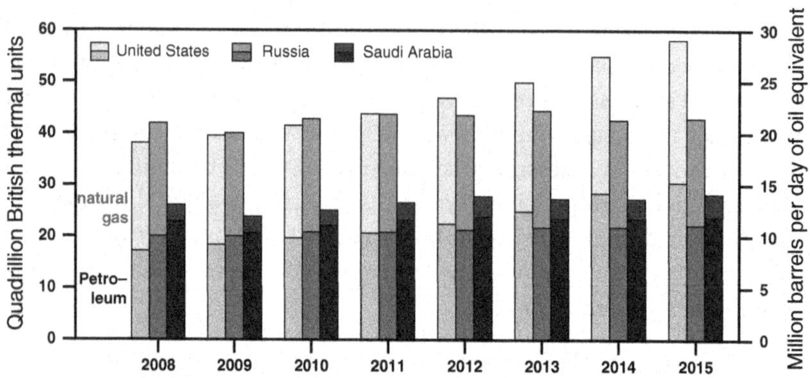

Figure 5.12 Petroleum production from the United States, Russia, and Saudi Arabia from 2008 to 2015. The United States became the world's leading producer of petroleum in 2012 because of hydraulic fracturing (fracking) and horizontal drilling. (Figure from the EIA 2019e).

producers as their profit margins shrank. (For more on natural gas pricing, see https://geology.com/articles/natural-gas-prices/).

Imagine you own 200 acres of land in western Pennsylvania that sits on top of the Marcellus Shale. You have watched your neighbors sell mineral rights to their land and seen obvious signs of an improved standard of living in terms of houses, cars they drive, private schools their children attend, and so on. You have also seen some of the downsides of natural gas drilling, as massive trucks bring water and equipment along farm roads and turn tiny crossroads into Wild West–style boom towns.

Now, you have been approached by a firm to obtain rights to drill on your property. They are promising to pay 200,000 dollars over six years for exclusive rights. They say this number comes from the expected flow rates over six years and 12.4 percent royalties on the wellhead price, which would be locked in at 3.35 dollars per 1,000 cubic feet. As recently as 2008, the wellhead price for 1,000 cubic ft of natural gas was 10 dollars, and the company says that since the price so low right now (3.35 dollars, because of increased supply, as mentioned earlier) (EIA 2019a), the 12.5 percent is the highest royalty they can offer (EIA 2019d) (Figure 5.13).

Generally, the price that a buyer pays for natural gas is most influenced by three factors: (1) the quantity of gas being purchased, (2) the amount of processing that has been done to prepare the gas for the buyer, and (3) the type of transportation and transit distance required to deliver the gas to the buyer. Although there are plenty of buyers for natural gas in the Midwest, of which western Pennsylvania could be seen as a part, the company might get a higher price for selling the natural gas to Europe.

Continued

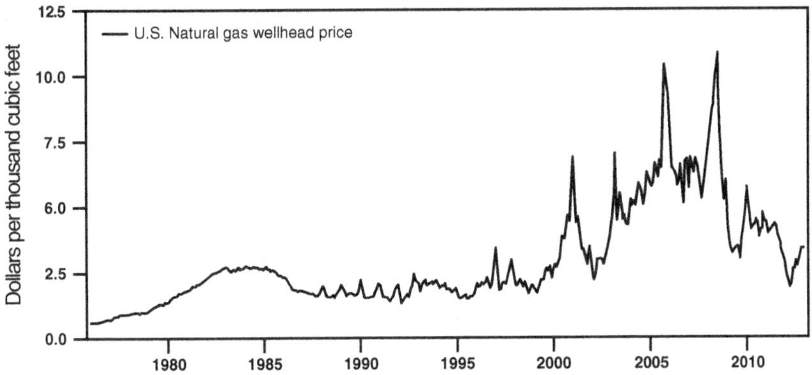

Figure 5.13 US natural gas price per 1,000 cubic feet at the wellhead. Residential consumers will pay two or three times the wellhead price after accounting for refining, delivery, and transportation among other factors. (Figure from the EIA 2019d).

You have no way of knowing what the ultimate buyer will pay for the gas and so must take the company's price at face value. You did know that the wellhead price is two to three times lower than what a typical homeowner pays, but it is customary for royalties to be based on wellhead and not final purchase price.

In any case, using on the following data and equation 5.1, evaluate the company's offer based on your own assessment of the value of your natural gas deposit:

1. Use a 12.4 percent royalty rate.
2. Use the US Department of Energy wellhead price per 1,000 cubic feet = 3.35 dollars per 1,000 cubic feet of gas (latest data, which are from 2012: https://www.eia.gov/dnav/ng/hist/n9190us3M.htm).
3. Use three different production rates and time periods. Production rates vary greatly with deposit, global price, and age of the well. In the Marcellus Shale, flow rates are usually between 1 and 2 million cubic feet per day for the first few months (there are a thousand times a thousand cubic feet in a million cubic feet). Production can then drop dramatically (see https://geology.com/royalty/production-decline.shtml). Based on your neighbor's experience, you can expect the first six months' average yield to be 1.1 million cubic feet per day (cfd), the second six months' average yield to be 0.5 million cfd, and the next five years' average yield to be 0.2 million cfd.
4. Use a property size of 200 acres, and assume you are the exclusive owner so that the royalties will not have to be subdivided.

Continued

Multiple the variables together to estimate the projected royalty amount:

$$\text{Royalty} = r * p * f * t \quad (5.1)$$

where

r = royalty rate (decimal, not percentage)
p = price
f = flow rate per day (in thousands of cubic feet)
t = time in days

Divide the periods for evaluation in to blocks based on the flow rates above: two six-month blocks and a five-year block. What are the expected royalties? Should you reject the offer? What would the royalties be if the price per 1,000 cubic feet rose to the 2010 level of 5.70 dollars? Assuming that your annual income is equivalent to the US Department of Agriculture forecast for 2019/2020, or about 76,800 dollars (USAD 2020), would you continue exploring fracking on your property?

NOTES

1. The report calls for "unabated" extraction, meaning without the benefit of carbon capture and storage, a process discussed in Chapter 7, which reduces the damage of fossil fuel extraction but at a high cost.
2. Exhaust gas from combustion power plants is called flue gas.
3. Amines are organic compounds with a side chain of NH_3.
4. Weathering is the loss of materials over the long term from exposure to the atmosphere. It is the natural process that makes soils from minerals in rocks.
5. Photovoltaic refers to the production of electrical current between two semiconductors in the presence of light.

REFERENCES

Boyle, Godfrey, ed. 2012. *Renewable Energy: Power for a Sustainable Future.* 3rd ed. *Oxford University Press and Open University, Oxford.*

Energy Information Administration (EIA). 2011. "Emissions of greenhouse gases in the US." Accessed June 23, 2020, at https://www.eia.gov/environment/emissions/ghg_report/ghg_overview.php

Energy Information Administration (EIA). 2013. "Further sensitivity analysis of hypothetical policies to limit energy-related carbon dioxide emissions." Accessed June 23, 2020, at https://www.eia.gov/outlooks/aeo/supplement/co2/

Energy Information Administration (EIA). 2015. "Despite recent closures, US nuclear capacity is schedule to increase by 2020." Accessed June 12, 2019, at https://www.eia.gov/todayinenergy/detail.php?id=23592

Energy Information Administration (EIA). 2016a. "Transportation sector energy consumption." Accessed June 23, 2020, https://www.eia.gov/outlooks/ieo/pdf/transportation.pdf

Energy Information Administration (EIA). 2016b. "Recent increases in global nuclear capacity led by Asia." Accessed June 12, 2019, at https://www.eia.gov/todayinenergy/detail.php?id=28112

Energy Information Administration (EIA). 2017. "EIA forecasts growth in world nuclear electricity capacity, led by non-OECD countries." Accessed May 28, 2019, at https://www.eia.gov/todayinenergy/detail.php?id=33672

Energy Information Administration (EIA). 2018a. "Carbon dioxide emissions by fuel." Accessed April 19, 2019, at https://www.eia.gov/environment/

Energy Information Administration (EIA) 2018b. "US battery storage market trends." Accessed May 24, 2019, at https://www.eia.gov/analysis/studies/electricity/batterystorage/

Energy Information Administration (EIA) 2018c. "Biomass explained." Accessed May 30, 2019, at https://www.eia.gov/energyexplained/index.php?page=biomass_home

Energy Information Administration (EIA). 2019a. "US natural gas consumption" Accessed May 15, 2019, at https://www.eia.gov/dnav/ng/hist/n9140us2a.htm

Energy Information Administration (EIA). 2019b. "US renewable electricity generation has doubled since 2008." Accessed May 27, 2019, at https://www.eia.gov/todayinenergy/detail.php?id=38752&src=email

Energy Information Administration (EIA). 2019c. "Annual energy outlook 2019 with projections to 2050." Accessed June 10, 2019, at www.eia.gov/aeo

Energy Information Administration (EIA). 2019d. "Natural gas wellhead price." Accessed June 23, 2020, at https://www.eia.gov/dnav/ng/hist/n9190us3M.htm

Energy Information Administration (EIA). 2019e. "The US leads global petroleum and natural gas production with record growth in 2018." Accessed June 23, 2020, at https://www.eia.gov/todayinenergy/detail.php?id=40,973#

Energy Information Administration (EIA). 2019f. "Short term energy outlook, January 2019" Accessed June 23, 2020, at https://www.eia.gov/todayinenergy/detail.php?id=38053#

Energy Information Administration (EIA). 2020. "Petroleum and other liquids." Accessed December 2, 2020, at https://www.eia.gov/dnav/pet/hist/LeafHandler.ashx?n=PET&s=MCRFPUS1&f=M

Fung, Brian. 2017. "Tesla's enormous battery in Australia, just weeks old, is already responding to outages in 'record' time." *Washington Post.* Accessed December 15, 2020.

Gislason, Sigurdur R., and Eric H. Oelkers. 2014. "Carbon storage in basalt." *Science* 344:373–374.

Haszeldine, Stuart R. 2009. "Carbon capture and storage: How green can black be?" *Science* 325:1647–1652.

Hubbert, M. K. 1956. "Nuclear energy and fossil fuel." Spring Meeting of the Southern District, American Petroleum Institute, San Antonio, Texas, March 7–9, 1956. Publication No. 95, Shell Development Co.

Intergovernmental Panel on Climate Change (IPCC). 2018. IPCC, 2018: Global Warming of 1.5°C. An IPCC Special Report on the impacts of global warming of 1.5°C above pre-industrial levels and related global greenhouse gas emission pathways, in the context of strengthening the global response to the threat of climate change, sustainable development, and efforts to eradicate poverty [Masson-Delmotte, V., P. Zhai, H.-O. Pörtner, D. Roberts, J. Skea, P.R. Shukla, A. Pirani, W. Moufouma-Okia, C. Péan, R. Pidcock, S. Connors, J.B.R. Matthews, Y. Chen, X. Zhou, M.I. Gomis, E. Lonnoy, T. Maycock, M. Tignor, and T. Waterfield (eds.)]. Geneva Switzerland

International Energy Agency (IEA). 2017. "Energy, technology perspectives."

International Energy Agency (IEA). 2018. "Renewables 2018: Market analysis and forecast from 2018 to 2023."

International Energy Agency (IEA). 2019. "Shapshot of global PV markets. International Energy Agency Photovoltaic Power Systems Programme."

International Energy Agency (IEA). 2019. World energy outlook. https://www.iea.org/reports/world-energy-outlook-2019. accessed December 15, 2020

International Energy Agency (IEA). 2020a. *Global Energy Review 2020.* Accessed June 23, 2020, at https://www.iea.org/reports/global-energy-review-2020

International Energy Agency (IEA). 2020b. Coal Information Overview. World total coal production, 1971-2019 provisional. https://www.iea.org/data-and-statistics/charts/world-total-coal-production-1971-2019-provisional

International Energy Agency (IEA). 2020c. World energy outlook. Accessed December 15, 2020. https://www.iea.org/reports/world-energy-outlook-2020

Keith, David W. 2009. "Why capture CO_2 from the atmosphere?" *Science* 325:1654–1655.

King RR, DC Law, KM Edmonson, CM Fetzer, GS Kinsey, H Yoon, RA Sherif, NH Karam. 2007. 40% efficient metamorphic GaInP/GainAs/Ge multijunction solar cells. *Applied Physics Letters* 90 (18). https://doi.org/10.1063/1.2734507

Lars, Scott R., Tom Will, and Peter P. Marra. 2013. "The impact of free-ranging domestic cats on wildlife of the United States." *Nature Communications* 4:1396.

Le Quere, Corinne, Robert B. Jackson, Matthew W. Jones, Adam J. P. Smith, Sam Abernethy, Robbie M. Andrew, Anthony J. De-Gol, David R. Willis, Yuli Shan, Josep G. Canadell, Pierre Friedlingstein, Felix Creutzig, and Glen P. Peters. 2020. "Temporary reduction in daily CO_2 emissions during the COVID-19 forced confinement." *Nature Climate Change* 10:647–653. DOI: 10.1038/s41558-020-0797-x

Mufson, Steven, and Salwan Georges. 2019. "A climate change solution slowly gains ground." *Washington Post*, April 22.

National Academy of Science. 2016. *Climate Intervention: Carbon Dioxide Removal and Reliable Sequestration.* National Academies Press, Washington, DC. DOI: 10.17226/18805

NEO. 2018. "New Energy Outlook 2018." *Bloomberg New Energy Finance,* June 12. Accessed June 23, 2020, at https://bnef.turtl.co/story/neo2018?teaser=true

Nordic Energy Research. 2013. "The Nordics: Decoupling economic growth." Accessed June 12, 2019, at https://www.nordicenergy.org/figure/decoupling-economic-growth/

Rochelle, Gary T. 2009. "Amine scrubbing for CO_2 capture." *Science* 325:1652–1654.

Rigopoulos et al. 2018. Sustainable exploitation of mafic rock quarry wate for carbon sequestration following ball milling. *Resource Policy* 59:24-32.

NRDC 2013. America's (amazingly) good energy news. NRDC's First Annual Energy Report R:13-09-B. https://www.nrdc.org/sites/default/files/energy-environment-report-2013.pdf

Rubin, Edward S., John E. Davison, and Howard J. Herzog. 2015. "The cost of CO_2 capture and storage." *International Journal of Greenhouse Gas Control* 40:378–400.

US Department of Agriculture. 2020. "Farm household income forecast for 2019 and 2020—February 2020 update." Accessed May 25, 2020, at https://www.ers.usda.gov/topics/farm-economy/farm-household-well-being/farm-household-income-forecast/

US Department of Energy 2019. "Wind vision." Accessed May 28, 2019, at https://www.energy.gov/maps/map-projected-growth-wind-industry-now-until-2050

Wilberforce, Tabbi, Ahmad Baroutaji, Bassel Soudan, Hai Al-Alami Abdul, and Ghani Oladi Abdul. 2019. "Outlook of carbon capture technologies and challenges." *Science of the Total Environment* 657:56–72.

White House, 2015. "Statement by the President on the Keyston XL pipeline." Accessed January 30, 2020, at https://www.whitehouse.gov/the-press-office/2015/11/06/statement-president-keystone-xl-pipeline

Case Study 5.1

FRACKING, CLIMATE CHANGE, AND THE REGIONAL ECONOMY OF PENNSYLVANIA

DAN BARROW[1]

An Imaginary Debate in the 2026 Governor's Race

Fracking had an explosive effect on the rural economy of Susquehanna County, Pennsylvania (population 40,348), in 2009, as a home pipe explosion nearly killed Norma Fiorentino, a woman in Dimock Township. A local fracking operation failed to properly enclose a natural gas well, which allowed methane, a highly flammable compound, to enter one of her water wells. While Fiorentino was away from her house, a motorized pump ignited an explosion that destroyed a several-thousand-pound block of concrete. Officials believe that if the explosion had happened later, methane could have built up in her basement, creating an explosion large enough to kill her (StateImpact Pennsylvania 2020). Fiorentino is highly concerned that her other water well is unsafe, and she told the *Philadelphia Inquirer*, "I used to be able to drink that water. It smells and its white and foamy. I'm afraid it's gonna explode too" (Gensler 2017).

And at the time of the Dimock explosion, Susquehanna County was just at the beginning of a fracking boom. The struggling county in northeastern Pennsylvania (about 100 miles north of Allentown and 90 miles west of New York City) had a per capita income of 29,630 dollars, or about 3,000 dollars below the national average (US Census Bureau 2018). Since then, the unemployment rate in Susquehanna County has dropped from 8.5 percent in 2009 to 4.3 percent in 2019, before COVID-related unemployment drove it back up to 12.9 percent in April 2020 (Federal Reserve Bank of St. Louis 2020).

The explosion at Norma Fiorentino's home was part of a long list of complaints from residents that fracking operations were ruining Dimock Township's health and safety. The energy company involved, Cabot Oil and Gas, was fined $120,000 for failing to prevent the explosion, in addition to water well contamination and chemical spills (Shankman 2009). Fiorentino and other residents believed the fines did not go far enough and took legal action against Cabot Oil and Gas. In a federal civil litigation, the panel ordered the company to pay 4.2 million dollars to several families for being negligent toward water safety (Energy Justice Network n.d.).

Recalling these incidents, the scenario at the end of this Case Study will involve hypothetical democratic socialist Sean Greene, registered as the state's Green Party candidate for governor, who has proposed a ban on fracking. Having logged enough support to get on the 2022 gubernatorial race ballot, Greene agrees in October 2022 to a debate with hypothetical Democratic candidate Dara Jones and hypothetical Republican candidate Amani Soto. Greene has managed to get his proposed fracking ban into the political discussion. It has not gained much traction with the Republican candidate, and the Democratic candidate seems ambivalent. You will be assigned to take one of the three positions in the debate, and you will prepare for a discussion of the following question: "Should fracking be banned in Pennsylvania (and the United States more broadly)?"

The explosion at Norma Fiorentino's home isn't the only fracking-related incident of concern, either. A similar explosion that occurred in Ohio blew a house off its foundations and left a neighborhood with no drinkable water. And in September 2009, 8,000 gallons of fracking fluid leaked from faulty supply pipes into wetlands poisoning streams and killing fish. Drinking water turned brown and corrosive and would ignite when a match was held to it as it came out of the tap. People reported dizziness, headaches, and skin sores from showering. In October 2009, the PA Department of Environmental Protection shut down water wells in the area due to major contamination of the aquifer (Source Watch 2020).

Countless residents of Pennsylvania are understandably concerned over fracking's local environmental and health effects, but the national debate over fracking mostly surrounds its impact on climate change. Many environmentalists who advocate for immediate and substantial reductions in emissions view fracking as irreconcilable with a responsible climate policy and seek to ban fracking outright. Proponents of US industry view fracking as an opportunity to reduce emissions by replacing coal, since the natural gas produced by fracking is likely a lower emitting alternative. While coal emits 214 to 228 pounds of CO_2 per million BTUs, natural gas emits 117 pounds of CO_2, per million BTUs, or roughly half the CO_2 of coal (Energy Information Administration [EIA] 2019a).

As mentioned earlier in the chapter, the natural gas produced by fracking could serve as a bridge fuel to ease the transition to renewable energy. This means that instead of either immediately banning fracking, or encouraging it indefinitely, many policymakers believe it should be temporarily used until renewable energy becomes more economically viable. There is also intense disagreement over fracking's local environmental and health consequences. Opponents of fracking believe that it inevitably endangers local water supplies and increases health issues from smog and other emissions. Others believe these risks are extremely limited and greatly outweighed by the economic benefit of oil and natural gas production.

The Fracking Process

Fracking, or hydraulic fracturing, is a step within the process of unconventionally extracting oil and natural gas from shale or other impermeable rock formations (Figure 5.14). Extraction begins with vertical drilling from the surface down to the kickoff point, at which the layer of targeted shale begins. Then the well begins to be drilled horizontally across the shale layer for up to a mile. After drilling, the well is lined with steel and cement to facilitate movement of the oil or natural gas, and to prevent potential groundwater contamination.

Once the casing is lined, drillers create holes in the casing and shale by using perforating guns that detonate small amounts of explosives. These holes serve as pathways for oil and gas to flow through. Perforating is followed by fracking, which expands these pathways by injecting a mixture of water, sand, and chemicals at high temperature and pressure into the well. Due to underground pressure, fracking fluid returns to the surface with brine, heavy

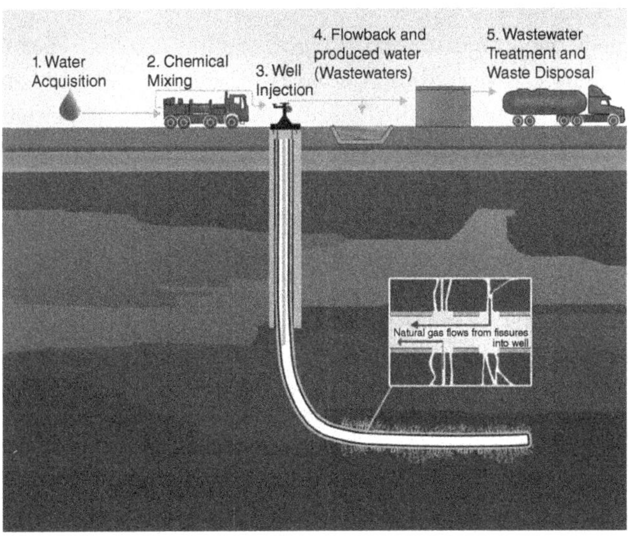

Figure 5.14 Hydraulic fracturing diagram. (United States Geological Survey. "What is hydraulic fracturing?" https://www.usgs.gov/faqs/what-hydraulic-fracturing?qt-news_science_products=0#qt-news_science_products)

metals, and other toxic compounds needing to be stored and later transported to treatment facilities. Finally, oil and gas flows to the surface to be transported via pipeline or truck.

Fracking has allowed energy companies to force natural gas and oil out of shale and other impermeable rock formations. These processes have drastically increased accessible natural gas and oil reserves and allowed the United States to increase natural gas production from 24 million cubic feet in 2000 to 40 million cubic feet in 2019 (EIA 2020a). Likewise, domestic oil production has jumped from 5.8 million barrels a day to over 12 million barrels a day over the same period (EIA 2020b).

Fracking has been described as a solution to the dwindling supply of fossil fuels. This method of extracting oil and natural gas is common in states across the nation, such as Pennsylvania, West Virginia, Louisiana, Texas, and North Dakota. Indeed, fracking in the United States is said to have turned the nation from a net importer of energy to a net exporter in only a few short years. During the COVID-19 pandemic, drops in the price of oil from 55 dollars per barrel to 19 dollars per barrel have reduced fracking by making it less profitable (EIA 2020c). However, as the economy settles after the pandemic, the debate of whether fracking is a cost-effective way of reducing Greenhouse Gas Emissions and deferring "peak oil" (the year when the demand for fossil fuels will outstrip supply) will continue. This is expected to happen around 2030, according to the IIEA (2019a).

In the United States, where it is most common, fracking produced about 44 percent of domestic energy consumption in 2019. In 2020, as mentioned, fracking fell due to the COVID-induced oil glut, but experts predict that fracking will surge in the long run. According to the EIA (2020d), natural gas produced by fracking is expected to increase by 81 percent, and oil from fracking will increase by 38 percent. This will cause natural gas to generate 36 percent of US electricity in 2050, while electricity generated by coal will fall to 13 percent (EIA 2020d). Fracking is poised to increase in the United States, but it is highly unclear whether it will be common throughout the world in the coming decades. Prior to 2017, the only nations with commercial fracking operations were the United States, Canada, and China (Yan et al. 2017). The global rise of fracking depends on whether national governments encourage it, and whether geographic conditions in other countries allow the practice to be profitable.

Economic Impact

Fracking has undoubtedly driven economic growth in Pennsylvania by expanding the energy sector, increasing tax revenue, lowering energy prices, and creating well-paying jobs. Pennsylvania sits atop one of the largest shale plays in the United States, so fracking has caused the state's natural gas production to skyrocket. In 2008, Pennsylvania produced nearly of 200 billion cubic feet, but production soared to about 7 trillion cubic feet in 2019 (EIA 2020e). Pennsylvania now extracts the second most natural gas in the United States, behind only Texas, and produces 20 percent of natural gas in the nation (Rand Corporation 2017). From 2010 to 2018, the share of electricity generated from natural gas in Pennsylvania doubled, from 18 percent to 36 percent, while electricity from coal was been cut in half (EIA 2018a). The Marcellus Shale region produces 300 million dollars in annual revenue from fracking in Pennsylvania, West Virginia, and Ohio (Rand Corporation 2017). Overall, fracking serves 37 percent of the nation's energy needs in the form of natural gas and oil combined, ahead of conventionally drilled oil and gas, which serves 29 percent; coal, with 14 percent; renewables, at 11 percent; and nuclear, at 8 percent (American Geosciences Institute 2017).

Pennsylvania's state legislature capitalized on the fracking boom by enacting a natural gas tax in 2012 that has generated hundreds of millions of dollars. The tax yields about 200 million dollars annually and has produced a total of 1.7 billion dollars in tax revenue so far. These funds have been used to improve infrastructure, emergency preparedness, and water security (Hurdle 2019). Many landowners in Pennsylvania have also reaped the benefits of fracking's rise, because they can receive massive payments by leasing their land to natural gas companies. An estimated 1.2 billion dollars is paid to Pennsylvania landowners per year to rent parts of their property to launch fracking operations. This income has been life-changing to residents who were previously living paycheck to paycheck (Begos 2013).

Fracking has created thousands of jobs in Pennsylvania that pay more than most rural jobs. Outside of fracking, economic opportunity is relatively limited in Susquehanna County. In 2009, the county had a per capita income of 29,630 dollars, or about 3,000 dollars below the national average (US Census Bureau 2018). Because of this, jobs created by fracking have had a profound impact on employment opportunities in the region. Jobs involving fracking and natural gas extraction itself increased from 1,700 to 6,600 from 2005 to 2015 (Rand Corporation 2017).

Furthermore, fracking creates more midstream and downstream jobs, because more workers must be employed to transport and refine the newly produced natural gas. When including pipeline and service workers, fracking produces 32,000 jobs throughout the state (Eavis 2020). Natural gas–related jobs in Pennsylvania pay up to 110,000 dollars per year. This is considerably higher than the state average wage of 51,000 dollars, and fracking jobs tend to be located in rural areas, where wages from other jobs are significantly lower (US Bureau of Labor Statistics 2020). In Susquehanna County, median household income is about 53,000 dollars, so jobs in natural gas are even higher compared to alternative jobs in the area (US Census Bureau 2018). Fracking also improves the wages of workers within industries outside of oil and gas, from the indirect spillover effects of fracking's revenue. It is estimated that every 1 million dollars of oil and gas revenue creates 257,000 dollars in local wages and 286,000 dollars in business income. Fracking has generated 198 million dollars in revenue in Pennsylvania, so the state likely experienced 51 million dollars in wage gains and 56 million dollars in increased business income (Feyrer et al. 2017).

Transitioning from coal to natural gas has also benefited Pennsylvania by decreasing the economic damage from emissions. The social cost of carbon is an estimation of the economic damage of an energy source's pollution. For instance, the emissions from both coal and natural gas produce smog, which increases health costs and reduces tourism, among countless other externalities (indirect costs that are often not factored in). According to one economist, producing 1.00 dollar of coal results in 2.20 dollars of total economic damage, while the social cost of natural gas is only 34 cents (Raimi 2017). Fracking has allowed Pennsylvania to replace a fuel source that has a net negative economic impact with an energy source that has manageable external costs. Fracking has also lifted the regional economy of northeastern Pennsylvania, but environmentalists fear that this growth comes at the cost of climate change due to increased long-term greenhouse gas emissions.

Fracking's Impact on Climate Change

Hydraulic fracturing has expanded natural gas production in the United States and contributed to reduced coal production. Many policymakers are optimistic that this transition will help establish a low-carbon economy. The first question in accessing fracking's impact on climate change, however, is whether the overall production of natural gas is less detrimental to global temperatures than coal. This largely depends on its level of methane leakage, which is highly uncertain. Another important consideration surrounding fracking's impact on climate change is its indirect impact of prolonging a shift to renewable energy. Supporters of natural gas as a bridge fuel argue that using natural gas as opposed to coal could provide modest benefits to global emissions, while critics of this notion claim that these benefits could be outweighed by reducing the rise of cleaner renewable energy.

Methane Leakage

In terms of emissions at combustion, the point at which fuel is converted into energy, natural gas is less harmful to global emissions. CO_2 is the most common greenhouse gas, and natural gas emits roughly half as much CO_2

as coal does at combustion. Natural gas combustion also produces less of the minor contributors to climate change, such as SO and NO_x. According to the US Environmental Protection Agency (2014), coal combustion emits about 16 times more NO_x than natural gas. However, natural gas is potentially more damaging to global temperature than coal due to methane emissions.

Prior to combustion, natural gas emits more methane in storage and transportation from venting, flaring, and fugitive emissions. Venting is when natural gas is intentionally released into the air in the form of methane. Flaring is another way to dispose of excess gas on site; when a well is unable to transport all the natural gas it produces, companies often flare the excess gas by burning it on the site without generating any electricity. Flaring contributes less severely to climate change than venting, however, because methane is converted into CO_2, a far less potent greenhouse gas (Cheremisinoff 2015). Natural gas also emits methane in the form of fugitive emissions, when either pipes or storage tanks leak methane into the atmosphere.

Flaring and venting are more common in fracking wells than at conventional drilling sites, meaning that fracking likely increases methane emissions. When the Bakken Shale region in North Dakota experienced an increase in fracking operations, the amount of natural gas emitted as methane increased from 5 to 30 percent in just two years; furthermore, total methane emissions in the region increased by 25 times (Union of Concerned Scientists 2016). As mentioned, methane is also released into the atmosphere as it is transported by pipelines from leakages (fugitive emissions). A variety of studies have attempted to estimate the amount of natural gas released as methane from all of these processes, but their findings have been extremely conflicting.

If methane emissions from natural gas reach a certain threshold, then natural gas production is more detrimental to global temperatures than coal. It is estimated that if two percent of natural gas is emitted as methane, switching from coal to natural gas will result in increased long-term global temperatures (Wigley 2011). Estimates of methane emissions in the United States vary, but studies find that less than two percent of natural gas is emitted as methane. The Environmental Protection Agency's most recent study found that about 1.4 percent of natural gas is released into the atmosphere as methane on average (Environmental Defense Fund 2018). However, climate scientists have recently challenged these findings, and claim that studies have underestimated methane emissions associated with natural gas. They believe that flaring is less efficient than originally predicted, because it fails to completely convert methane into CO_2. A study by the National Academy of Sciences estimated that natural gas production's methane emissions are about 1.5 times higher than predicted by the Environmental Protection Agency (Zavala-Araiza et al. 2015).

Reducing Methane Leakage

Many supporters of fracking concede that methane leakage is a serious concern but believe it can be dramatically reduced by applying modern technology and practices. Improved monitoring software of methane leakage across pipelines could enable natural gas companies to detect and stop major leaks earlier (Cheremisinoff 2015). This could both reduce fugitive emissions and enable more accurate measuring of methane emissions. Improved technology can also reduce methane emissions by utilizing methane instead of flaring. As stated earlier, shale operations burn natural gas that cannot be stored or transported through flaring. Instead of either transporting natural gas to the market or flaring it, some companies have begun to use produced natural gas to fuel their own equipment and heating (Cheremisinoff 2015). The National Resource Defense Council believes that implementing such technologies could reduce methane leakage by up to 90 percent (Green 2014). Supporters of natural gas are also optimistic that efficiency will improve over time, since newly constructed plants tend to leak much less methane than older sites (Raimi 2017).

A Bridge to Renewables

Those who support fracking as a bridge fuel argue that transitioning from coal to natural gas is the best way of simultaneously lowering greenhouse gas emissions while keeping energy costs low. Advocates of the bridge fuel approach believe that the United States must replace fossil fuels with renewable energy in the long term, but that doing so immediately would damage the economy. The United States could achieve short-term emissions reductions by replacing coal with natural gas, because it emits about half the CO_2. The transition from coal to gas is already underway and has reduced observed US greenhouse-gas emissions by 11 percent (EIA 2020d). It is important to note, however, that fracking could also be increasing unobserved greenhouse emissions in the form of methane, as discussed earlier. Proponents of natural gas respond to these concerns by arguing that methane emissions can be reduced by applying modern technologies and practices, as described in the previous section.

One of renewable energy's largest shortcomings is the need for energy storage. Renewable energy is intermittent, since solar panels only generate power during the day and wind turbines only generate power when there is enough wind. To access power at all times, renewable energy must be stored, but doing so is currently expensive. Although the cost of energy storage has consistently decreased in the United States, it remains an obstacle.

As described in the chapter, the cost of storing renewable energy has dropped from 200 dollars per kilowatt-hour to 70 dollars per kilowatt-hour within just six years. These cost reductions likely fall short in making solar energy cost-competitive, however, even in regions where it is unusually efficient, such as Texas and Arizona. It is estimated that the cost of energy storage would have to drop to about 20 dollars per kilowatt-hour for solar energy to outcompete fossil fuels, which is considerably lower than the 70 dollars per kilowatt-hour it currently costs (Ziegler et al. 2019). While the COVID-19 pandemic–related oil shock seems to have further lowered the cost of fossil fuels, supporters of natural gas believe it should be produced until the cost of storing renewables is dramatically reduced.

In addition to the regional economic benefits described earlier, the US national economy has greatly benefited from fracking in the form of reduced energy costs and greater energy security. Fracking has allowed the United States to become a net exporter of crude oil. In 2005, US energy imports were 12.5 million barrels per day greater than exports, but the United States currently exports 772,000 barrels per day more than it imports (EIA 2020f, 2020g). This has resulted in massive savings amongst American consumers. The United States consumed 2.7 billion more gallons of oil in 2015 compared to 2014, but it paid 188 billion dollars less because of its declining price. Similarly, US consumers have also benefited from cheaper natural gas. The United States saved 70 billion dollars in natural gas spending in 2012 compared to 2008 despite consuming more (Raimi 2017). The surge in domestic oil production has also reduced fluctuations in energy prices, contributing to less uncertainty within the private sector (Kersting 2016). The United States is not completely energy independent, however, because domestic energy costs still shift based on global oil and gas supply, but most economists agree that increased oil and gas production has made domestic prices far more stable.

Transitioning Directly to Renewables

Critics of the bridge fuel approach argue that the emissions improvements in transitioning from coal to natural gas are insufficient. Because of this, policymakers should immediately facilitate a transition from all fossil fuels to renewable sources of energy. Even if natural gas produces slightly less damaging greenhouse gases than coal, these benefits may be outweighed by prolonging a shift to renewable energy. Natural gas and renewable energy compete for investment funds, and increased investment in natural gas could restrict investment in renewable energy, leading to less research, development, and expansion of renewables. Cheaper natural gas also reduces incentives for companies to improve energy efficiency, and for individuals to reduce their energy consumption. For instance, if oil and gas are cheaper, driving a more fuel-efficient car results in less savings, making it less likely for consumers to make the switch. Opponents of using natural gas as a bridge fuel are also skeptical that natural gas will only be used temporarily. They fear that increases in natural gas consumption will only cause companies to become more dependent on fossil fuels and increase the cost of switching to renewables over time (Staddon & Depledge 2015).

Furthermore, although natural gas may provide economic benefits in terms of job creation and lower energy costs, but it is important to consider its external costs from a climate change vantage point as well. The economic costs associated with climate change in the form of damage to infrastructure, ecosystem services, and public health could offset the direct economic benefits of fracking (Staddon & Depledge 2015). Climate change is the primary source of national debate regarding natural gas, but local residents of Pennsylvania tend to perceive its local environmental risks as a more pressing policy issue.

Local Environmental Risks

Residents of Susquehanna County fear that fracking is detrimental to local environmental and community health because it contaminates water supplies and harms air quality. Fracking poses significant risks of groundwater contamination, because improper casing can allow methane to enter groundwater stores. Although the natural gas and oil contained within shale rarely enters underground water sources through rock pathways, methane can travel upward alongside the well. This methane is meant to be sealed off by cement casing, but if the casing fails, methane migrates into aquifers and can enter local water supplies. Methane is naturally occurring, but fracking can elevate the risk of water sources containing unsafe levels of methane.

Fracking was likely the culprit for the flaming faucets in Dimock (Raimi 2017). The National Academy of Arts and Sciences studied methane levels in water wells within the Marcellus Shale region, which encompasses Susquehanna County as well as most of Pennsylvania and West Virginia. They determined that water wells within one kilometer of fracking sites were more likely to have elevated methane levels than water wells farther away from fracking (Osborn et al 2011). In 2015, the Earth and Environmental Institute also studied water methane levels in the Marcellus Shale region and determined that 0.5 percent of shale gas wells contained unnaturally high levels of methane (Wen et. al 2018). This percentage of wells may sound miniscule, but if these numbers are reflective of methane levels throughout the state, then 342 of the 68,421 Pennsylvanian natural gas wells could potentially be increasing methane counts in water supplies (EIA 2019b). Health risks from heightened methane exposure include loss of consciousness and acute respiratory distress (Jo et al. 2013).

Fracking is even more likely to contaminate water sources from the surface spillage of either fuel or wastewater. As stated earlier, fracking produces an extreme amount of wastewater that contains fracking fluid, brine, heavy metals, organic compounds, and even radioactive materials in some circumstances. This water is usually stored on site in pits, and if the pits are improperly lined, these toxic compounds can enter groundwater stores through the surface. Surface spillage is also a concern when wastewater is transported to treatment facilities by truck or pipeline. In northeastern Pennsylvania, surface spills caused fracking fluid to enter at least 32 water wells. (Raimi 2017). Likewise, natural gas and oil can enter water supplies in the event of a truck crash or pipe breakage.

In addition to the threat of groundwater contamination, fracking may compromise public health through exposure to harmful emissions. Volatile organic compounds (VOCs) are naturally occurring, but human activities can release additional VOCs. Drilling and fracking releases VOC's stored underground, causing oil and natural gas to be the greatest industrial contributor of VOCs (Cheremisinoff 2015, Nyberg et al. 2018). From 6 to 12 percent of VOC emissions in the Marcellus area are from fracking operations (Anirban 2014). Fracking operations also produce NO_x, which interacts with VOCs and other chemicals in the atmosphere to form smog. Fracking in the Marcellus region produces an estimated 129 tons of NO_x a day, or about 12 percent of NO_x emissions in the region (Anirban 2014). Smog can result in many forms of respiratory dysfunction, such as worsened asthma, and often requires hospitalization. Residents of northeastern Pennsylvania are more likely to develop asthma and related respiratory issues if they are exposed to emissions from fracking sites (Rasmussen et al. 2016). There have also been widespread accounts of nosebleeds and eye irritation in areas near fracking developments, which may be the result of increased smog and other emissions associated with shale development (Lourdes 2016).

Unsurprisingly, stakeholders and policymakers in Pennsylvania have responded differently to the uncertainties in groundwater contamination from fracking. Many environmentalists support a precautionary approach, meaning fracking should only be allowed when the risks have been fully understood. They encourage Pennsylvania lawmakers to follow the lead of neighboring New York state and ban fracking. Opponents of this outlook believe that the local risks of fracking are minimal, and that they can be reduced further by proper regulation and applying the most environmentally responsible technologies and procedures.

Fracking is not regulated under the Safe Drinking Water Act or Clean Air Act, which allows the Environmental Protection Agency to set standards for CO, $NO_{x,}$ VOCs, and other emissions. Hydraulic fracturing is only subject to federal regulations if it occurs on federally owned land, but most fracking occurs on privately owned land. Therefore, fracking is predominantly regulated by state and local governments. This lack of federal regulation has contributed to a variety of regulations among states toward containment of both stray gas and wastewater. Pennsylvania's regulation of fracking is relatively extensive, since water must be tested before and after drilling. The state also enacted a policy of presumptive liability, which mandates that oil and gas companies must prove that methane levels have not changed in order to continue their operations. These policies have helped lower the amount of methane levels in water wells from 0.5 to 0.2 percent (Raimi 2017).

Conclusion

Hydraulic fracturing has allowed oil and natural gas companies to extract fossil fuels from previously inaccessible rock formations. This not only has allowed US natural gas production to skyrocket, but has also expanded long-term oil and gas reserves in the United States. The fracking boom has unleashed several local environmental concerns, most of which pertain to groundwater contamination, climate change emissions, and air quality. Improper well casings can allow methane in the form of stray gas to enter aquifers, and surface spillage can cause wastewater or fuel to infiltrate groundwater as well. Although fracking's impact on air quality is less detrimental than coal, hydraulic fracturing heavily contributes to smog and other local air pollutants, negatively impacting health. Many environmentalists believe these effects and uncertainties should lead policymakers to ban fracking, while industry advocates believe these issues can be mitigated by proper regulation and modern practices. Similar disagreements are held over fracking's impact on climate change. Those who support fracking as a bridge fuel believe that, if methane emissions are reduced, natural gas can serve as a bridge to a lower carbon economy. On the other hand, those who are in favor of sharp and immediate emissions reductions believe that natural gas will only prolong a shift to a renewables-based economy.

EXERCISE

SCENARIO FOR CLASSES

After reading the case study and carefully considering the advantages and disadvantages of fracking, divide into three teams. Each team will support one of the hypothetical candidates in the 2022 Pennsylvania gubernatorial debate. Democratic socialist and Green Party candidate Sean Greene's team will argue in favor of an immediate ban on fracking. This team asserts that fracking poses severe risks to local environments and should be replaced by renewable energy to combat climate change.

Continued

Democrat Dara Jones's team is in favor of using natural gas in the near term as a bridge fuel, but also improving regulations to reduce its environmental effects. Republican Amani Soto team's will argue that fracking should be used indefinitely, because its economic benefits outweigh its environmental effects.

STARTING POINT QUESTIONS:

1. Should Pennsylvania ban fracking until its risks on water safety and air quality are fully understood?
2. Do the methane emissions from natural gas production prevent its ability to serve as a bridge fuel?
3. Should Pennsylvanian and federal policymakers immediately transition to renewable energy instead of using natural gas as a bridge fuel?

CASE STUDY 5.1 NOTES

1. Dan Barrow is an undergraduate student at American University majoring in political science and economics. He is interested in pursuing economic and environmental policy surrounding climate change.

CASE STUDY 5.1 REFERENCES

American Geosciences Institute. 2017. "What are the major sources and users of energy in the United States." Accessed June 23, 2020, at https://www.sightline.org/2017/10/30/is-your-natural-gas-actually-fracked/#:~:text=Nationally%2C%20fracking%20produces%20two%2Dthirds,percent%20of%20the%20nation's%20oil

Anirban, Roy A., Peter J. Adams, and Allen L. Robinson. 2014. "Air pollutant emissions from the development, production, and processing of Marcellus shale natural gas." *Journal of the Air & Waste Management Association* 64(1):19–37.

Begos, Kevin. 2013. "Pennsylvania fracking royalties could top $1 billion as private landowners rake in cash." *Associated Press*, January 28. Accessed June 23, 2020, at https://www.businessinsider.com/fracking-royalty-payments-in-pa-2013-1

Cheremisinoff, Nicholas P. 2015. *Hydraulic Fracturing Operations: Handbook of Environmental Practices*. Scrivener Publishing, Beverly, MA.

Eavis, Peter. 2020. "Fracking once lifted Pennsylvania. Now it could be a drag." *New York Times*, March 31. Accessed June 23, 2020, at https://www.nytimes.com/2020/03/31/business/energy-environment/pennsylvania-shale-gas-fracking.html

Energy Information Administration (EIA). 2018. "Pennsylvania state profile and energy estimates." Accessed June 23, 2020, at https://www.eia.gov/state/?sid=PA#tabs-1

Energy Information Administration (EIA). 2019a. "How much carbon dioxide is produced when different fuels are burned?" Accessed June 23, 2020, at https://www.eia.gov/tools/faqs/faq.php?id=73&t=11

Energy Information Administration (EIA). 2019b. "Pennsylvania state energy profile." Accessed June 23, 2020, at https://www.eia.gov/state/print.php?sid=PA

Energy Information Administration (EIA). 2020a. "US dry natural gas production." Accessed June 23, 2020, at https://www.eia.gov/dnav/ng/hist/n9070us2A.htm

Energy Information Administration (EIA). 2020b. "US field production of crude oil." Accessed June 23, 2020, at https://www.eia.gov/dnav/pet/hist/LeafHandler.ashx?n=PET&s=MCRFPUS2&f=M

Energy Information Administration (EIA). 2020c. "Short-term energy outlook." Accessed June 23, 2020, at https://www.eia.gov/outlooks/steo/report/prices.php

Energy Information Administration (EIA). 2020d. "Annual energy outlook." Accessed June 23, 2020, at https://www.eia.gov/outlooks/aeo/

Energy Information Administration (EIA). 2020e. "Pennsylvania natural gas marketed production." Accessed June 23, 2020, at https://www.eia.gov/dnav/ng/hist/n9050pa2a.htm

Energy Information Administration (EIA). 2020f. "Today in energy." Accessed June 23, 2020, at https://www.eia.gov/todayinenergy/detail.php?id=42735

Energy Information Administration (EIA). 2020g. "Petroleum and other liquids." Accessed June 23, 2020, at https://www.eia.gov/dnav/pet/hist/LeafHandler.ashx?n=pet&s=mttntus2&f=a

Energy Information Administration (EIA). 2020h. "US natural gas gross withdrawals" Accessed June 23, 2020, at https://www.eia.gov/dnav/ng/hist/n9010us2m.htm

Environmental Defense Fund. 2018. "New study finds US oil and gas methane emissions are 60 percent higher than EPA reports." Accessed June 23, 2020, at https://www.edf.org/media/new-study-finds-us-oil-and-gas-methane-emissions-are-60-percent-higher-epa-reports-0

Energy Justice Network. n.d. "Dimock families win water contamination case against Cabot Oil and Gas." Accessed June 23, 2020, at https://www.energyjustice.net/content/dimock-families-win-water-contamination-case-against-cabot-oil-gas

Federal Reserve Bank of St. Louis. 2020. "Unemployment rate in Susquehanna County, PA" Accessed June 23, 2020, at https://fred.stlouisfed.org/series/PASUSQ5URN

Feyrer, James, Erin T. Mansur, and Bruce Sacerdote. 2017. "Geographic dispersion of economic shocks: evidence from the fracking revolution." *American Economic Review*107(4):1313–1334.

Gensler, Howard. 2017. "As fracking lawsuits draw to a close, a look back at what happened in Dimock." *Philadelphia Inquirer*, September 28. Accessed June 23, 2020, at https://www.inquirer.com/philly/news/pennsylvania/what-happened-in-dimock-pa-fracking-drilling-natural-gas-20170928.html

Green, K. P. 2014. "Managing the risks of hydraulic fracturing." *Fraser Research Bulletin*. Fraser Institute. https://www.fraserinstitute.org/sites/default/files/managing-the-risks-of-hydraulic-fracturing-an-update.pdf

Hurdle, Jon. 2019. "Gas impact fee revenue rises to 7-year high, boosted by stripper wells." StateImpact Pennsylvania, June 27. Accessed June 23, 2020, at https://stateimpact.npr.org/pennsylvania/2019/06/27/gas-impact-fee-revenue-rises-to-7-year-high-boosted-by-stripper-wells/

International Energy Agency (IEA). 2019. "World energy outlook 2019." Accessed June 23, 2020, at https://www.iea.org/reports/world-energy-outlook-2019/oil

International Energy Agency (IEA). 2020. "Global energy review 2020." Accessed June 23, 2020, at https://www.iea.org/reports/global-energy-review-2020

Jo, Jun Yeon, Yong Sik Kwon, Jin Wook Lee, Jae Seok Park, Byung Hak Rho, and Won-Il Choi. et al. 2013. "Acute respiratory distress due to methane inhalation." *Tuberculosis and Respiratory Diseases* 74(3):120–123.

Legere, Laura. 2017. "Eight years later, Dimock's water is still under scrutiny." *Pittsburgh Post-Gazette*, August 4. Accessed June 23, 2020, at https://www.postgazette.com/business/powersource/2017/08/04/Dimock-methane-Cabot-Agency-for-Toxic-Substances-Marcellus-shale-natural-gas-water-contamination/stories/201708030135

Le Quere, Corinne, Robert B. Jackson, Matthew W. Jones, Adam J. P. Smith, Sam Abernethy, Robbie M. Andrew, Anthony J. De-Gol, David R. Willis, Yuli Shan, Josep G. Canadell, Piette Friedlingstein, Felix Creutzig, Glen P. Peters. 2020. "Temporary reduction in daily CO_2 emissions during the COVID-19 forced confinement." *Nature Climate Change* 10:647–653. DOI: 10.1038/s41558-020-0797-x

Lourdes, Vera. 2016. "Community health impacts from oil and gas development in Texas: The perspective of Sharon Wilson." *Journal of Environmental and Occupational Health Policy*26(3):496–507.

Nyberg, Daniel, Christopher Wright, and Jacqueline Kirk. 2018. "Dash for gas: Climate change, hegemony and the scalar politics of fracking in the UK." *British Journal of Management*29(2):235–251.

Osborn, Stephen G., Avner Vengosh, Nathaniel R. Warner, and Robert B. Jackson. 2011. "Methane contamination of drinking water accompanying gas-well drilling and hydraulic fracturing" *Proceedings of the National Academy of Sciences of the United States of America*108(20):8172–8176.

Raimi, Daniel. 2017. *The Fracking Debate: The Risks, Benefits, and Uncertainties of the Shale Revolution*. Columbia University Press, New York.

Rand Corporation. 2017. "Developing a skilled workforce for the oil and natural gas industry." Accessed June 23, 2020, at https://www.rand.org/pubs/research_reports/RR2199.html

Rasmussen, Sara G., Elizabeth L Ogburn, Meredith McCormack, Joan A Casey, Karen Bandeen-Roche, Dione G Mercer, and Brian S Schwartz. 2016. "Association between unconventional natural gas development in the Marcellus shale and asthma exacerbations." *JAMA Internal Medicine* 176(9):1334–1343.

Shankman, Sabrina. 2009. "Pennsylvania tells drilling company to clean up its act." *ProPublica*, November 6. Accessed June 23, 2020, at https://www.propublica.org/article/pennsylvania-tells-drilling-company-to-clean-up-its-act-1106

Source Watch. 2020. "Marcellus shale." Accessed June 23, 2020, at https://www.sourcewatch.org/index.php/Marcellus_Shale

Staddon, Philip L., and Michael H. Depledge. 2015. "Fracking cannot be reconciled with climate change mitigation policies." *Environmental Science and Technology* 49(14):8269–8270.

StateImpact Pennsylvania, "Dimock, PA: 'Ground zero' in the fight over fracking." Accessed June 23, 2020, at https://stateimpact.npr.org/pennsylvania/tag/dimock/

Szyszkowicz, M., R. Shutt, T. Kousha, and B. H. Rowe. 2014. "Air pollution and emergency department visits for epistaxis." *Clinical Otolaryngology* 39(6):345–351.

Union of Concerned Scientists. 2016. "The truth about tight oil." Accessed June 23, 2020, at https://www.ucsusa.org/resources/truth-about-tight-oil

US Bureau of Labor Statistics. 2020. "State occupational employment and wage estimates Pennsylvania." Accessed June 23, 2020, at https://www.bls.gov/oes/current/oes_pa.htm

US Census Bureau. 2018. "Susquehanna County, Pennsylvania." Accessed June 23, 2020, at https://www.census.gov/quickfacts/susquehannacountypennsylvania

US Environmental Protection Agency (EPA). 2014. "Emissions factors for greenhouse gas inventories." Accessed June 23, 2020, at https://www.epa.gov/sites/production/files/2015-07/documents/emission-factors_2014.pdf

Wen, Tao, Xianzeng Niu, Matthew Gonzales, Guanjie Zheng, Zhenhui Li, and Susan L. Brantley. 2018. "Big groundwater data sets reveal possible rare contamination amid otherwise improved water quality for some analytes in a region of Marcellus shale development." *Environmental Science & Technology* 52(12):7149–7159.

Wigley, Tom. 2011. "Coal to gas: The influence of methane leakage." *Climatic Change* 108(3): 601–608.

Yan, Beizhan, Martin Stute, Reynold A. Panettieri Jr., James Ross, Brian Mailloux, Matthew J. Neidell, Lissa Soares, Marilyn Howarth, Xinhua Liu, Pouné Saberi, and Steven N. Chillrud. 2017. "Association of groundwater constituents with topography and distance to unconventional gas wells in NE Pennsylvania." *Science of the Total Environment*. 195(201). doi: 10.1016/j.scitotenv.2016.10.160.

Zavala-Araiza, Daniel, D. Lyon, Ramón A. Alvarez, K. J. Davis, R. Harriss, S. Herndon, A. Karion, E. Kort, B. Lamb, X. Lan, A. Marchese, S. Pacala, A. Robinson, P. Shepson, C. Sweeney, R. Talbot, A. Townsend-Small, T. I. Yacovitch, D. Zimmerle, and S. Hamburg. 2015. "Reconciling divergent estimates of oil and gas methane emissions." *Proceedings of the National Academy of Sciences of the United States of America* 112(51): 15597–15602.

Ziegler, Micah, Joshua M.Mueller, Gonçalo D.Pereira, JuhyunSong, Marco Ferrara, Yet-Ming Chiang, and Jessika E.Trancik. Mueller, 2019. "Storage requirements and costs of shaping renewable energy toward grid decarbonization." *Joule* 3(9): 2134–2153. doi.org/10.1016/j.joule.2019.06.012

6

Beyond the Energy Sector

Technological Transformations on the Way to "Decarbonization"

By 2019, about 50 percent of consumers had the option of receiving their home electricity from renewable sources, but only two percent made use of that option (Cummins 2020). Why is that? For one, utility companies do not necessarily tell consumers this is an option, so many people just don't know about it. Opting for renewable electricity in recent years has also been a bit more expensive (10 dollars extra per month on the average utility bill), although now that wind energy is even cheaper than natural gas, this might change (Cummins 2020). In Chapter 5, we addressed fossil fuels, which indeed are the most direct cause of climate change. But broader behavioral changes, such as in the sourcing of utility bills, will be needed if we are to reach the Paris Agreement goal of keeping climate change to 2°C by the end of this century. We will have to rely on a combination of technological fixes and behavioral changes to get us where we need to be.

This chapter examines some areas other than energy production where reducing carbon dioxide (CO_2) emissions is a challenge. We start with recognizing that reducing the rate of extraction and development of fossil fuel resources is absolutely necessary if we are at all serious about meeting the goal of limiting warming this century to 2°C or less. We also examine the issues the energy grid infrastructure faces as more alternative sources of electricity come online and as electric cars become more mainstream. Government policy around tax incentives drives both consumer interest and costs for energy "at the pump" or switch. Therefore, policy influences the market for energy, and this idea is addressed. Cities are where most point sources of CO_2 emission occur, so we also look at carbon emissions in urban environments, where LEED (Leadership in Energy and Environmental Design)-accredited buildings and other green infrastructure hold promise for reducing overall emissions by lowering energy demands. Finally, we examine what agricultural practices and human dietary preferences have to do with emissions as we move forward.

The Concept of "Stranded Oil"

If we take a step back and look at where all the CO_2 we are pouring into the atmosphere comes from, we have to appreciate that it represents hundreds of millions of years of plant productivity, sequestered by photosynthesis over oceans of time (think of it as tree and plankton biomass, captured and concentrated). If we continue the way we are headed, insisting that all the petroleum we can find must be burned (as most economic resource use models suggest we will), then we will be releasing to the atmosphere these hundreds of millions of years' worth of carbon within only a few hundred years. There is zero doubt that, over the next 100 years or so, this will (1) acidify the oceans, wiping out the foraminifera and related calcareous plankton that form the base of the ocean food chain; (2) trap enough heat to recreate the atmosphere of the Jurassic Period (150 million years ago), when atmospheric CO_2 concentration was five times that of today (Van Der Meer et al. 2014); and (3) accelerate global sea level rise, leading to trillions of dollars of damages in what ultimately will be futile efforts in coastal cities to adapt (Roston 2019, 4). In turn, these changes will cause massive shifts in global human migration, as crop failures, increased droughts, severe weather, and expansion of uninhabitable equatorial areas drive people to search for more hospitable climates, along with other dramatic civilization- and even planet-altering impacts over the long term. Without a technological fix (e.g., managed sequestration), a dramatic shift to carbon emissions–free energy sources, or some enhanced biological drawdown, all these undesirable outcomes will come to pass. Given that we are not exactly making leaps and bounds of progress in any of the three mitigation possibilities, it is useful to consider leaving some of the fossil carbon in the ground.

"Stranded oil" is a term used to describe petroleum (or coal, by extension) that is left in the ground deliberately even though it could be economically extracted. "Stranded" is a loaded word, and it is very deliberately used in this context. Being "stranded" is never a good thing, so the term connotes a sense of loss or fear. And the term "stranded oil" was coined by public relations firms working for petroleum interests to engender a feeling of loss at the abandonment of potential wealth and resources (Rotherham et al. 2014). It is similar to why development interests insist on calling the Arctic National Wildlife Refuge "ANWR" rather than using the actual name. The public instinctively cares less about drilling in some vague acronym versus drilling in something called a "Wildlife Refuge." In fact, people instinctively balk at the idea of drilling in a "wildlife refuge"! The vast majority of Americans, regardless of political leanings, love national parks and seashores, so if ANWR is viewed as one of those, then drilling is much less popular.

Likewise, the term "stranded oil" is intended to make the public sympathetic to the petroleum companies that have been forced to abandon the oil just waiting to be extracted. The narrative implies that those who want to abandon the "stranded" oil are those environmental people who only seem to care about the climate rather than the oil, and the wealth it represents, right there for the taking. Even if we continue on our path of maximum exploitation, however, the oil will eventually run out. In 2017, the International Energy Agency (IEA) estimated that there would be adequate liquid hydrocarbon until 2050, with the caveat that there is substantial uncertainty regarding demand, future efficiencies changing the ratios of use versus supply, and possible new reserves becoming available (IEA 2017). In any case, the petroleum will run out, and society will still confront the same problem of decarbonizing energy sources—but will do so amidst much altered planet.

As you may have surmised, we do not think that "stranded oil" is the proper view of petroleum resources that may be left intact or saved for the future. Just because a resource is economically extractable does not mean it should be extracted right now. Consider the US National Forests, which could be more extensively harvested but are not, or the cod fisheries in the North Atlantic, which could be fished out but are not because they would collapse (Worm et al. 2006). Causing the collapse of a resource is guaranteeing that the next generation will lose any economic benefit that resource may have provided. This would obviously be unsustainable and also incredibly selfish. The fact that excessive harvesting, fishing, or mineral extraction can destroy a resource and the environment around it is not new information.

We know that long-term thinking about resource use leads to more wealth for more people over time. In regard to petroleum resources, the "stranded" oil can still be harvested if necessary (consider the US National Petroleum Reserve in Alaska). The petroleum is not "lost" if it is not taken quickly. The prudent thing is to extend the life of the resource by decreasing its use and allowing natural systems (and massive mitigation efforts by governments and institutions) to absorb or sequester the great amount of excess CO_2, thereby avoiding at least some of the harmful externalities associated with burning fossil fuels. We have to curb our extraction, not stop it completely. Petroleum has other uses, after all, besides burning it for transportation or electricity. These include plastics, construction material, pharmaceuticals, lubricants, industrial chemicals, and a host of others that will necessitate continued extraction and refining of petroleum. What we argue is that the rate of extraction and wasteful burning, when there are alternatives, is unsustainable, dangerous, and foolish—not to mention cruel to the most vulnerable who will (or already are) feeling the brunt of the nonreversible impacts of climate change. In Chapter 8, we take this logic even further by discussing the proposal by policy analysts to launch a "Fossil Fuel Non-Proliferation Treaty."

In earlier chapters, we described dramatic changes occurring in the electrical utility sector as carbon-free or natural gas–generated electricity has become cheaper than coal or petroleum. While this trend portends a large increase in fuel with lower carbon emissions, significant challenges (and opportunities) are associated with bringing the "clean" energy to market.

Changing the Energy Grid System

As noted in Chapter 5, there has recently been a large increase in clean, reduced-carbon energy supply worldwide. In the future, most developed nations (and advanced developing nations) will be producing more electrical power from alternatives and renewables. Just as a reminder of recent trends in the United States, 25 gigawatts (GW) of power capacity were added in 2017, of which 12 GW were renewable and 3.5 GW were small-scale solar (e.g., rooftop panels) (Energy Information Administration [EIA] 2018). Furthermore, as of 2016, 29 states plus the District of Columbia had mandated renewable energy goals, and the federal government is aiming for 30 percent renewable electricity for all its agencies (Tran & Smith 2017). Worldwide, the number of new coal-fired power plants has dropped by 84 percent since 2015, and the decline has been steeper in recent years (Carrington 2019). With some exceptions in Asia (where new coal-based power plants came online in 2018- 2020, the decline of new coal plant production will continue, as renewable solar and wind energy have decreased in cost and are now cheaper than coal and even natural gas (Carrington 2019).

The cheapest renewable energy is solar, with wind not far behind. These energy-harvesting pathways, however, present challenges within existing power grid infrastructure. That infrastructure was built on a model for centralizing power and then distributing it to end users through high-voltage lines and substations. Most of that electricity delivery structure was located in urban areas or near the populations that consumed the energy. Since a traditional fossil fuel power plant can be built anywhere as long as there is a supply of fuel for heat and water for cooling, there are relatively few constraints regarding where they go. Power plants favor being close to their fuel sources, of course, since that saves fuel transportation costs, but the infrastructure for moving coal via rail or natural gas via pipeline in developed countries is robust. So proximity to sources will not make or break a potential traditional power plant.

In contrast, solar power is largely generated in smaller amounts relative to 500-megawatt, coal-fired power plants, and the energy created is often used locally. Solar power often is "decentralized" or "distributed" rather than centrally located and pushed outward. Decentralized power can be beneficial, both because it cuts the need for expensive and difficult-to-maintain transmission lines (which can also be a fire hazard, as California's 2019 fire season demonstrated) and because this puts power generation closer to the hands that use that power (i.e., more control). But it also has drawbacks, which range from energy distribution planning to extensive point-source infrastructure that must be maintained by a homeowner or non-utility entity (who may or may not be regulated).

Creating transmission lines that can upload solar power from individual locations as well as distribute energy to those locations will require some modification of infrastructure. An additional wrinkle exists, however, because the modeling used to project power needs at different times of the day and then modulate accordingly has been based on a single source. Once you have thousands, or tens of thousands, of additional sources feeding into the lines, the models break down. They were not developed to deal with decentralized, intermittent power. It is possible that the models could incorporate data on megawatt-scale solar energy deployed for various regions and then use weather forecasting to project when those units might be performing optimally or not. And in the future, these models will have to, since solar power is so much cheaper than other sources of electrical power. These challenges cannot be dismissed.

Implementing the Infrastructure for Electric Vehicles and Sustainable Power

All alternative energies require high-tech engineering and infrastructure that concentrates that energy into a form dense enough to use (except for hydropower, which is concentrated power generation for the most part). In contrast, fossil fuels exist because natural processes create the energy-rich molecules and concentrate them. The creation of the alternative infrastructure (both to harvest and to deliver) will require high-tech engineering and extensive manufacture, and a lot of energy (probably derived from fossil fuel energy) will be necessary to smelt the metals, machine the parts, prepare the sites, pour and punch silicon wafers, and so on. This point should not be ignored, and it needs to be understood that getting more carbon-neutral energy and the supporting infrastructure will itself have a substantial carbon footprint.

According to the *Better Growth, Better Climate* report (New Climate Economy 2014), 94 trillion dollars will need to be invested on new infrastructure projects through 2030 in order to replace the carbon-intensive infrastructure we have built over the last 100 years. The cost is so astonishingly high because 70 percent of all emissions are linked to our existing carbon-intensive infrastructure. As mentioned also in the Chapter 8 discussion of zero-carbon emissions, an ambitious but highly elusive goal, almost all emissions should be replaced if we are going to break out of carbon dependency. It does not have to be done all at once, of course. And it is highly likely that costs will come down as technology improves, and as best practices reveal lower-cost paths to efficiency. But we are going to have to show a more robust effort than the anemic one that exists now.

The costs of doing nothing are also extreme. The cost of weather-related episodes has been climbing dramatically in recent years. Globally, 450 billion dollars were spent on disaster recovery over the 2015–2018 period which amounted to 150 billion dollars a year. This amount has been rising as well. In 2014, the amount spent was closer to 100 billion dollars (New Climate Economy 2014), and just a few years later, it was up by 50 percent. As storm events intensify and once-rare events such as 500-year floods become more common, these costs will continue to rise. Although it is recognized that billions are needed in infrastructure upgrades, governments are grossly underinvesting. In the United States, throughout the years of the Trump administration. there was even a push to incentivize higher emissions industries (e.g., coal) and create impediments for alternatives. For example, the Trump administration actively thwarted progress by denying previously granted permits for wind-power development or fighting California's clean air regulations. Indeed, even the mitigation ambitions of the most "climate progressive" nations in the world, the United Kingdom are only half as stringent as what is needed to deliver on the Paris Agreement (Anderson et al. 2020).

Another hindrance to smoothly adopting alternatives into the energy infrastructure is the infrequency of "drop-down" replacements for current high-carbon technologies. Take electric cars, for example. The electricity to run the car can be derived by solar or wind, of course, but the very expensive challenges (in addition to gathering the wind or solar power) include (1) large-scale battery production, (2) deploying an extensive network of recharging stations, and (3) creating new electrical transmission facilities to meet with additional demand from electric cars. It is no wonder that an estimated 20 years will be required to scale up a promising new technology to reach any level that impacts overall trends. For example, as of 2018, there were 3.2 million electric cars worldwide—out of 70.5 million total cars (not counting commercial vehicles) (OICA 2018). If infrastructure to produce and charge them can keep up with current growth (a phenomenal 76 percent annually!), then the electrical market could equal that for conventional combustion cars by 2029 (Heinberg & Fridley 2016). It is unlikely, however, that the current growth rates can be sustained. The infrastructure supporting the growth is not matching pace.

As with electric vehicles (EVs), electric alternatives to existing furnaces, physical plants, and internal combustion engines of various kinds (not to mention batteries for utilities and cars) cannot be dropped into place because of expense, project size, and the fact that our existing electrical infrastructure is not capable of taking on all that added load. This is why most efforts in reducing CO_2 emissions so far have been made by improving the efficiency of existing fossil fuel infrastructure.

It is worth reminding ourselves that the extensive network of gasoline filling stations was a massive infrastructure undertaking that took decades to develop.

The first filling station was in Pittsburg in 1913, and as of 2017, there are approximately 156,000 in the United States (the number was higher 20 years ago, but consolidation and expenses dropped the total number). Each of these locations has large underground tanks holding the extremely hazardous, explosive, and carcinogenic fuel. They have to be monitored frequently to make sure the fuels are not leaking into groundwater (which, unfortunately, happens a lot). These stations were a major investment, and maintaining them continues to be costly.

Tax Incentives, Externalities, and the Market Place

Developing the infrastructure for EVs will likely take decades and be costly. Almost without the public noticing, however, more than 64,000 individual public charging stations have been installed in the United States (almost one-third of those in California) according to the US Department of Transportation (n.d.). Many of these stations provide power for free, so consumers can save a substantial amount of money if they choose the EV route (plus there may be federal or state tax incentives to purchase, depending on the year and laws passed by Congress). The existence of tax incentives and free fuel encourages EVs to gain market share, however, is decried by some as the government "tipping the scales" in the marketplace. And this brings us to a discussion of tax incentives in the energy industry and related fields. It is indeed true that "free" electric hook-ups and direct incentives such as the 2,500- to 7,500-dollar tax credit the United States offers encourage individual investment in alternative vehicles.

TAX CREDITS FOR ALTERNATIVES

Additional incentives for investing in alternative vehicles exist as production tax credits (PTCs) and investment tax credits for some alternative energies. In the United States, PTCs are for new wind, geothermal, and closed-loop biomass plants, and companies/utilities can obtain a up to a 24-dollar credit per megawatt-hour for generation as long construction starts by 2020 (EIA 2019). Other renewable or alternative technologies are eligible for a up to a 12-dollar credit per megawatt-hour (as long as construction starts before 2020) These credits would be in effect during the first 10 years of operation. Additionally, there is an investment tax credit in the United States (started in 2018) that will cover 30 percent of all solar construction costs for projects started before 2020 and brought online by 2024 (projects starting later receive less credit in increments down to 22 percent) (EIA 2019). Governments can also incentivize innovation by loaning money or guaranteeing loans for companies.

Loan programs are very common for governments interested in encouraging innovation, but they also attract attention when a loan fails and the taxpayers are left "holding the bag."

Some US readers may recall the collapse of a solar company called Solyndra back in 2011 after it had been granted 535 million dollars in loan guarantees. The loans had been made by the Department of Energy as part of the 40 billion dollars in guaranteed federal loan stimulus associated with the 2008 financial crisis. The company was experimenting with tubular solar (not panels) and was not using polysilicon, the common base material of solar panels. Unfortunately, the tubular solar was not readily compatible with household panel photovoltaics (see Chapter 5 for a discussion of photovoltaics) or large utility arrays. Additionally, the price of polysilicon dropped from 400 dollars per kilogram to 50 dollars in 2008. Both of these factors wiped out any cost benefit of Solyndra's tubular product, and the company went bankrupt (Fehrenbacher 2015).

Although these incentives were obviously created to encourage rapid investment in alternatives, we remind the reader that the petroleum industry gets billions in tax deductions from "intangible drilling and development costs" alone, which can be 60 to 80 percent of the cost of drilling a well (American Petroleum Institute 2016). By using the word "intangible," any costs associated with getting to a potential drill area, exploring that area, drilling test wells, and extracting the oil could conceivably be deducted. These deductions as business expenses were created to spur investment in petroleum and natural gas early in the industry's history when it was riskier than today and was viewed as a matter of national security. Today the industry is well developed and less risky from a development perspective, but riskier from a climate impact perspective. Incentivizing more petroleum development may actually increase risk in the long run because of uncontrolled CO_2 production and the resulting effects. Tax incentives have always been used to support behavior or industry that the government wants to encourage. Given the costs of climate change (adaptation, natural disasters), it may be more appropriate and farsighted at this time to continue inspiring the use of alternative energy and ceasing to fund fossil fuel development.

On top of tax breaks to petroleum on the exploration/development side, there is also an externalization of costs associated with waste produced. Although petroleum companies are increasingly paying damages for large oil spills, the other damaging impacts of their products are not accounted for on any balance sheet. Harms from cancer-causing volatile organic hydrocarbon emissions, particulates ($PM_{2.5}$ especially) and smog aside, rising CO_2 emissions are clearly having impacts on human infrastructure and health, yet the costs are borne by individuals or governments.

EXTERNALIZED COSTS

Overall, the true costs of petroleum products are not reflected in the price people pay when purchasing the product. Downstream costs have been "externalized." Some would say the petroleum industry is shielding its true costs with tax breaks (or tax avoidance) on the extraction side, and with avoidance of externalities on the consumption side. Some could also say that government policy encourages the development of alternative energy, which is also undoubtedly true. When complaints are made regarding unfair tax breaks for renewable energy, there are frequent counter-accusations of hypocrisy. Energy companies, like all other companies that can afford to, attempt to influence government policy regarding their industry. They use lobbyists to bend lawmakers' ears and help write legislation that favors their positions, and they make campaign contributions to all sides in an effort to get a seat at the policymaking table. Such is modern capitalism.

LEED Buildings and Urban Population Density

Buildings represent about 40 percent of energy use worldwide and are point sources for almost 50 percent of CO_2 emissions. Designing buildings with an eye toward limiting the energy use and carbon footprint of new or retrofitted buildings is a growing trend. In discussions around urban sustainability, the subject of LEED-accredited buildings often comes up, so this section investigates how LEED buildings can reduce carbon emissions from urban areas in some surprising ways, plus provide lower operating costs and enable ancillary environmental benefits.

LEED is an acronym for Leadership in Energy and Environmental Design. It is a US Green Building Council rating system to encourage building designs that save on energy, water, and materials. LEED-certified buildings also have low impact on their building sites, reduce CO_2 and methane emissions, and attempt to improve indoor environmental quality. Buildings are ranked based on how well they address efficiency and sustainability according to specific criteria. Rankings are Certified, Sliver, Gold, and Platinum. The buildings are ranked on a range of criteria, including indoor air quality, energy and atmosphere (efficiency), and innovation and design.

While the LEED buildings are designed to accomplish more sustainable operations, few independent studies have quantified their effectiveness. A 2006 study by the New Buildings Institute found among a group of 121 LEED buildings, there was a 25 to 30 percent energy savings relative to conventional buildings. The study was criticized, however, because the energy data were volunteered, because the median (i.e., the middle score of all listed) energy intensity for LEED was compared to mean (i.e., the numerical average of all scores) energy intensity for other buildings, and because there was "little correlation between simulated and measured energy savings for individual buildings" (Shofield 2015).

In 2012, a study examining the US Navy's efforts to reduce energy use by 30 percent between 2008 and 2015 showed that only two of the 11 LEED buildings built to meet the goal achieved the hoped for 30 percent energy usage reductions (Menassa et al. 2012). In fact, most actually increased their electricity use (Menassa et al. 2012). As of now, LEED does not require electricity to be sourced from CO_2-free renewables, which is surprising considering the ability to choose is becoming more widespread. And it seems that while LEED buildings provide benefits for cities in multiple important ways (reduced water use, healthier landscaping, more productive work environment, etc.), there are improvements that could be made for energy use/source criteria and more consideration for reducing CO_2 external emissions (LEED is very good at examining internal atmospheric conditions). Facing regional climate shifts that will affect urban subclimates dramatically, LEED could play a more dominant role.

The Nexus Between Land Use, Climate, and Food

About 50 percent of the terrestrial land mass is used for growing crops or raising livestock. Just looking at grasslands, the most productive lands for growing crops, the conversion to pure agriculture is close to 90 percent. Ninety-seven percent of North America's tallgrass prairie was converted to croplands between 1800 to 1950, for example (Primack 2008). Given that land use dramatically impacts the carbon cycle, it stands to reason that the way we use land to grow food is going to have an impact on climate-changing emissions.

When thinking about linkages between climate and human food, we can break them down into two that seem divergent at first glance, but that also impact each other. The first is food security, because productivity of some regions and some crops will decrease, which could lead to humanitarian crises and mass migration. The second is sources of food for the growing population, because depending on the amount of animal protein (and type of animal) that we consume, there are energetic costs and greenhouse-gas (GHG) emissions.

Climate change is predicted to undermine food security for much of the world, as shifts in crop hardiness zones, migration patterns, pollinator distribution, and heat stress to plants and workers (among other factors) accelerate. As we have reviewed in earlier chapters, many of these shifts will be asynchronous (i.e., do not occur at the same time), and so seasonal crop yields will be harder to predict (hence, less food security). Here, we show the aggregate impacts of climate change on food production (Figure 6.1). In Chapter 9, we discuss how individuals might change their diets to minimize these impacts.

LAND USE CHANGE AND GHG EMISSIONS

Land can be both a source of and a sink for CO_2. Through photosynthesis and an increase in living biomass, the biosphere draws down atmospheric CO_2 and can sequester it in woody plant tissue and cellulose. This captures atmospheric CO_2 and removes it from the carbon cycle for at least a short period of time (years to tens of decades). As discussed in Chapter 2, the soil itself is also a major sink for carbon, containing almost three times the carbon in the

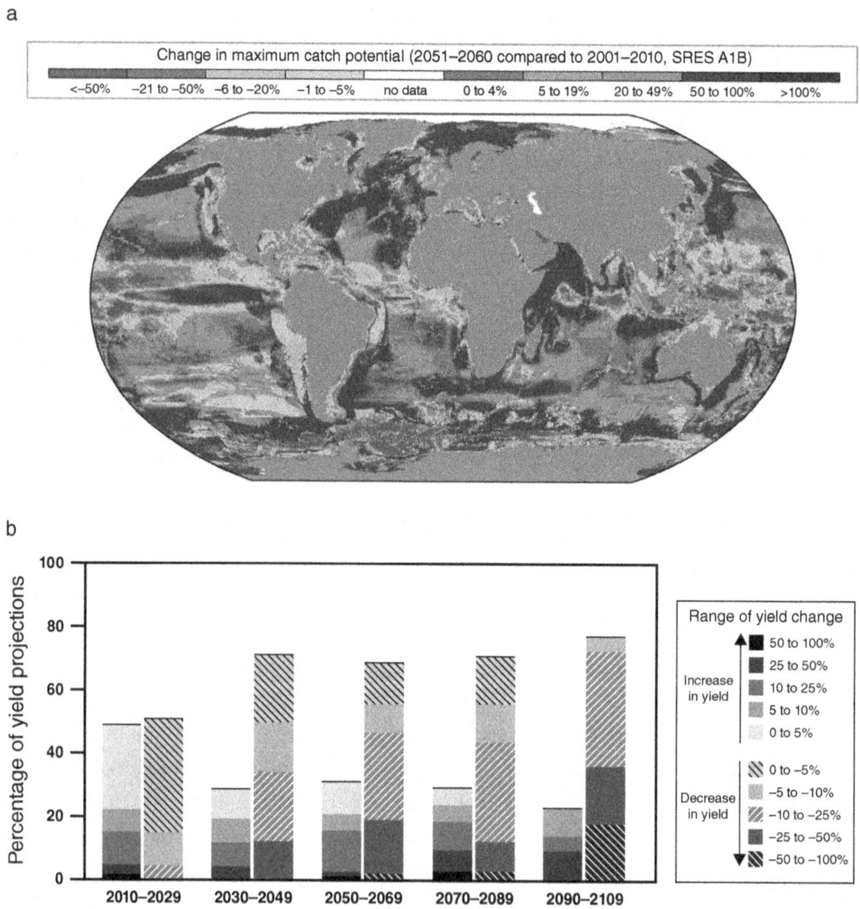

Figure 6.1 Panel (a) shows changes expected in maximum catch potential for the world's fisheries between 2051 and 2060 relative to those of the period 2001–2010. Projections are based on a single climate model under a moderately high warming scenario. Note that the greatest losses would be in tropical and Antarctica waters. Panel (b) shows a summary of expected yield changes for wheat, maize, rice, and soy (mostly) as time moves forward to 2109. While there will be some areas that experience greater yields because of longer growing seasons and the possibility of multiple harvests, for most crops in most areas, yields could be substantially decreased. (Figure from IPCC 2014a).

atmosphere (Schlesinger & Bernhardt 2013). The land can emit CO_2 as well as absorb or sequester it. Soils that are warmer release CO_2 as microbial respiration increases (respiration results in CO_2 release from metabolic activity). Anywhere there is organic litter or organic soil plus oxygen, there will be decay (though microbial activity) and CO_2 release.

A classic study in England and Wales examined carbon content of 5,662 soil plots in 1978 and then revisited them in 2003. It was found that the soils lost 13 teragrams (10^{12} grams, or about 14.3 million tons) each year on average over the 21 years (Bellamy et al. 2005). The losses were independent of local weather, day-to-night temperature flux, or rainfall, indicating that the entire southern UK region was affected. They found that regional climate variation in both precipitation and temperature on the decade scale was the driver of the losses. Another interesting finding was that the amount of organic carbon in the soil was significantly correlated with the amount of carbon lost; that is, the more carbon there was initially, the more was lost (so peat soils, the richest organic soils, lost the most carbon). Frustratingly, the losses of soil carbon were almost exactly equivalent to the carbon emissions reductions the UK accomplished through the Kyoto Protocol (12.7 teragrams C per year)!

In addition to carbon losses from warming soils, CO_2 is also released from soils when surface vegetation is disturbed or there is a change in land use. For example, converting forest to cropland or grassland reduces the carbon content of the land/soils. Anytime there is plowing or planting activity on a grassland or cleared forest, there will be carbon losses through emission to the atmosphere (CO_2 from decay and respiration) and from water percolating through the soil (carbon compounds dissolve in the water and are transported from the surface to groundwater and, from there, to oceans). Although 65 percent of all GHG emissions are from fossil fuels and industrial processes releasing CO_2, about 11 percent of all CO_2 emissions are from forestry and land use practices (IPCC 2014a).

Globally, humans have actually enhanced the removal of 11.2 ± 2.6 gigatons of CO_2 per year from the atmosphere by (1) increasing nitrogen in rainfall, (2) climate change resulting in earlier spring and longer growing seasons, and (3) "extra" CO_2 causing a carbon fertilization effect (see the discussion in Chapter 2) (IPCC 2019). During that same period, it is estimated that 5.2 ± 2.6 gigatons of CO_2 per year have been emitted from terrestrial landscapes from land-use change alone. It is expected that as soils warm, the emissions of CO_2 will increase and shrink the extra 6.0 ± 2.3 gigatons of CO_2 that terrestrial systems have absorbed (IPCC 2019).

Of particular concern is how organic carbon locked in permafrost will react to warming. Given the CO_2 release from warming, organic-rich soils (e.g., bogs) in England and Wales observed in the Bellamy et al. (2005) paper, it is assumed that the atmospheric flux will be substantial. Recall here that annual emissions from fossil fuel combustion alone in 2014 were approximately 10.1 gigatons of carbon (which is about 37 gigatons when it combines with oxygen to produce CO_2, and remember that oxygen is quite a bit more massive than carbon!) (IPCC 2014a). While it is encouraging that terrestrial systems are still sequestering carbon globally, we have to keep in mind that emissions from burning are six times the amount sequestered in recent years.

Changing land use not only impacts CO_2 release from those soils, it can also result in the emission of other GHGs. Land managed for agriculture, forestry, or other modified uses accounts for 44 percent of methane and 82 percent of nitrous oxide (N_2O) emissions, both of which are GHGs. In terms of total warming energy, methane contributes about 16 percent and N_2O about 6 percent (IPCC 2014a). Total methane emissions between 2007 and 2016 were 162 ± 48 million tons (4.5 ± 1.4 gigatons of CO_2 equivalents, to put things in a CO_2 perspective). Prior to 2007, growth in methane emissions was small and actually paused between 1999 and 2006. Biological sources are far more important for methane emissions than industrial processes, and it is thought that increased cattle production (and production of other ruminants) plus rice cultivation could be responsible for increases since 2007 (IPCC 2019). Most N_2O emissions are a result of inefficient nitrogen-fertilizer application. N_2O is a trace gas in the atmosphere, but also an important one among the GHGs. It is approximately 300 times more potent at trapping heat radiation than CO_2. Between 2007 and 2016, 8.3 ± 2.5 million tons per year were emitted (equivalent to 2.3 ± 0.7 gigatons of CO_2 per year). Most N_2O is emitted from oxygen free-environments as inorganic nitrogen compounds are converted into N_2 gas by bacteria.

It is thought that increased numbers of livestock on managed pastures (growth of the cattle industry, again) have led to more manure deposited to soils, which then decays, removing oxygen in local soils and causing N_2O production (IPCC 2019). All three GHGs emitted from soils following altered land use arise from organic material being consumed or otherwise metabolized by bacteria. Clearly, development projects or agricultural planning should make efforts to limit exposure of the soil to air and water, both of which encourage decay and exacerbate GHG emissions.

Human use directly affects approximately 70 percent of all ice-free land (IPCC 2019). As human populations continue to grow, greater amounts of land are converted to cropland and commercial forests. These changes result in the land becoming an ever-greater source of CO_2 to the atmosphere. Climate change itself contributes to soil loss through increased rainfall intensity in some areas and increased dryness in others. The soil losses contribute to agricultural expansion, because new croplands are needed to replace what was lost. Additionally, climate change is increasing the expansion of arid zones (borders of deserts are expanding), further straining productivity on existing croplands. Drylands[1] experiencing drought increased by one percent each year on average between 1961 and 2013. During the same period, the population in these same areas increased 200 percent. The added population places increasing strain on soil resources (not to mention water), and there has been an increase in dust storms (which obviously remove soil) in Central and East Asia, Australia, and sub-Saharan Africa.

DIET AND GHG EMISSIONS

In recent years, more attention has been placed on human diet and GHG emissions. Initially, concern in shifting human diets was related to health, as populous nations became wealthier (e.g., China). This is because a host of diseases associated with highly saturated fats, red meat consumption, and processed foods correlate with the diet shift. For example, in 1980, less than one percent of China's population suffered from type II diabetes, but by 2008, that had risen to 10 percent (Hu 2011). The trend is not only in type II diabetes, however, and not only in China. The trend is worldwide. A normal omnivorous diet with red meat as well as a mix of other food types shows higher incidences of cancer, coronary mortality, and mortality overall (Tilman & Clark 2014). When researchers compared the incidence of these causes of mortality (including type II diabetes) with Mediterranean diets (vegetables, fruit, seafood, grain, sugar, oil, eggs/dairy, and moderate poultry, pork, lamb, and beef), pescatarian diets (vegetables and seafood), and vegetarian diets (one serving of meat or less per month, plus vegetables and dairy), they found dramatically lower mortality rates relative to the omnivorous diet (Figure 6.2).

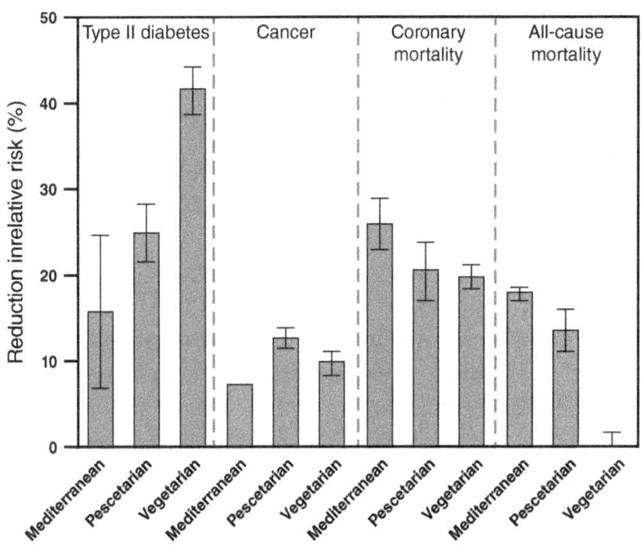

Figure 6.2 Reduction of risk of mortality associated with different diets versus an omnivorous diet, which is the 2009 global average diet that includes all food groups. (Figure from Tilman & Clark 2014).

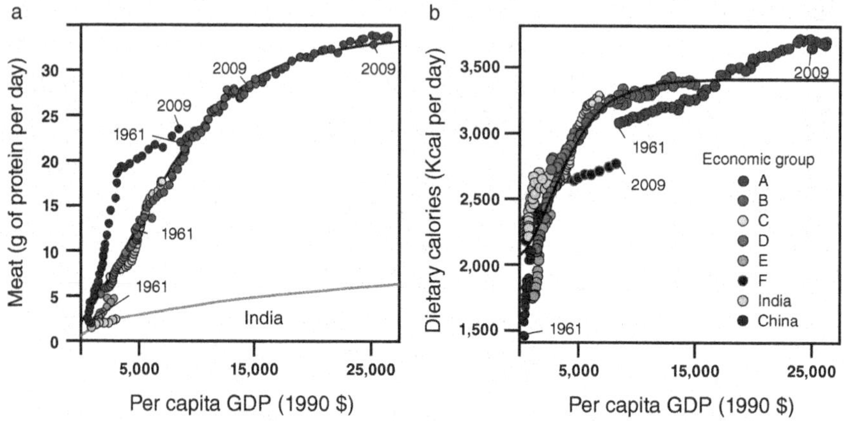

Figure 6.3 Relationship between per capita income and meat protein consumed (a) as well as calories consumed (b). Note that while grams of protein per day per capita (a) starts decoupling from GDP growth around 15,000 dollars per capita, the total calories consumed (b) continues to increase until GDP is 25,000 dollars per capita. The extra calories are derived from increases in processed foods plus sugar and "empty calories." The economic groups are based on GDP for 100 countries. Group A is highest 15, group B is the next 15, and so on; group F is lowest 25, with China and India standing alone. Group A: Australia, Austria, Canada, Denmark, Finland, France, Germany, Ireland, Japan, The Netherlands, Norway, Sweden, Switzerland, United Kingdom, and United States. Group B: Argentina, Chile, Greece, Israel, Italy, Malaysia, Mauritius, New Zealand, Portugal, Saudi Arabia, South Korea, Spain, Trinidad and Tobago, Uruguay, and Venezuela. Group C: Botswana, Brazil, Colombia, Costa Rica, Ecuador, Guatemala, Iran, Jordan, Mexico, South Africa, Syria, Thailand, Tunisia, and Turkey. Group D: Algeria, Bolivia, Cuba, Dominican Republic, Egypt, El Salvador, Indonesia, Jamaica, Lebanon, Morocco, Paraguay, Peru, Philippines, Sri Lanka, and Swaziland. Group E: Bangladesh, Pakistan, Benin, Cameroon, Cote d'Ivoire, Ghana, Honduras, Libya, Mozambique, Myanmar, Nicaragua, Nigeria, North Korea, Senegal, and Vietnam. Group F: Burkina Faso, Central African Republic, Chad, Democratic Republic of the Congo, Eritrea, Ethiopia, Gambia, Guinea, Haiti, Kenya, Madagascar, Malawi, Zambia, Zimbabwe, Mali, Nepal, Niger, Rwanda, Burundi, Sierra Leone, Sudan, Tanzania, Togo, and Uganda. (Figure from Tillman & Clark 2014).

This trend toward a diet higher in meat and fat has been closely correlated with a rise in disposable income in all countries except India, where cultural practices strongly discourage meat consumption. Tracking per capita gross domestic product (GDP) and meat consumption per day shows a strong positive correlation until 15,000 dollars per capita GDP is reached (in 1990 US dollars), after which the relationship decouples (Figure 6.3a). Between 1961 and 2013, the number of people considered by the United Nations as overweight increased by 80 percent globally (IPCC 2019). Obviously, from a health care cost perspective, the planetary diet change has been monitored closely (and with a bit of alarm). This is discussed further in Chapter 9.

Aside from health consequences, higher animal consumption among the most populous area of the world has caused some concern, because land used to grow feed for livestock could be used to grow crops for humans instead. For every 10 kg of beef raised, 100 kg of grains or grasses must be consumed, which is a roughly 10 percent efficiency (90 percent of the food mass is lost as waste or metabolic activity). Usually when we think of red meat, we are referring to cattle, sheep, ox, bison, yak, and antelope (among others). These are all ruminants (herbivores that have three- or four-chambered stomachs and chew cud), and their diet efficiencies are all about 10 percent, as observed in cattle. Efficiencies are higher for pork and, in particular, poultry, which can be 50 percent efficient as measured by mass fed versus mass produced. As the world gets richer (in the long term, most models assume it will, regardless of recessions and short-term fluctuations) and the year 2050 approaches, food consumption models show that protein in diets will increase by 11 percent, consumption of beef and other ruminates will rise by 31 percent, consumption of pork and poultry will rise by 23 percent, and vegetable consumption will drop by 18 percent (Tilman & Clark 2014).

If our goal is to use cropland primary production efficiently, with a goal of feeding everyone adequately—which is not being done now by any stretch of the imagination; indeed, 815 million people are undernourished worldwide (UN 2019)—then growing food for animals with very low efficiency probably is not the way forward. It is not effective land-use policy to encourage growing crops for food animals that are extremely inefficient at converting calories of feed into calories of animal protein (nothing against cows, they are just very inefficient). Beef and other ruminants produce meat high in protein and nutrients per unit mass. However, consuming them frequently can be unhealthy, and other species (e.g., swine and poultry) are much more efficient sources of high-calorie and nutrient-rich meat. Of course, in some countries, cattle are kept on rangelands for grazing, and rangelands are too dry to grow crops. Studies have shown that grazing ruminants sustainably on noncrop lands while feeding them crop residue can produce high-quality foods, improve food security for communities, and enhance nutrient cycling (Eisler et al. 2014, Smith et al. 2013). As with most agriculture, the method used to raise any specific crop or stock has a large impact on the sustainability of the practice, the GHG emissions, and the impact on local human populations. While grazing on rangelands or noncrop lands has drawbacks, the issue of diverting cropland production to feed the cattle is not an issue.

It has long been known that cattle (the world's dominant food ungulate) not only require an enormous amount of food to maintain growth (again, only about 10 percent of feed calories become cow calories), but they also expel 600 liters of methane a day on average (Schlesinger & Berhardt 2013, Williamson-Rea 2018). Methane is 25 times more potent as a GHG than CO_2, but it is (thankfully) far less abundant than CO_2 and does not remain in the atmosphere for long periods of time. Global methane is rising, however (Figure 6.4), and of all anthropogenic sources, ruminate emissions make up about one-third of the total each year.

On top of the feed issue and the methane issue is the fact that producing ruminant meat is by far the most GHG-intensive agricultural practice on the planet. Protein production per unit CO_2 equivalents is 250 times higher in ruminates than is it is for legumes (Tilman & Clark 2014). In terms of grams of animal protein and kilocalories produced, growing ruminates produces, respectively, six times and five times more CO_2 equivalents than either pork or poultry (Figure 6.5a,c). In terms of "servings," ruminates also produce about six times more CO_2 equivalents than either pork or poultry (Figure 6.5b). Interestingly, farmed pork and poultry produce less CO_2 than many types of seafood products, including trawled fisheries (where weighted nets are dragged along the ocean bottom), non-recirculating aquaculture, and recirculating aquaculture (in a recirculating system, the water is run through a biofilter before being released back into the tank). In a non-recirculating system, the water flushes with the surroundings, and fish are kept from escaping with nets or other barriers. Non-recirculating systems produce 50 percent or less CO_2 equivalents per gram protein or calorie when compared to recirculating systems (Figure 6.5).

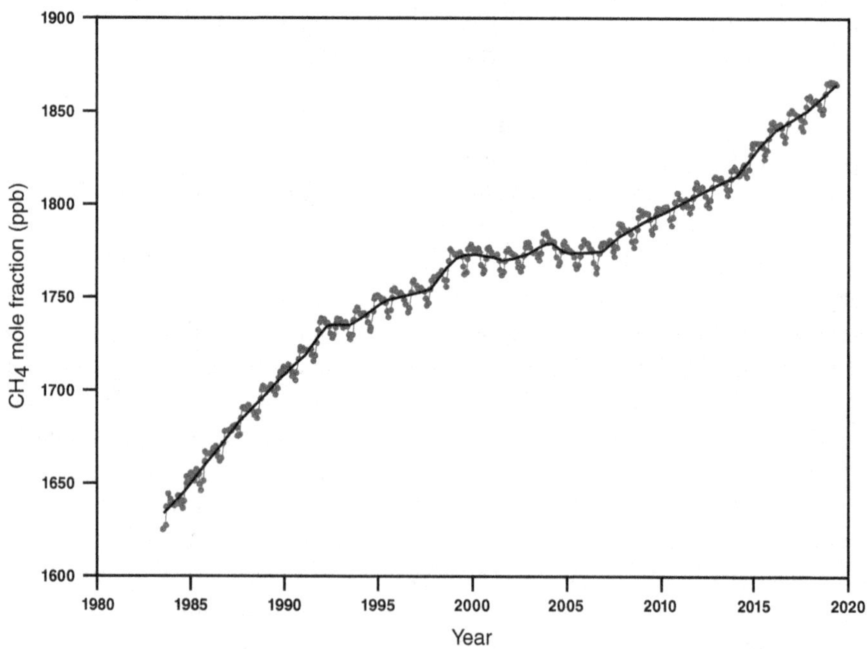

Figure 6.4 Atmospheric methane (CH_4) concentration versus time as measured from marine surface sites. The gray circles and lines are globally averaged monthly mean values (centered on the middle of the month). The black line shows a long-term trend where the average seasonal cycle has been removed. (Figure from Global Monitoring Laboratory 2020).

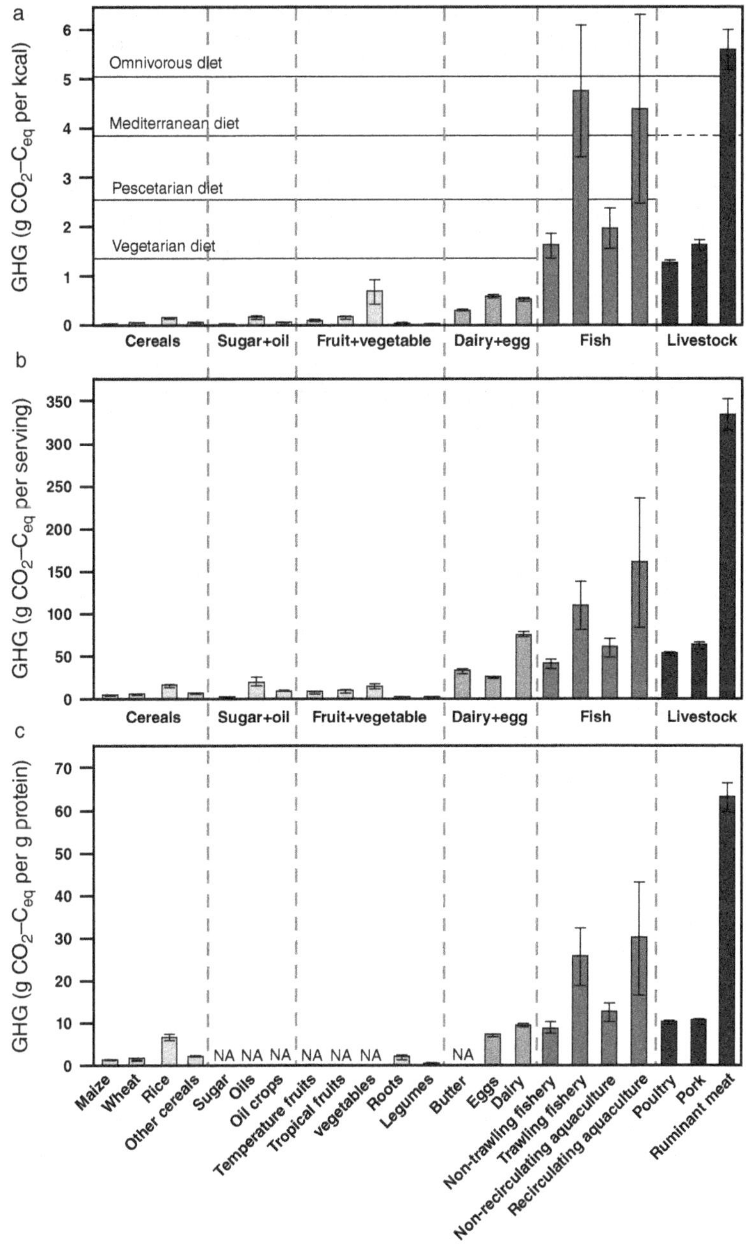

Figure 6.5 Life-cycle GHG emissions in CO_2 equivalents (CO_2-C_{eq}) for 22 different food types used in 555 food production systems. (a) CO_2-C_{eq} per kilocalorie, (b) CO_2-C_{eq} per serving (as defined by US Department of Agriculture), and (c) CO_2-C_{eq} per gram protein. (Figure from Tilman & Clark 2014).

Raising land animals for food products but not directly eating them (dairy or eggs) has the least GHG impact of any animal agriculture. For eggs, approximately nine grams of CO_2 equivalents are produced per gram protein, versus 11 grams of CO_2 equivalents for pork or poultry or 69 grams of CO_2 equivalents for ruminant meat (Tilman & Clark 2014). For dairy (mostly milk and cheese), 10 grams of CO_2 equivalents are released per gram of protein. Looking at fisheries, the results are mixed, depending on whether trawling is used or open-water (non-recirculating) aquaculture is used. Per gram protein and kilocalorie produced, trawling fisheries and recirculated aquaculture produce two to three times the CO_2 equivalents as pork and poultry and one-half to three-fourths the CO_2 that raising ruminant meat does (Figure 6.5).

Worldwide, the consumption of meat has been growing between five and six percent per year and of dairy between 3.4 -3.8 percent per year since the 1960s (Bruinsma 2003). However, the livestock supplying the meat has been shifting. Although beef consumption dominated the meat market in the 1970s and 1980s, chicken overtook beef as the most consumed animal protein in 2005. Today, the consumption of poultry (chicken and turkey) and of fish/seafood is rising, and beef has declined from its peak of 89 pounds per person to about 54 pounds. Pork consumption has remained fairly steady in the United States since the 1970s (and was 50 pounds per capita in 2016), but it was overtaken by chicken in the early 1990s (Figure 6.6). The US population has risen, of course, even as its beef consumption per person has decreased, so while individuals are consuming less beef, the total amount consumed has been rising (Figure 6.6). The increase is largely driven by chicken consumption, however. For example, in 2016, the average American consumed 194 pounds of beef, pork, or chicken versus 184 pounds in 2012 (Hill 2016).

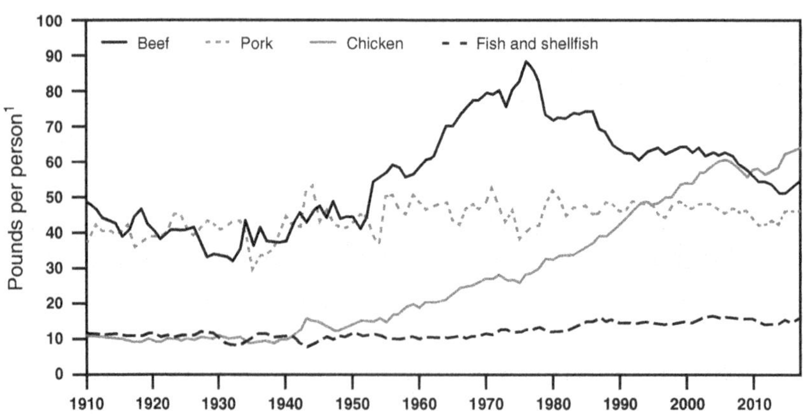

[1]Calculated on the basis of raw and edible meat in boneless, trimmed (edible) weight. Excludes edible offals, bones, viscera, and game from red meat. Includes skin, neck, and giblets from chicken. Excludes use of chicken for commercially prepared pet food. Source: USDA, Economic Research Service, Food Availability Data.

Figure 6.6 Mean consumption trends in the United States for the last 100 years. In 2017, edible pounds per capita for chicken increases to 64.1 pounds versus beef's 54.3 pounds. (Figure from US Department of Agriculture 2019).

The shift in type of meat consumed is probably driven by the cheaper prices for chicken (less feed and land needed) and by public awareness that red meat can be unhealthy. The shift is not driven by any public awareness about the climate impacts of diet, although a minority of Americans are aware of the link and do factor it into their lifestyle choices, including diet.

Exercises

As with earlier chapters, exercises appearing at the chapter closing are meant to be standalone sections that complement the material covered but are not directly dependent on the chapter contents. For this chapter, we present two problems that address different aspects of electrical power generation and use. The first exercise is a critical evaluation of electrical energy use in the United States and why it is so much higher than in other developed economies. The second presents a series of electrical power generation scenarios for hydroelectrical power. Alternative methods of generating electricity, such as hydropower, are often derided as being unreliable, since they depend on wind, sun, or water, all of which are in variable supply depending on weather, climate, and time of day. This is true, of course, which is why improvement in batteries or other methods of power storage must be discussed along with alternatives. In any case, it is helpful to examine how electrical power is actually generated and to consider ways it may be used more efficiently. In these two exercises, readers may (1) consider how energy could be used more efficiently in the United States and (2) use some relatively simple power generation problems to gain an appreciation of how electrical power is generated using renewable hydropower.

Exercise 6.1 Examining Electrical Power in the United States

The United States uses far more energy per capita than other developed (postindustrial) countries, such as the UK or Japan (although the Middle Eastern countries use far more energy per capita than the United States). In 2015, the United States consumed over 12,000 kWh per capita, while Japan consumed roughly 7,500 kWh per capita and the UK less than 5,000 kWh per capita. Both the UK and Japan, however, have been maintaining a relatively robust economy while trimming energy use at the same time (Figure 6.7). If the United States could alter its electrical energy consumption to more closely align with changes the UK, for example, has done, billions of dollars could be saved, and substantial reductions in CO_2 emissions could be achieved.

Continued

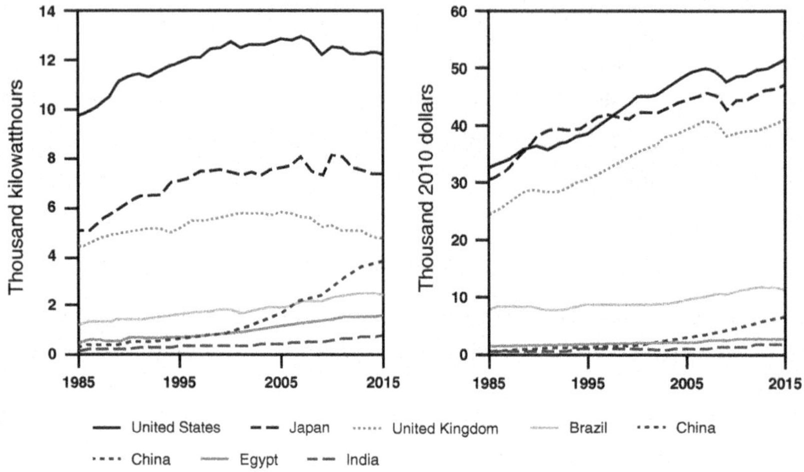

Figure 6.7 Electricity use per capita for three developed and four developing countries between 1985 and 2015 (left). Gross domestic production (GDP) per capita for the same countries over the same time period (right). (Figure from EIA 2017).

Here are some research questions related to this topic.

1. What are some reasons why US electrical power consumption is so much higher than other post-industrial economies?
2. What are some policy changes the United States could implement that would curb electrical power consumption?
3. If the United States reduces its electrical consumption, what do you think will occur to the natural gas sector with regard to demand?
4. If alternative energy sources continue to be cheaper than natural gas for electrical power generation, can you project what percentage of the total US energy would be derived from alternative sources by 2050?

Exercise 6.2 Calculating Hydroelectric Power

Since hydropower facilities are largely influenced by their local geography, their power capacity is highly variable. If you want to build a 500-megawatt natural gas power plant, you can just build one using industry standards. For a hydropower facility, however, you will have to know (1) the height your water will fall from as it moves to the turbine (called hydraulic head), (2) the volume of water moving through the turbine, (3) the type of turbine, (4) the number of turbines, and (5) the efficiency (i.e., how much of the water's potential energy can be converted to electrical energy).

As with the other "alternative energies," output from a hydroelectric power plant is variable, depending on environmental factors. There are times when power capacity exceeds demand and times when low flow rates or reduced hydraulic head can decrease capacity. Hydroelectric power, therefore, is another power source that would need to be combined with very good battery storage or pumped storage. Understanding and being able to estimate power capacity under different conditions is vital for having hydrological power incorporated into an energy portfolio. Considering that almost 20 percent of all electrical power is generated from it, examining the basics of calculating power from hydropower can be useful.

We start with highly perched water falling through space and into a turbine. The energy of the falling water is found by calculating its potential energy (PE):

$$PE = M * g * H \tag{6.1}$$

where M is mass (1 m^3 of water is 1,000kg), g is acceleration of gravity (9.81 m s^{-2}), and H is hydraulic head in meters (the distance the water "falls").

Power (P) is usually given in watts, which is joules per second. To calculate power, you just make a few modifications to equation 6.1:

$$P = 1{,}000 * \mu * Q * g * H \tag{6.2}$$

where μ (pronounced "mu") is efficiency after losses from friction or drag (as a decimal, not a percentage), Q is discharged per second (converted from l^3 s^{-1} to m^3 s^{-1}; there are 1,000 l^3 in a m^3), g is acceleration of gravity (9.81 m s^{-2}), and H is hydraulic head. The 1,000 is just a scaling multiple. If you are using m^3 s^{-1} for discharge and want output in watts, you need to multiply by 1,000; if you want kilowatts, do not multiply by 1,000.

Equation 6.2 works for propeller turbines (usually used in low-hydraulic-head situations) or Francis turbines (the most common type found at large dams like Hoover in the United States or Three Gorges in China). A third type of turbine is a Pelton impulse turbine. It is used in situations where H is very large but water volume may be limited. For a Pelton turbine, water is shot into the turbine with jets, and the flow rate (or discharge rate) is dependent on the area of the jet and the hydraulic head. The potential energy of the free-falling water is turned into kinetic energy using the following equation:

$$M * g * H = 0.5 * M * v^2 \tag{6.3}$$

and

$$2 * M * g * H = M * v^2 \tag{6.4}$$

Continued

so

$$v^2 = 2 * g * H \tag{6.5}$$

then

$$v = \sqrt{2 * g * H} \tag{6.6}$$

Since a Pelton turbine requires jets to deliver the water, jet nozzle area is needed to find discharge rate (Q):

$$Q = A * \sqrt{2 * g * H} \tag{6.7}$$

Then, Q is plugged into equation 6.2 to find the power. Interestingly, for Pelton turbines, additional jets can be added to a single wheel. To find the new power output, you just multiply the power generated for one jet by the number of jets. Some Pelton arrangements can have eight jets or more! To find the amount of power, simply multiply the Q from equation 6.7 by the number of jets (provided the jet nozzle area is the same).

The quantitative objective of this exercise is to show how variable hydroelectric power can be, depending on the height of the dam, the type of turbine, the number of turbines, the efficiency, and seasonal precipitation (hence, flow rate). How adding additional turbines to existing power plants can add to capacity is a second objective. Thus, using the appropriate equations outlined here, calculate the power capacity for the following hydropower stations:

1. *Galloway, Scotland, hydroelectric power station:* Hydraulic head (H) = 25m, Francis turbine = 88 percent efficiency, flow rate (Q) = 6,00 l^3/min, g= 9.81 m s^{-2}. Convert to cubic meters per second, and your answer will be in kilowatts.
2. *Niagara Falls hydroelectric power station:* Hydraulic head (H) = 100 m, Francis turbine = 83 percent efficiency, flow rate (Q) = 360,000,000 l^3/min, g = 9.81 m s^{-2}. Convert to cubic meters per second, and your answer will be in kilowatts.
3. *Pelton hydroelectric power station (using H from Three Gorges):* Hydraulic head (H) = 181 m; efficiency = 90 percent. What is the output in kilowatts from one jet, A = 1.3 m^2? What if you had three jets on the same Pelton wheel?
4. *Itaipu power station:* Shared by Paraguay and Brazil, it supplies 95 percent of Paraguay's power and up to 80 percent of Brazil's power! The flow of water into it is seasonal, however, so output can be variable. What are low- and high-power possibilities for Itaipu? Flow depends on season, Q = 9,000–36,000 m^3/sec, H = 118.4 m, Francis turbine = 90 percent efficiency. Your answer will be in kilowatts.

NOTES

1. Drylands are defined by the United Nations Environmental Programme as areas where the ratio of average annual precipitation and potential evapotranspiration is less than 0.65. See http://www.fao.org/dryland-forestry/background/what-are-drylands/en/

REFERENCES

Anderson, Kevin, John F. Broderick, and Isak Stoddard. 2020. "A factor of two: the mitigation plans of 'climate progressive' nations fall far short of Paris-compliant pathways." *Climate Policy* 20(10):1290–1304. DOI: https://doi.org/10.1080/14693062.2020.1728209

American Petroleum Institute. 2016. "Eliminating the ability to expense intangible drilling and development costs US investment." Accessed December 28, 2020 at https://www.api.org/~/media/Files/Policy/Taxes/16-February/API-IDC-2016.pdf

Bellamy, Pat H., Peter J. Loveland, Ian R. Bradley, Murray R. Lark, and Guy J. D. Kirk. 2005. "Carbon losses from all soils across England and Wales 1978–2003." *Nature* 437(7056):245–248.

Bruinsma. 2003. *World Agriculture: Towards 2015/2030. An FAO Perspective*. Earthscan Publications Ltd. London England. ISBN 1844077

Carrington, Damian. 2019. "Global 'collapse' in number of new coal-fired power plants." *The Guardian*, March 27. Accessed December 28, 2020, https://www.theguardian.com/environment/2019/mar/28/global-collapse-in-number-of-new-coal-fired-power-plants

Cummins, Eleanor. 2020. "Here's how to switch to renewable energy." *MSN Lifestyle*. Accessed December 28, 2020, https://www.msn.com/en-us/lifestyle/home-and-garden/here-s-how-to-switch-to-renewable-energy/ar-BB15vD8t

Energy Information Administration (EIA). 2017. "Link between growth in economic activity and electricity use is changing around the world." Accessed December 2, 2020, at https://www.eia.gov/todayinenergy/detail.php?id=33812

Energy Information Administration (EIA). 2018. "Nearly half of utility-scale capacity installed in 2017 came from renewable." Accessed June 23, 2020, at https://www.eia.gov/todayinenergy/detail.php?id=34472

Energy Information Administration (EIA). 2019. "Levelized cost and levelized avoided cost of new generation resources in the *Annual Energy Outlook 2019*." Accessed December 16, 2020. https://www.eia.gov/outlooks/aeo/pdf/electricity_generation.pdf

Eisler, Marc C., Michael R. Lee, John F. Tarlton, Martin B. Graeme, John Beddington, J. A. Dungait, Henry Greathead, Jianxin Liu, Stephen Mathew, Helen Miller, Tom Misselbrook, Phil Murray, V. K. Vinod, R. Van Saun, and Michael Winter. 2014. "Agriculture: Steps to sustainable livestock." *Nature* 507(7490):32–34.

Fehrenbacher, Katie. 2015. "Why the Solyndra mistake is still important to remember." *Fortune*, August 27. Accessed at https://fortune.com/2015/08/27/remember-solyndra-mistake/

Global Monitoring Laboratory. 2020. "Trends in Atmospheric Methane." Accessed December 2, 2020, at https://esrl.noaa.gov/gmd/ccgg/trends_ch4/

Heinberg, Richard, and David Fridley. 2016. "Transportation: The Substitution Challenge." In *Our Renewable Future*. Island Press, Washington DC. https://doi.org/10.5822/978-1-61091-780-3_5

Hill, Catey. 2016. "This chart proves Americans love their meat." *Market Watch*, December 1 Accessed June 23, 2020, at https://www.marketwatch.com/story/this-chart-proves-americans-love-their-meat-2016-08-15

Hu, F. B. 2011. "Globalization of diabetes: the role of diet, lifestyle and genes." *Diabetes Care* 34:1249–1257.

International Energy Agency (IEA) 2017. World Energy Outlook. Accessed December 15, 2020 at https://www.iea.org/reports/world-energy-outlook-2017

Intergovernmental Panel on Climate Change (IPCC). 2014a. *Climate Change 2014: Synthesis Report. Contribution of Working Groups I, II and III to the Fifth Assessment Report of the Intergovernmental Panel on Climate Change*, edited by R. K. Pachauri and L. A. Meer. *IPCC*, Geneva, Switzerland.

Intergovernmental Panel on Climate Change (IPCC). 2014b. *Climate Change 2014: Mitigation of Climate Change. Contribution of Working Group III to the Fifth Assessment Report of the*

Intergovernmental Panel on Climate Change, edited by O. Edenhofer, R. Pichs-Madruga, Y. Sokona, E. Farahani, S. Kadner, K. Seyboth, A. Adler, I. Baum, S. Brunner, P. Eickemeier, B. Kriemann, J. Savolainen, S. Schlömer, C. von Stechow, T. Zwickel, and J. C. Minx. *Cambridge University Press*, Cambridge, UK.

Intergovernmental Panel on Climate Change (IPCC). 2019. *Special Report on Climate Change, Desertification, Land Degradation, Sustainable Land Management, Food Security, and Greenhouse gas fluxes in Terrestrial Ecosystems.* Accessed December 28, 2020 *https://www.ipcc.ch/srccl/*

Menassa, Carol, Seth Mangasarian, Mounir El Asmar, and Carl Kirar. 2012. "Energy consumption evaluations of US Navy LEED-Certified buildings." *Journal of Performance of Constructed Facilities* 26(1):46-53.

New Climate Economy. 2014. *Better Growth, Better Climate.* World Resources Institute, Washington, DC.

OICA. 2018. "International Organization of Motor Vehicles Manufactures." Accessed June 23, 2020, at http://www.oica.net/category/sales-statistics/

Primack, Richard B. 2008. *A Primer of Conservation Biology*. 4th ed. Sinauer Associates, Sunderland, MA.

Roston, Eric. 2019. "The massive cost of not adapting to climate change." *Bloomberg,* September 9. Accessed December 28, 2020 at https://www.bloomberg.com/news/articles/2019-09-09/the-massive-cost-of-not-adapting-to-climate-change

Rotherham, Duncan, James D. Brown, and Jennifer Suke. 2014. "Stranded oil and gas assets—Is the wolf at the door or waiting in the forest?" ICF International, Inc. Fairfax, VA.

Schlesinger, William H., and Emily S. Bernhardt. 2013. *Biogeochemistry: An Analysis of Global Change*. 3rd ed. Academic Press, Cambridge, MA.

Shofield, John H. 2015. "Are US 'green buildings' really saving energy? The facts may surprise you." *Brinknews*, June 10. Accessed June 23, 2020, at http://www.brinknews.com/are-u-s-green-buildings-really-saving-energy-the-facts-may-surprise-you/

Smith, Jimmy, Keith Sones, Delia Grace, Susan MacMillan, Shirley Tarawali, and Mario Herrero. 2013. "Beyond milk, meat and eggs: role of livestock in food and nutrition security." *Animal Frontiers* 3:6–13.

Tilman, Dave, and Michael Clark. 2014. "Global diets link environmental sustainability and human health." *Nature* 515:518–522.

Tran, Thomas D. T., and Amanda D. Smith. 2017. "Evaluation of renewable energy technologies and their potential for technical integration and cost-effective use within the US energy sector." *Renewable and Sustainable Energy Reviews* 80:1372–1388.

United Nations. 2019. *Sustainable Development Goals*. Accessed June 23, 2020, at https://www.un.org/sustainabledevelopment/hunger/

US Department of Agriculture. 2019. "Per capita available of chicken higher than that of beef." Accessed December 2, 2020, at https://www.ers.usda.gov/data-products/chart-gallery/gallery/chart-detail/?chartId=58312

US Department of Transportation, n.d. "Electric vehicle charging stations." Accessed December 2, 2020, at https://afdc.energy.gov/fuels/electricity_stations.html

Van Der Meer, Douwe G., Richard E. Zeebe, Douwe J. J. van Hinsbergen, Appy Sluijs, Wim Spakmana and Trond H. Torsvikd. 2014. Plate tectonic controls on atmospheric CO2 levels since the Triassic. *Proceedings of the National Academy of Science* 111(12): 4380–4385.

Williamson-Rea, Jacob. 2018. "Want to reduce emissions? Start in the gut of a cow." *Penn Today,* October 26. Accessed December 28, 2020 at https://penntoday.upenn.edu/news/limiting-amount-methane-produced-cattle

Worm, Boris, Edward B. Barbier, Nicole Beaumont, Emmet Duffy, Carl Folke, Benjamin S. Halpern, Heike K. Lotze, Fioenza Micheli, Stephen R. Palumbi, Enric Sala, Kimberley A. Selkoe, John J. Stachowicz, and Reg Watson. 2006. "Impact of biodiversity loss on ocean ecosystem services." *Science* 314:787–790.

7

The Economics of Carbon Taxes and Emissions Trading, and Who Will Foot the Bill?

In 2017, one of the authors of this book visited Bangladesh's Kitubdia Island, where cement ocean embankments were to be built years earlier to protect residents from massive storm waves during the cyclone (hurricane) season. The embankments had instead been built of sand, and the head of the local *upazila* (county government) told our group that there should not have been any complaints, as 70 percent of the funds allocated for recent embankment construction had been spent on that task.[1] No explanation was given about the other 30 percent, or for why the embankments had been built of sand, thus endangering hundreds of people when they broke and sent saltwater streaming over what had been fertile rice-growing lands. After this, the brackish water–sated lands were suitable only for shrimp farming (in land-based saltwater pools) and salt harvesting (from evaporation in shallow pools). The incident offers a cautionary tale about opening financial spigots for climate finance (for both mitigation and adaptation), at the national and international levels, without better safeguards for accountability by politicians.

Bangladesh is particularly susceptible to climate change. Every several years, severe flooding damages infrastructure, and cyclones have caused as many as 300,000 to 500,000 deaths (Sovacool & Linner 2016, 36). Authorities anticipate that some 15 percent of the country will be inundated by 2050, generated more than 25 million climate migrants (nearly one-sixth of the country's population), and to resolve this, the government has undertaken massive construction projects at a cost of billions of dollars (Sovacool & Linner 2016, 46). Still, Sovacool and Linner (2016) affirmed our observation, in that construction "siphons off resources, through corruption, to workers and laborers associated with local elites" (46).

As billions of dollars have started flowing into climate projects around the world over the last three decades, transparency has become an issue, although far from the only one. The Bangladesh Water Board, which dictates construction of ocean and river embankments in that flood-vulnerable nation, is known among policy analysts for building projects in the political districts of board administrators, but not always where these are most needed.[2] More broadly, Bangladeshis—and citizens from most of the rest of the world—seem more willing to try to resolve natural disasters than to "solve" climate change. Government officials are extremely willing to "call out" natural disasters like monsoons, cyclones, floods, and drought. Governments provide short-term relief for concrete, short-term problems like disasters, and if they are organized, this can help improve politicians' stock among voters. But climate change is an amorphous, distant, and foreboding challenge. How do we bring it to the fore?

Flood victims in Bangladesh believe that while the government manages the country's frequent emergencies, its overall level of service provision may be lower than those of other governments. Mohammad Abul Kasheem, a lifelong resident of Panjar Bhanga, near Rangpur, said: "When there is no disaster, the government forgets us" (Eisenstadt and Haque 2017). The more than a dozen people we questioned in Panjar Bhanga were mostly stoic, insisting they themselves were responsible for restoring order and recovering from yet another natural disaster (Eisenstadt & Haque. 2017).

This unfortunate story from Bangladesh, where it bears mentioning that many noble public servants have raised awareness of climate change to a higher level than currently exists in most countries, is that it may be emblematic of a worldwide problem. In general, people seem unwilling to face climate change, and especially adaptation (accepting that climate change has arrived and vulnerable nations must be retrofitted to cope, as mentioned in Chapter 6). Chapter 2 considered the disorder that climate change has already brought to the natural world. This chapter looks at the human cost of fixing climate change, as estimated by economists.

First, we consider overall estimates (which vary) based on economic models and how particular expenditures can be expected to diminish the problem. Next, we examine some of the most cost-effective strategies for reducing emissions in the energy sector, promoting alternative energy, energy efficiency, and the move to low-energy lifestyles. Then, we look in more depth at a particularly important component of the problem generating extensive debate internationally: the viability of emissions trading systems (ETS) versus carbon taxes, and how to measure emissions to thus ensure transparency in international and domestic "carbon accounting." Following this, we explore hitherto-failed efforts to catalyze large-scale public and private funding for "climate finance" (a new form of international assistance over the past few decades specifically targeting mitigation and occasionally adaptation), and the fragmentation of this new stream of funding. We also address the issue of "climate risk" and how financial instruments are being established to address this problem.

Modeling the Financial Costs of "Fixing" Climate Change

Few economists have been bold enough to offer monetary estimates for the cost of "fixing" climate change ("deep decarbonization" to drastically reduce the use of carbon-based fuels), and almost no one includes the cost of adaptation (adjusting to the 1.5°C in average global temperature increase we have accepted as inevitable). This is probably because of the realization they would be subjected to a lot of the "climate denial" arguments briefly discussed in Chapter 1, as well as to a slew of new arguments against the assumptions they made in their models, which are inexact but necessary in any model of possible future events. Economist William Nordhaus, however, won the Nobel Prize in 2019 for his level-headed efforts to put a price on carbon emissions, which he and other economists like Nicholas Stern have viewed as the central pillar in adapting markets to "bake in" the costs of climate change by recognizing the costs of high emissions-producing goods. While Nordhaus was receiving the Nobel in Stockholm, yellow-vest protesters shut down the Louvre and Eiffel Tower in Paris, contesting gasoline taxes. Indeed, as noted by Nordhaus, "Climate change may be an existential crisis, but in their day-to-day lives, many people are more worried about the problems created by the most obvious solution than by climate change itself" (quoted in Leonhardt 2019).

In contrast to Nordhaus, Costa Rican diplomat Christiana Figueres, the director of the United Nations Framework Convention on Climate Change (UNFCCC) during the Paris negotiations, agreed that:

> an economist would probably argue that the most efficient way to reduce greenhouse gases is to put a price on carbon. But efficient is not always what can be attained from a political perspective. I would rather move now on what we can do than wait for economists' perfection. (quoted in Leonhardt 2019).

Indeed, waiting for "economists' perfection" may have also deterred many from offering concrete numbers, which are readily critiqued, and often with tedious precision, as critics pick apart each assumption in any model. But Nordhaus, in his masterful 2013 book, *The Climate Casino: Risk, Uncertainty, and Economics for a Warming World,* updates cost projections he and colleagues had been working on for over a decade, but also situates them within a discussion of how to interpret risk scenarios like the ones he generates, which are also familiar to readers of documents from the Intergovernmental Panel on Climate Change. Nordhaus' (2013, 177–181) best estimates can contain climate change within the UN 2°C limit codified in Paris with expenditure of approximately 1.5 percent of world income (gross national product aggregated to cover all countries) per year over time (presumably decades).

Much more sobering, however, is Nordhaus' "likely" scenario, where imperfect policy implementation leads to a 3°C limit on temperature rise by the end of the 21st century, with a cost upward of 3 percent of world income (Nordhaus 2013, 177–181). Nordhaus writes: "To the extent that countries do not participate, or that inefficient measures are used, or that the timing of actions is inefficient, then we cannot realistically attain an ambitious temperature target such as that adopted at Copenhagen [the 2°C]" (181). The inadequacy of present UN pledges in member-state Nationally Determined Contributions (NDCs) are evident in real time on Climate Interactive's vivid pledge thermometer at https://www.climateinteractive.org/programs/scoreboard/. In the next section, we consider policies for addressing mitigation over time (and, indeed, some of these are addressed as "sectoral policies" in Chapter 3).

Here, the unit of analysis we consider is cost of a policy as well as emissions reduction, whereas in Chapter 3, we considered mostly emissions reductions. If Chapter 3 was about the "low-hanging fruit" in the climate policy orchard, in this chapter we recognize, as anyone who has gone apple picking in the fall knows, that we must pay the orchard grower an admission fee and, often, a cost per pound of the fruit we take (our own labor picking the fruit is not the orchard owner's problem, and thus an externality as far as the owner is concerned).

We may propel ourselves to needed emissions reductions by incentivizing reductions and internalizing the negative externalities of greenhouse-gas (GHG) emissions by establishing the proper social cost of carbon. The economist Robert Pollin (2015, 125–131) claims we should be able to achieve such emissions reductions at a cost of about 1.5 percent of world economic output over the next 20 years, by investing in green technologies (principally renewable energy and improving the energy efficiency of existing facilities). Indeed, Pollin argues that needed reductions of global carbon dioxide (CO_2) emissions (by about 40 percent within 20 years, from 33 billion to 20 billion tons) "is a realistic goal" (92). If investments are made in clean renewable energy (excluding nuclear energy, which he considers dangerous in the wake of the Fukushima nuclear reactor meltdown in 2011 and more recent controversies about weapons uses of nuclear reactors), energy efficiency is improved, and use of oil, coal, and natural gas is reduced by one-third over this same period, global per capita CO_2 emissions will fall by the needed 40 percent.

Several policy proposals need to combine to produce these outcomes, according to Pollin (2015, 95–110). High prices need to be set on carbon emissions, but for this economist, these prices serve the function of making fossil fuels more expensive and renewables relatively cheaper. (In the next section, we discuss further the social cost of carbon and the creation of emissions trading markets.) Pollin's emissions savings accrue from substituting renewables for fossil fuels. Other factors needed to promote this change include "feed-in tariffs," which are contracts that require utilities to buy renewable electricity at fixed prices with long-term contracts. Additionally, he writes that accessible financing is necessary to allow many borrowers to participate in the market. Pollin also cites the importance of alternative forms of utility ownership, noting that publicly owned oil and gas companies control some 75 percent of fossil fuel production worldwide (99). On the whole, Pollin's view might be considered somewhat optimistic compared to that of Nordhaus, as described in the next section, which is driven by the market for emissions rather than by deploying renewables and energy efficiency.

The Social Cost of Carbon: The Most Important Number in the Green Economy

Bear in mind that the social cost of carbon represents the most important number to economists, who believe—along with increasing numbers of others, like your humble authors—that a carbon tax or mandatory ETS would be the single best way to address climate change swiftly and effectively. Including the cost of mitigating emissions from carbon produced would bring an overwhelmingly expensive negative externality into pricing models, and thus bring what companies pay to generate fossil fuel–based energy closer to its real cost (the direct cost of producing the energy, which they pay, plus the indirect cost of "cleaning up after themselves" later, which no one presently pays). But other means also exist for mitigating and adapting to climate change, based on regulating carbon-intensive industries like automobile transportation (i.e., the Obama administration's fuel efficiency requirements on new cars which were reversed by Trump) and spending on relevant projects (e.g., building ocean embankments, but out of concrete, in Bangladesh; subsidizing substitution of solar power for coal-fired electric plants in the United States). In this chapter, however, we first address pricing carbon, as this has been the sharp focus of economists for decades.

The social cost of carbon represents the true economic cost associated with climate change posed from the emission of each ton of CO_2 into the atmosphere. In other words, economists reason, if we could find the actual cost required to dispose of that ton of CO_2 and mitigate the damages it causes (which until recently were never factored in to costs at all), then governments could use funds collected—via carbon taxes or emissions bonds issued in an ETS—to offset the impacts of future emissions (although past and existing emissions, which linger in the atmosphere for about 150 years, would still need to be addressed). Worldwide estimates of the social cost of carbon ranged between 10 and 1,000 dollars per metric ton of CO_2 (Ricke et al. 2018), with the authoritative Stern-Stiglitz Report placing the cost between 40 and 80 dollars per ton of CO_2 by 2020 and 50 to 100 dollars by 2030, provided a supportive policy environment exists (Carbon Pricing Leadership Coalition 2017, 3).

Ricke et al. (2018, 897) estimated the median global cost at approximately 417 dollars per metric ton of CO_2 emissions. This is the per ton cost worldwide of the economic damages from CO_2 emissions (within statistically given confidence intervals), and it is considered by some as "the most important number in the global economy" (Rotman. 2019).

Ricke et al. (2018, 896) also distilled the country "winners and losers" using their equation for the social cost of carbon, which included variables accounting for the evolution of national economies (e.g., gross domestic product [GDP], population, and CO_2 emissions), climate impacts from those emissions and collateral damage from other nations' emissions, country-level climate damages, and the discounting of future damages into present=day economic values. The researchers found that the nations exacting the highest portion of the global cost are India, with a social cost of carbon of 86 dollars per ton of CO_2 (21 percent of the world total); the United States, at 48 dollars per ton; Saudi Arabia, at 47 dollars per ton (11 percent of total); and then Brazil, China, and the United Arab Emirates, with each at 24 dollars per ton (6 percent of total). The good news from this model for Canada, countries in northern Europe, and the former Soviet Union is that as their present temperatures are below the economic optimum, they have negative social costs of carbon, at least for the time being (Ricke et. al. 2018, 897).

Note that while this calculation does consider economic damages from climate change as part of the equation, many other calculations do not. Recall that most of the policy discussion in Chapter 3 was about emissions reduction, or mitigation. Adaptation, or retrofitting climate-affected areas to better cope with the "new normal" of life under changing weather and climate conditions, is less commonly addressed. Recall that most of the most vulnerable nations in Chapter 4 (Table 4.3) contributed few emissions to the worldwide total, making them climate change victims rather than perpetrators. We enter further into the discussion with the caveat that most of the solutions considered by economists to date assess mitigation; that is, they strive to find a cost of reducing emissions to the point of keeping climate change within prescribed limits. The costs of adaptation, which are born primarily by the most climate-vulnerable nations (all tropical latitude countries in the Global South), is discussed in Chapter 8. While the blunt tools of economists may be of less use in understanding the ill-defined parameters of adaptation, which is even harder to measure than mitigation, lawyers are entering the fray to sue for loss and damage.

Nordhaus does distinguish between abatement, which we will liken to mitigation, and what he calls damages, which we will liken to adaptation. He argues that if policies are efficient and properly designed, and if future costs are the same as present ones (an assumption we question later), then an economically feasible target for maximum warning (i.e., the goal we can pay for with some sacrifice) is a 2.25°C increase, with a cost of about one percent of global income (Nordhaus 2013, 208). However, recalling that by 2012 only some 20 percent of worldwide emissions were covered in the Kyoto regime, Nordhaus assumes only a 50 percent participation rate by emitting nations in abatement (mitigation) policies, and the affordable temperature rise target then becomes 3.8°C. Relaxing the efficiency and/or stringency of policies, in addition to their "coverage" (the number of nations that comply), worsens the temperature rise even more, especially when Nordhaus introduces tipping points—that is, likely inflection points in temperature rise that prompt increases in the rate of temperature change. He warns that the prediction of tipping points is difficult because of the "empirical problem that comes from our inability to predict the impacts of the threshold damages reliably" (214). In other words, there is a point where models can continue to predict outcomes based on inputs, but where those outcomes then extend to uncharted territory.

The Discount Rate of Uncertainty: A Second Important Number in the Green Economy

Let us return to the assumption in Nordhaus's model cited above that future costs are the same as present ones. The problem, as Nordhaus and many other economists recognize, is that future costs are never the same as present ones. We know that if you borrow 1,000 dollars today, it will not be able to purchase the same number of goods in 2035. But how much lower will its purchasing power be in some 20 to 25 years when compared to its value today? The question is a speculative one, but we do know that, given the power of compounding (which benefits those with interest-bearing, long-term savings), a 1,000-dollar investment that compounds at three percent per year will be worth 4,400 dollars in 50 years, whereas one that grows at seven percent per year will be worth over 29,000 dollars.[3] According to Irwin (2019), the Obama administration's moderate three percent discount rate put the social cost of carbon at 42 dollars per metric ton. That rate was a little low, but still within the accepted range of the most recent comprehensive study, the so-called Stern-Stiglitz Report, which puts the cost of carbon needed to keep temperature changes within Paris targets at 40 to 80 dollars per metric ton of CO_2 by 2020 and 50 to 100 dollars per metric ton of CO_2 by 2030 (Carbon Pricing Leadership Coalition 2017). And hey, at least the Obama administration, sandwiched between the "ostrich" (head-in-the-sand) climate policies of Bush and Trump, placed a social cost on carbon, even if nothing further really came of it at the national level. Expect the Biden administration to pick up where Obama left off.

The Obama policy was established in the aftermath of perhaps the first alarmist climate change report by an economist, the 2006 release of the Stern

Review. The British economist Sir Nicholas Stern argued that climate change, if unchecked, could cost up to 20 percent of world GDP (the aggregate measure of all the domestic production in a country's economy in a given year but here aggregated worldwide). Nordhaus took a more conservative approach to compounding than Stern, arguing for a 5.5 percent discount rate for the first half of the 21st century and a 4.0 percent discount rate over the entire 21st century. Stern's discount rate was only 1.4 percent, meaning that to him there would be a lower return on current investments and that the "gap" between expenses to mitigate and adapt to climate change would continue to expand even as current spending continued to increase, because he anticipated future needs to be so much greater (Carey 2011).

Discount rates are important for other reasons in addition to the mere power of compounding, which accelerates the increase in value of resources over time. In our example, to Stern the compound impact is less, because he assumes that the costs of fixing climate change will increase dramatically. But beyond this straightforward issue of saving more to compound more (i.e., gain interest on top of interest over time), there is the opportunity cost of investing in climate change. For example, when saving for a big purchase (e.g., college tuition, a car, or a house), one wants to save just the right amount, rather than too much or too little. And while saving too little is a greater problem than saving too much, saving too much may still be inefficient, because you cannot receive the full utility on funds that you are saving for a car if that means you are not investing those same funds in a mutual fund that offers a higher rate of return, or spending the money on a fun activity like attending a Rolling Stones concert with a friend (the authors did admit to being Gen Xers—and perhaps we could harness energy via Mick Jagger's choreography). So "tying up funds" yields an opportunity cost—namely, the cost of not using them in other ways.

Additionally, a fixed sum of money like our 1,000 dollars has a lot more relative value to a poorer person than to a richer one. Indeed, this argument is used often in the North-South Debate. If the developed nations have such a surplus, then what amounts to small sums of money to those in the North can make meaningful improvements in the lives of those in the Global South. And as we have also seen, there are still "unknown unknowns"[4] regarding climate change, such as if and when the process might initiate a feedback loop that "compounds" temperature increases, and whether the financing to adapt to climate change will be needed gradually over time or might reach a point where much liquidity is needed all at once. While we will discuss adaptation, especially by poor and vulnerable nations (which tend to be the same ones), in Chapter 8, it is evident here that liquidity (i.e., cash on hand to meet immediate needs, such as after a climate-induced hurricane) may be even more critical—given their relative poverty and also their geographical tendency to be on the front lines of climate change—to nations in the Global South.

Having now considered the social costs of capital and the key discounting assumptions in two leading estimates, we show how the social cost of carbon has been integrated into policies (carbon taxes) and markets (ETS), but with limited successes.

Carbon Taxes and ETS as Lynchpins of a Solution

Chapter 3 briefly considered what economists like Nordhaus view as the foundation of a lasting solution to climate change: mitigation and adaptation (or, in his terms, abatement and damage). We introduced the concepts of carbon taxes (e.g., the one the United Kingdom introduced at a national level) and ETS, under which companies, utilities, and other public and private entities are issued permits by a central authority—presumably the government—to emit quotas of CO_2 and other GHGs. Permit-receiving entities can buy and sell the right to emit under this system, which was implemented in the European Union (EU) in the 1990s and in North America and China (among others) at the national and subnational levels more recently. Nordhaus, and others like Stiglitz and Stern, all argue that "putting a price on carbon represents a societal decision about the priority of reducing CO_2 emissions . . . that emissions are harmful and should be reduced" (Nordhaus 2013, 222).

The problem mentioned previously is that if the "true" or social cost of emissions to a state is 40 to 50 dollars per ton of emissions (usually given in CO_2), and the price charged to refineries to purchase the rights for a ton of emissions is three dollars, then a wide gap exists. Polluters are "getting away" with paying much less than the true cost of emissions. Table 7.1 below illustrates this pattern of excessively low prices set on a ton of emissions for the score of nations and subnational entities that have set a price for carbon. As can be seen from the second column (bolded), the price of a metric ton of CO_2 emissions offsets ranges from 36 cents in Ukraine to over 129 dollars in Sweden. The actual cost seems to reside somewhere in the middle, but only nine entities (eight countries plus British Columbia in Canada's subnational system) even price carbon at over 25 dollars, and some 175 countries (the ones that do not even have carbon taxes and hence did not even make the table) do not even have a carbon tax.

At the individual level, pricing carbon increases the cost of carbon-intensive activities (e.g., electricity use, driving, and air flight) relative to non–carbon intensive activities. For example, Nordhaus (2013, 229–230) estimates that a low carbon tax (25 dollars per ton of CO_2) would increase the cost of petroleum about 11 percent, natural gas by 30 percent, electricity by 31 percent, and coal by 134 percent (as you can see, coal emits many more units of CO_2 per million BTU, which is the unit of measurement). At the level of the household, such a carbon tax would increase spending per year on "big-ticket items"

Table 7.1 **Traits of Nations That Price Carbon**

	Percent of Global Emissions (2016)	*Price of Carbon ($US per metric ton, 2019)*	*Year of Implementation*	*Fossil Fuel Consumption (% of total, 2014)*	*Renewable Consumption (% of total final energy consumption, 2014)*	*Fossil Fuel over Renewable (divide two columns to left)*
Argentina	0.60	10.00	2019	87.72	10.90	8.05
Canada	1.62	15.00	2019	73.19	22.02	3.32
Chile	0.25	5.00	2017	73.25	26.72	2.74
Colombia	0.24	5.27	2017	76.69	24.44	3.14
Denmark	0.11	26.94	1992	67.70	30.27	2.24
Estonia	0.05	2.29	2000	14.49	25.33	0.57
Finland	0.13	71.12	1990	42.18	41.22	1.02
France	0.99	51.16	2014	46.23	13.35	3.46
Iceland	0.01	32.19	2010	10.93	76.34	0.14
Ireland	0.12	22.94	2010	84.44	8.52	9.91
Japan	3.47	2.65	2012	94.41	5.63	16.78
Latvia	0.02	5.16	2004	56.72	40.24	1.41
Lichtenstein	0.00	96.68	2008		62.59	
Mexico	1.34	3.01	2014	90.28	9.76	9.25
Norway	0.13	60.27	1991	57.98	57.20	1.01
Poland	0.92	0.08	1990	90.09	11.57	7.79
Portugal	0.15	14.61	2015	72.81	30.46	2.39
Singapore	0.18	3.70	2019	90.58	0.62	145.71
Slovenia	0.04	19.84	1996	59.66	22.30	2.68
Sweden	0.13	129.74	1991	29.77	49.69	0.60
Switzerland	0.11	96.67	2008	48.74	23.42	2.08
United Kingdom	1.12	23.51	2014	82.72	7.40	11.19
Ukraine	0.69	0.36	2011	75.35	3.50	21.53

Notes and sources:

Percent of Global Emissions: "Each country's share of global carbon dioxide (CO_2) has been calculated by Our World in Data based on annual national emissions data published by the Global Carbon Project (GCP). This is calculated as each country's share of the sum of all country emissions. This is calculated as each country's emissions divided by the sum of all countries' emissions in a given year. This doesn't include emissions from international aviation and shipping (so-called 'bunkers') and 'statistical differences.'" Source is https://ourworldindata.org/co2-and-other-greenhouse-gas-emissions

Price of Carbon: Source is https://carbonpricingdashboard.worldbank.org/map_data. Note from data source: Nominal prices on February 01, 2019. Prices are not necessarily comparable between carbon pricing initiatives because of differences in the number of sectors covered and allocation methods applied, specific exemptions, and different compensation methods. Due to the dynamic approach to continuously improve data quality and fluctuating exchange rates, data of different years may not always be comparable.

Year of Implementation: Source is https://carbonpricingdashboard.worldbank.org/map_data

Fossil Fuel Consumption: There is no fossil fuel per capita. What the World Bank has is overall/ total energy use per capita. For fossil fuel, what exists is a percentage of the total, and this is same for renewables consumption as well. Data here are for fossil fuel energy consumption (percent of total); fossil fuel comprises coal, oil, petroleum, and natural gas products Source is https://data.worldbank.org/indicator/eg.use.comm.fo.zs

Renewable Consumption: Renewable energy consumption is the share of renewable energy in total final energy consumption. Source is https://data.worldbank.org/indicator/eg.fec.rnew.zs

as follows: electricity costs by 19 percent, driving costs by 8 percent, and airplane flight by 6 percent (measured per flight, as opposed to "per year" as the other costs are measured) (Nordhaus 2013, 229–230).

How carbon pricing is applied, however, can vary. Such taxes are applied directly in the case of carbon taxes and indirectly in the case of ETS, which regulate the amount of emissions an entity (typically a utility or any carbon-polluting company) can undertake without penalty. Beyond the government-allowed emissions, the entity must buy "polluting rights" from entities that do not use their entire quota of CO_2 pollution. ETS went into more widespread use before carbon taxes, and an ETS consisting of the entire EU was established in 2005. Indeed, by 2017, some 12 percent of global emissions were covered by ETS. These systems are complicated, requiring "cap setting, permit allocation, trading guidelines that avoid carbon leakage, regulation of offsets, high compliance, transparent and continuous monitoring, and careful collaboration between systems" (Riehl et. al. 2016, 227).

But which is better, a carbon tax or ETS? It seems either one can serve the purpose, depending on how stringent it is and how well it covers the full gamut of emitters, rather than just certain countries or economic sectors. Still, as French President Emmanuel Macron's experiences with the yellow-vest movement convey, taxes, while more transparent and easier to implement (especially if they are "flat," across-the-board taxes like sales taxes in the United States or, in Macron's case, a tax on gasoline purchases), are politically unpopular, as they come directly from consumer pockets. ETS tend to make oil refineries and electricity utilities pay for emitting more than their allowed quotas of contaminants and then pass these taxes on to consumers over time, in the form of rate hikes. And then there is the issue of transparency. While carbon taxes are levied against the provision of carbon-intensive goods and guarantee collection of funds, which are concrete and fungible (unless and until Bitcoin dominates the currency market), ETS regulates emissions rather than collects funds. Emissions are diffuse, hard to measure, and subject to measurement error and even corruption. But no one wants more taxes, so ETS are often more palatable. And indeed, they are more numerous.

Critiques of ETS: Marketization of Carbon Pricing, but with Offsets and Other Complications

ETS seeks to marry the private sector's faith in the marketization of products (like the benefits of reduced emissions) with the imperative for climate protection. In 2005, a dozen years after the EU failed to implement a carbon tax in 1992, all 28 member-states (plus Iceland, Liechtenstein, and Norway) adopted the ETS as a more politically palatable alternative. The ETS was also embraced by private interests represented by the

International Emissions Trading Association after 1999, and advocated by the more capitalist economies of the developed world, led by the United States (until the House-approved American Clean Energy and Security Act [ACES], or Waxman-Markey Bill, failed to reach Senate consideration in 2009). Several dozen nations (and subnational entities) beyond the EU now participate in ETS, which is gaining momentum after early years of failure. While the European ETS remains the largest (said to price some 45 percent of EU emissions, although the claim is disputed), national systems exist in countries including China, Korea, New Zealand, and Switzerland, and several subnational ETS exist in Canada and China. In the United States, over a dozen eastern states have one.

The basic model for ETS is through the enactment of "cap-and-trade" legislation. Carbon rights are purchased from the government, either as permits (the right to emit a certain quota of GHGs) or as emissions reductions (emissions "avoided" or offset through sinks, as per our discussion of offsets in Chapter 1). Often, the cap-and-trade systems in developed nations are linked to offset programs in developing nations, perhaps the largest being REDD (Reducing Emissions from Deforestation and Forest Degradation) and REDD+ (more broadly, the conservation and sustainable management of forests as carbon stocks), which has experienced some success, but also controversy, in countries like Brazil and Indonesia. Article 6 of the Paris Agreement reiterated the Kyoto Protocol's engagement with ETS by allowing that NDCs could meet mitigation commitments through carbon pricing.

Pearse and Bohm (2014) elaborate several critiques of ETS. They state that, in addition to the EU's failure to reduce emissions using a region-wide ETS for over 20 years, offsets can allow "developed countries to appear to be reducing emissions, while passing on the abatement task to the developing world" (329). Additionally, Pearse and Bohm argue that while carbon traders have pledged a new mechanism for sustainable development from a portion of the proceeds, the distribution of Clean Development Mechanism finance (subsidies to developing nations still in effect from the Kyoto Protocol) continues to approximate "the historical flow of foreign direct investment to middle-income countries such as China, Brazil, and India" with sustainable development goals getting little more than "lip service" (330). Additional arguments have been made as well: that carbon markets are regressive in that adding costs to energy consumption, which makes up a greater relative portion of household spending of the poor, makes the most vulnerable pay more, whereas a carbon tax could more readily compensate for this; that fossil fuel lobbies have gotten their companies more permits than these need and thus their profits have been further subsidized; and that the opaqueness of allocation of permits and quotas has given rise to corruption (330–332). We elaborate on these in the next section.

Complications of Carbon Pricing: Transparency, Accountability, and "Climate Corruption"

The EU's ETS, which went into effect in 2005 covering the entity's then-15 countries (now nearly double that number) and was slated to reduce emissions 21 percent by 2020 from 2005 levels, nonetheless increased emissions 2.1 percent during its first two years, as it overallocated emissions permits to the 11,000 participating factories, power plants, and other emitting entities (UN Environment Program Global Environmental Alert Service [UNEP GEAS] 2013, 4–5). Corruption in that 134-billion-dollar trading arrangement over time has "enabled and facilitated the re-sale and misreporting of used carbon offsets, sophisticated computer hacking schemes for theft from national carbon emission registries and continuing value-added tax fraud" (UNEP GEAS 2013, 4–5).

It seems that some opportunities for corruption, or at least for ambiguity in the data, emerge from the inadequate nature of measuring, reporting, and verifying emissions. GHG emissions are usually calculated by an emissions factor, which is the factor of GHG emissions used by one unit of a commonly tracked activity, such as fuel consumption. For example, Rotman (2019) offers forth gasoline consumption by a normal-mileage auto in a direct trip from Los Angeles to Chicago as an approximate equivalent of consuming one ton of CO_2. At the 2010 UNFCCC meeting, delegate nations agreed to submit biennial reports relating national GHG inventories, mitigation actions taken, national needs, and support received to address these needs (called climate finance). However, according to Transparency International Bangladesh (2013, 113), funds have been spent "to contract experts to prepare single reports rather than invested in establishing permanent data collection processes and programmes" (113). This failure to harmonize data collection and reporting, and of climate finance accounting more generally, is discussed later. For its part, the Paris Agreement addressed this issue by recognizing the need to create a new "enhanced" transparency framework with annual reporting of emissions subject to technical review (Verkuijl & van Asselt 2019, 13). Still, the agreement did not specify the degree of "differentiation" of these responsibilities (i.e., whether developing countries lacking technical capacity had to perform to the same level as the developed countries), and subsequent Conference of the Parties (COP) meetings did not manage to agree on detailed modalities for designing and implementing a new transparency framework.

In general, international donors have argued that "large influxes of resources coupled with an imperative to spend can create conditions ripe for corruption . . . [which] can seriously compromise development outcomes, diverting funds away from intended beneficiaries and undermining development of local knowledge, skills, governance, and institutional capacity"

(UN Development Programme 2010). This is more of an issue in spending on adaptation, which is even more ill-defined and poorly measured than mitigation. With regard to the Bangladesh ocean embankment mentioned at the beginning of this chapter, questions have been raised, both in Bangladesh (Transparency International Bangladesh 2011, 2013) and beyond, about whether funds are reaching those most in need (Sovacool& Linner 2016). Imprecisions over the "downstream" use of climate finance (in development projects on the ground) extend from questions at the aggregate level (the UN) over what even constitutes climate finance. So we turn now to perhaps the biggest imprecision in climate finance, which is also one that could be fixed with standardization of the means of addressing how climate funds are defined and spent.

The Debate over the Climate Finance Architecture and its Fragmentation

While climate change has been a source of international lending at least since the Rio Conference extensively recognized the problem in 1992, "climate finance" has become an issue in the international landscape particularly since the 2009 COP meeting in Copenhagen. At that meeting, the developed nations agreed to pay 100 billion dollars per year by 2020 to address climate change (both mitigation and adaptation) in the developing world. But activists, governments of developing countries, and environmentalists have spent the last decade asking the pledging nations to "show me the money." Weikmans and Roberts (2019) show that the controversy extends way beyond the funding shortfall; they claim that "the lack of internationally agreed modalities to account for climate finance has given rise to . . . widely contrasting statements" (97). Furthermore, they show that the lack of consensus on an accounting system means that UN member-states trying to assess compliance with NDCs and other pledges are forever comparing apples and oranges, as a definition has yet to be decided for what even constitutes climate finance.

Prior to the 2015 COP meeting in Paris meeting, the Organization for Economic Cooperation and Development (OECD) and the NGO Climate Policy Initiative reported on funding from the developed world to the developing world, claiming those sums, in the run-up to the pledged100 billion dollars by 2020, had reached 52 billion dollars by 2013 and 62 billion dollars by 2014. Developing nation representatives questioned this sum, with the Indian Ministry of Finance claiming the true amount of funds raised to be just 2.2 billion dollars and the South African chief negotiator stating that "we don't now the veracity, credibility and the methodology of the report or who was consulted" (Weikmans & Roberts 2019, 98).

In addition to the methodology used to derive the figures, many questioned "additionality" in the OECD report, and even in the Paris Agreement's NDCs.

This refers to whether the funds counted represented "new and additional" donations to climate finance, or just "regifting" of development assistance already allocated that happened to contain a plausible climate "hook," as one negotiator told a coauthor of this book at the Paris COP. Weikmans and Roberts (2019, 100–102) raise additional concerns about accounting taken from the Rio marker methodology, adopted by OECD Development Assistance Committee countries in 1998, that continue to bedevil efforts to ascertain true levels of climate finance. By this methodology, development assistance from these nations is counted in several climate-related areas (mitigation, adaptation, biological diversity, and desertification), regardless of whether these climate areas are principal targets of funding or just minor ones. Additionally, according to Weikmans and Roberts (2019, 102), distinctions are not drawn between loans and grants.

"Costing" climate change and its possible solutions has thus far raised several questions in this chapter. Can climate finance processes be made transparent and accountable? Will the 100 billion dollars per year goal for developing country assistance be attained soon, and is that sum adequate to help vulnerable nations in the Global South adapt to climate change? Is it possible to arrive at a precise social price of carbon, and will the market "bear" that price, whether by carbon tax or ETS? Can "workarounds" can be found if the cost of carbon is too high to be politically palatable for citizens (as evidenced recently in France)? These questions do have a common assumption, however, that we should consider. They arise as part of an incremental strategy for solving climate change. But some, like the climate fatalist Dale Jamieson (2014), believe not only that incrementalism has already failed, but that:

> dusk has started to fall with respect to climate change and so the owl of Minerva can spread her wings. We can now begin the process of understanding why the global attempt to prevent serious anthropogenic climate change failed and begin to chart a course for living in a world that has been remade by human action. (ix)

This is a pessimistic view, albeit one that some scientists are starting to accept. Do know that most scientists believe it is not too late, that we are still in the late afternoon rather than dusk. But if Jamieson is right, do we have any options besides hoping to colonize other planets through terraforming?[5]

There is indeed a "Plan B," which has largely not been discussed publicly for reasons discussed in Chapter 8: geoengineering. However, just like radiation as a treatment for cancer involves radiating cancer cells and healthy ones, geoengineering requires the contamination of the atmosphere with sulfur dioxide (acid rain) to "reflect" heat and keep the heat from entering the earth's atmosphere. We will elaborate on this in Chapter 8, although first, we discuss other, more conventional escalations in level of effort that would seemingly help limit climate change without resorting to such drastic measures.

Before succumbing to consideration of such measures, let us consider another relevant economic issue in this chapter (and then some broader policy issues in Chapter 8) relating to risk pooling through climate insurance, preparing for adaptation damages and migration, and ways of conceiving decarbonization. After exhausting these alternatives, we will consider geo-engineering.

A Golden Age of Green Technology? New and Improved Energy Sources

Pricing carbon is viewed by economists as the workhorse of diminishing climate change, and it may be the most readily modeled means of finding an equilibrium between emissions and compensation for not emitting. Still, an unknown but possibly important aspect of the discount rate could be the advent of new technology or, at the very least, the use of existing technology to promote green growth. According to Fiorino (2018), green growth requires guiding growth away from dirty industries toward green economic sectors, which has the additional effect of improving social conditions, increasing incomes, and diminishing inequality. In addition, writes Fiorino, green growth seeks:

> complementary and synergistic linkages through new patterns of investment; institutions that integrate across policy sectors; methods for valuing ecosystem services and incorporating them into decisions; policies that deliver ecological and social as well as economic benefits, like eco-taxes; technologies that improve well-being while using resources efficiently; incentives that shift consumption in ecologically benign directions, via tax or pricing policies, and regulatory standards that promote technology innovation and diffusion. (27)

This is a comprehensive list, which essentially argues, in terms relevant to this book, that climate change mitigation need not always be costly. And that some mitigation technologies, like renewable energy sources and carbon capture and storage facilities, can—and are already—earning profits.

Technology breakthroughs are not formally modeled. In other words, the amount they diminish emissions in the future vis-à-vis the present cannot be readily calculated, as we do not know what solutions will be invented or discovered to reduce climate change. But we can try to think more holistically about green growth and the increasingly recognized water-energy-food nexus. And while we cannot fully factor unknowns into climate forecasting models, we can at least discuss some of the emerging energy technologies, as they are promising and this book could use a bit of good news.

THE PROMISE OF RENEWABLES

As discussed in Chapter 5, the use of renewables in electrical energy production has been growing rapidly in recent years. The combined growth of concentrated solar and photovoltaic (PV) solar will make up 37 percent of global electrical power by 2050 according to the International Energy Agency (IEA; 2018). Similar growth is being seen for wind power. In 2017, about four percent of global electricity was generated by wind power. The IEA projects that by 2050, wind will make up 18 percent of global electrical power (IEA 2018). Those projections are based on current policy and likely future projects; they do not include an assumption about dramatic policy changes that may favor the production of carbon-free power. What is driving the growth of the alternative energies is that their cost per kilowatt is now lower than that of natural gas for electrical power generation! (For a refresher on the surprising and hopeful trends in alternative electrical production, review Chapter 5.)

A REVOLUTION IN BATTERY STORAGE OF ENERGY: COST AND CAPACITY

The cost of batteries for storing excess solar or wind power, backing up conventional power supplies, or powering electric cars has been dropping in recent years and is projected to continue dropping into the future (Figure 7.1). The cost reductions will occur as large-scale battery storage capacity increases (Figure 7.2). At the moment, large-scale batteries are mostly used to provide backup power case of power failure in a region. As discussed in Chapter 5, however, large-scale batteries are increasingly being used to store excess power produced by solar or wind for later use at night or when the wind stops. Given the massive potential for wind energy, having a robust method for storing its power will go a long way toward making it a first choice for electrical

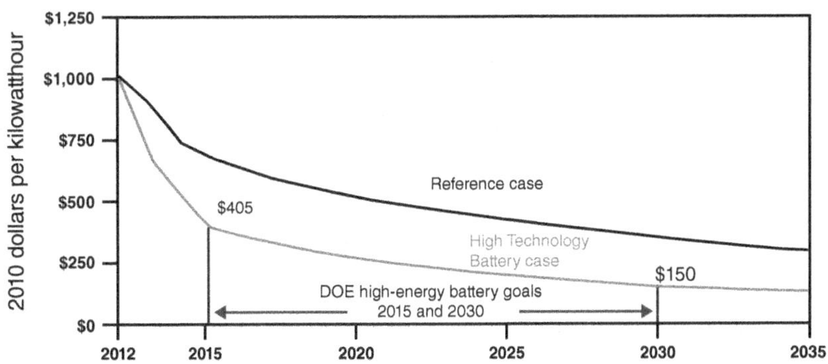

Figure 7.1 Trends for electrical vehicle battery storage projected to 2035 with the gray line representing the US Department of Energy's cost reduction goals. (Figure from EIA 2012.)

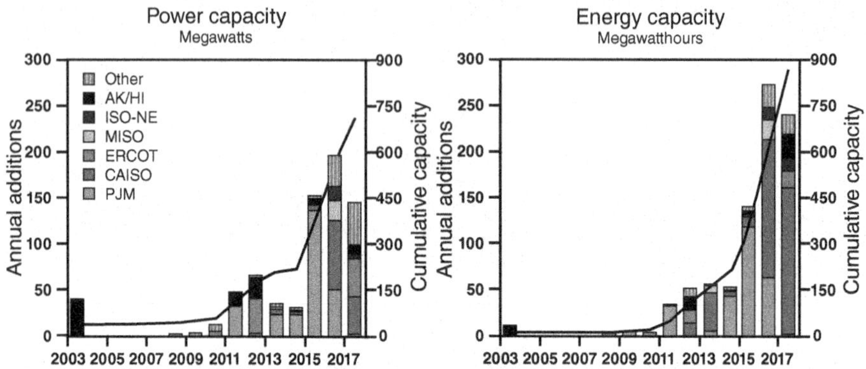

Figure 7.2 Increases in battery power and energy from 2003 to 2017. The different shades of gray represent regional organizations that share power supplies. For example, PJM is the Pennsylvania-New Jersey-Maryland Interconnection, which manages the power supply for 13 New England and mid-Atlantic states. (Figure from EIA 2020.)

power. As discussed in Chapter 5, wind energy is projected to increase from 4 percent of global power to 18 percent by 2050 (IEA 2018). And with advances in battery storage and changes in carbon reduction goals, it will likely surpass that projection.

DEVELOPMENT OF CARBON CAPTURE AND SEQUESTRATION

Another promising technological area is carbon storage and sequestration. It is becoming increasingly clear that, as we strive to reduce the CO_2 emission rate each year, we are going to have to start actively taking CO_2 out of the atmosphere. There are any number of ways to accomplish this, as discussed in Chapter 5. It is helpful to recall here, however, that pumping CO_2 into geological formation is a technology widely used since the 1970s, and that the capacity of underground storage could contain 76 years of emissions at projected rates (Energy Information Administration [EIA] 2018b). The reality is that we cannot capture all of our emissions, of course, but we do have the ability to capture large amounts. It is a matter of political will and allocating resources.

IMPROVED PROSPECTS FOR BIOFUELS

Although significant progress has been made in reducing CO_2 emissions from electrical production sources, the path for obtaining low-carbon energy for transportation is less clear. Petroleum is a high-density, easily transportable fuel, and while battery technology improves each year, the limited recharge rates and storage for personal vehicles makes switching to an electrical fleet quite a challenge. Biofuels are therefore being used as an attempt to transition away from using carbon from ancient plankton to using carbon from living plants or algae. Biofuels are diesel fuel or ethanol that is made from lipids (diesel) or carbohydrates (ethanol) obtained from corn, sugar cane, cellulose, vegetable oil, or even algae. Although a biofuel is still combusted to produce energy, the CO_2 that is emitted was part of the modern carbon cycle, so there is a net-zero increase in atmospheric CO_2.

Among the major criticisms of biofuel production are, first, that they are made from potential food crops or take up agricultural land that could be used to grow food crops and, second, that they are frequently fertilized with nitrogen, which means that methane was used to grow the biofuel (fertilizers are made by combining hydrogen from methane with nitrogen from the atmosphere: the Haber-Bosch process), not to mention the fuels used to operate the machines to harvest and process. Different countries approach biofuels differently. In the United States, biodiesel is mostly produced from soybeans, and 102 production plants collectively produce 2,553 million gallons per year (EIA 2019a). Most of the biodiesel consumed is in blends with conventional diesel, and its use has remained fairly steady since 2016 (Figure 7.3).

Ethanol is another commonly produced biofuel. In the United States, it is primarily made from corn, and in 2018, 14.4 billion gallons were consumed, with projections to reach 36 billion gallons by 2022 (EIA 2019c). The United States actually produces more ethanol than it consumes, and it ships a lot of ethanol to Brazil, which is the second-largest consumer (and producer). Brazil's production is seasonal and relies on sugar cane, but their vehicle fleet runs almost solely on ethanol and so they need a large supply at all times. Together, the United States and Brazil produce 85 percent of the world's ethanol (Fig 7.4).

As mentioned, nitrogen fertilizers are used to enhance growth in corn production (soybeans do not need nitrogen fertilizers, but are fertilized with phosphate). Modern agriculture uses manufactured nitrogen fertilizers created by the Haber-Bosch process, which combines nitrogen from the

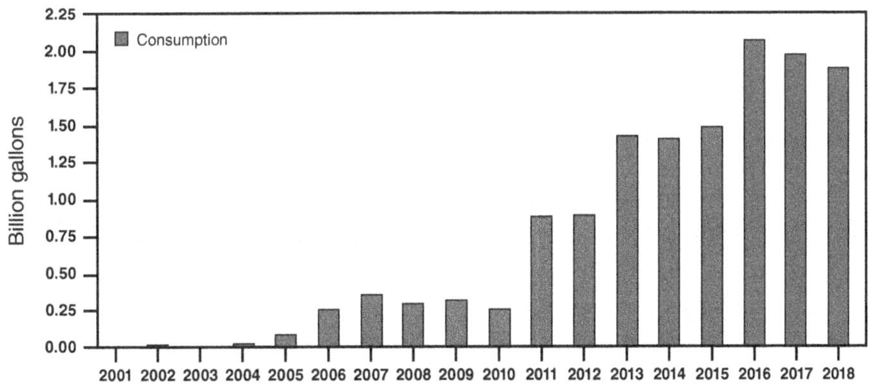

Figure 7.3 US biodiesel consumption is replacing billions of gallons of conventional diesel as blends with between 5 and 20 percent. (EIA 2019b.)

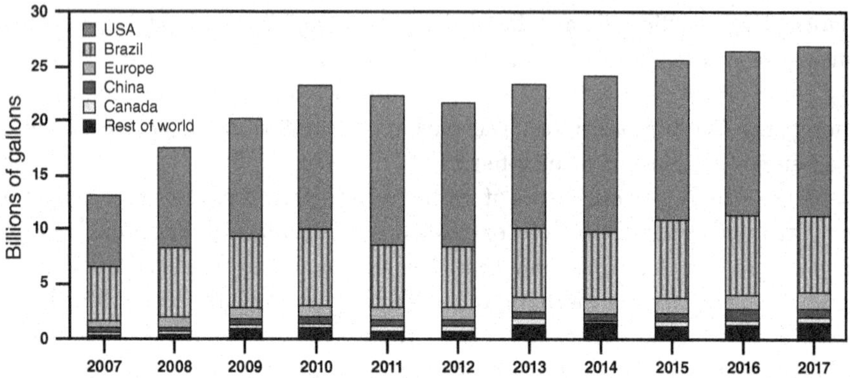

Figure 7.4 Production of ethanol worldwide by country. The United States is by far the largest producer, followed by Brazil. Production is on the rise but is controversial in the United States, because it uses corn as a feed stock, which takes away land that could be used to grow food. Additionally, the corn is often grown with fertilizers made from methane. (EIA 2019d.)

atmosphere with hydrogen to make ammonia (NH_3). The hydrogen used in the process is mostly derived from methane (CH_4), which is essentially natural gas. So you can see why there is some criticism of generating ethanol this way and then calling it a "green" substitute for fossil fuels.

Subsidizing New Technologies: Picking Winners Yes, but Greening Growth in the Process

Pollin (2015) makes the development of renewable energy sources the centerpiece of his book, arguing that green growth is the only kind we can encourage, and that the world needs a massive clean-energy transition. According to Pollin, however, even the cleanest transition will require that in 20 years, two-thirds of our energy still come from fossil fuels and nuclear energy (52). Pollin argues that solar energy is presently almost competitive with fossil fuel energy and has the potential for continued growth, whereas the other renewable energy sources are still maturing. In considering expense, Pollin figures in the costs of building new energy plants, the operational costs of generating electricity, and the cost of transmitting the electricity through the grid (54). Significantly, however, some sources of energy used near the source, such as PV solar and wind, do not need to have the energy produced uploaded to the grid. Radical improvements in battery storage and PV efficiency have made it possible for rural villagers in zones without grid infrastructure, which is still notably lacking in parts of Africa and Asia, to get electricity anyway. Not only does local sourcing of energy reduce costs, it diminishes the dependence of local villagers on capital-city bureaucrats and politicians who decide who gets connected to the grid and in what order.

A nine-fold increase in renewable energy expenditures (over 2012 levels) from 2016 to 2035 will be required to cut emissions worldwide by about 40 percent, which Pollin (2015, 59) estimates would cost about 1.2 trillion dollars per year over 20 years. The worldwide GDP is 87 trillion dollars in nominal terms (with the US GDP at about 21 trillion dollars and China's at about 14 trillion dollars). However, when we adjust for cost of living (i.e., how much basic items cost in each country), the relative value of GDP—what economists call purchasing power parity—yields a total GDP worldwide in 2019 of 143 trillion dollars, with China valued at 27 trillion dollars to the United States at 21 trillion dollars (other nations are measured relative to US GDP, which is the baseline value of one, with other national GDPs multiplied by a value higher or lower than one, depending on whether their costs of living are higher or lower).

Reductions in emissions from pricing carbon stem from economists' notion that factoring in what is presently the negative externality of GHG emissions will force polluters to pay the full price of burning coal and oil. The idea is that firms will emit fewer GHGs if the costs of abatement are added to the price. However, whether governments sell permits to polluters or tax carbon directly, the governments will receive new funds. Those funds need to be used to continue research on new and existing low-carbon sources of energy and to develop energy-saving devices, policies, and lifestyles. (In Chapter 8, we further study some of the policy innovations needed to adapt to life with climate change—from Bangladesh to Barbados—and how to accelerate the "deep decarbonization" of how we live. We will also discuss geoengineering.)

We end this chapter with one more cautionary note from Bangladesh, which has hardly contributed to GHG emissions but is suffering the brunt of climate vulnerability. "Much of what the Bangladeshi government is doing to stave off the coming deluge—raising levees, dredging canals, pumping water—deepens the threat of inundation in the long term," according to one long-time analysist (Harris 2014). When that retired coastal science professor told authorities that sea level rise there was 10 times faster than the global average, "the reaction among Bangladeshi government officials has been to tell me that I must be wrong" (Harris 2014). Given that cyclones in Bangladesh 40 years ago could kill half a million people, whereas nowadays the worst such hurricanes have claimed only some 300 lives, this attitude might be understandable, but domestic and international authorities are still failing to address the scope of the problem.

As many as 1.5 million of Dhaka's 5 million slum dwellers (total population of about 21 million people) may be coastal migrants forced to move as their agrarian livelihoods are washed away, to work as rickshaw drivers or textile factory workers in Bangladesh's capital city. It is time to recognize the connections of seemingly disparate phenomena to the climate problem, and to conceive of holistic, top-down solutions rather than forcing the vulnerable to solve their fates piecemeal, moving from adverse circumstances to slightly less adverse ones. Carbon taxes and ETS are concrete policies we could be implementing to ameliorate this crisis unfolding first in places we typically have not heard of, but which are quickly also moving closer to home. In Chapter 8, we present even more.

NOTES

1. Author interviews of Pekua officials in Cox's Bazaar district near Kitubdia with colleagues from North-South University in Dhaka on August 18, 2017.
2. Interview with Transparency International-Bangladesh director on April 13, 2019.
3. While this information could be gleaned from any actuarial table, Irwin (2019) offers this example and explains it well with regard to climate change.
4. President George W. Bush's Secretary of Defense Donald Rumsfeld said in a 2002 press conference relating to a different context (but prescient here): "There are known knowns; there are things we know we know. We also know there are known unknowns; that is to say we know there are some things we do not know. But there are also unknown unknowns—the ones we don't know we don't know. And if one looks throughout the history of our country and other free countries, it is the latter category that tend to be the difficult ones (Shermer 2005)."
5. While this is a fascinating idea, presented in science fiction, such as on one author's favorite television franchise, *Star Trek*, it is not a likely outcome upon which to hang our hopes. Still, in *Star Trek*, at least, this option is seemingly available as early as the late 21st century, albeit to alien cultures, when the Andorians turn a small planet near Vulcan into a habitable world by creating a livable atmosphere in "Cease Fire" (Season 2, Episode 15 of *Star Trek: Enterprise*). Another sci fi rendering of terra-forming can be drawn from *The Expanse*. For a discussion relating this project to macroeconomics and infrastructure spending, see Krugman's (2019) wonderful *New York Times* column.

REFERENCES

Carbon Pricing Leadership Coalition. 2017. *Report of the High-Level Commission on Carbon Prices*. World Bank Publications, Washington, DC. Accessed June 13, 2020, at https://static1.squarespace.com/static/54ff9c5ce4b0a53decccfb4c/t/59b7f2409f8dce5316811916/1505227332748/CarbonPricing_FullReport.pdf

Carey. John. 2011. "Calculating the True Cost of Global Climate Change," blog at YaleEnvironment360. Accessed January 5, 2021 at https://e360.yale.edu/features/calculating_the_true_cost_of_global_climate_change

Eisenstadt, Todd A., and Mahfuzul Haque. 2017. "Flooding in Bangladesh: Calling out climate change from the high ground." *New Security Beat*, August 28. Accessed May 9, 2019, at https://www.newsecuritybeat.org/2017/08/flooding-bangladesh-calling-climate-change-high-ground/

Energy Information Administration (EIA). 2012. "EIA's AEO2012 includes analysis of breakthroughs in vehicle battery technology" Accessed January 6, 2021, at https://www.eia.gov/todayinenergy/detail.php?id=6930

Energy Information Administration (EIA). 2018. "Carbon dioxide emissions by fuel." Accessed April 19, 2019, at https://www.eia.gov/environment/

Energy Information Administration (EIA). 2019a. "US biodiesel pant production capacity." Accessed January 13, 2020, at https://www.eia.gov/biofuels/biodiesel/capacity/

Energy Information Administration (EIA). 2019b. "Monthly energy review."May 2019. Accessed January 5, 2021 at https://www.eia.gov/totalenergy/data/monthly/archive/00351905.pdf

Energy Information Administration (EIA). 2019c. "Biofuels explained: Ethanol." Accessed January 13, 2020, at https://www.eia.gov/energyexplained/biofuels/ethanol.php

Energy Information Administration (EIA). 2019d. "International energy statistics." Accessed January 6, 2021, at https://www.eia.gov/international/data/world/biofuels/biofuels-production

Energy Information Administration (EIA). 2020. "Battery Storage in the United States: An Update on Market Trends." Accessed on January 6, 2021. https://www.eia.gov/analysis/studies/electricity/batterystorage/

European Commission. 2020. "Progress made in cutting emissions," blog accessed January 5, 2021 at https://ec.europa.eu/clima/policies/strategies/progress_en

Fiorino, Daniel J. 2018. *A Good Life on a Finite Earth: The Political Economy of Green Growth.* Oxford University Press, New York.

Harris, Gardiner. 2014. "Borrowed time on disappearing land: Facing rising seas, Bangladesh confronts the consequences of climate change." *New York Times*, March 8. Accessed July 31, 2019, at https://www.nytimes.com/2014/03/29/world/asia/facing-rising-seas-bangladesh-confronts-the-consequences-of-climate-change.html

High-Level Commission on Carbon Prices. 2017. Report of the High-Level Commission on Carbon Prices. Washington, DC: World Bank. Accessed January 5, 2021 at https://static1.squarespace.com/static/54ff9c5ce4b0a53decccfb4c/t/59b7f2409f8dce5316811916/1505227332748/CarbonPricing_FullReport.pdf

International Energy Agency (IEA). 2018. "Renewables 2018: Market analysis and forecast from 2018 to 2023." *International Energy Agency.* Accessed January 11, 2021 at https://www.iea.org/reports/renewables-2018

Irwin, Neil. 2019. "Climate change's giant impact on the economy: 4 key issues." *New York Times*, January 17. Accessed April 29, 2019, at https://www.nytimes.com/2019/01/17/upshot/how-to-think-about-the-costs-of-climate-change.html

Jamieson, Dale. 2014. *Reason in a Dark Time: Why the Struggle Against Climate Change Failed—and What it Means for Our Future.* Oxford University Press, New York.

Krugman, Paul. 2019. "The Macroeconomics of 'The Expanse," in New York Times, December 18. Accessed January 12, 2021 at https://www.nytimes.com/2019/12/18/opinion/the-macroeconomics-of-the-expanse.html

Leonhardt, David. 2019. "The problem with putting a price on the end of the world." *New York Times*, April 9. Accessed April 17, 2019, at https://www.nytimes.com/interactive/2019/04/09/magazine/climate-change-politics-economics.html

Nordhaus, William. 2013. *The Climate Casino: Risk, Uncertainty, and Economics for a Warming World.* Yale University Press, New Haven, CT.

Pearse, Rebecca, and Steffen Bohm. 2014. "Ten reasons why carbon markets will not bring about radical emissions reduction." *Carbon Management* 5(4):325–337.

Pollin, Robert. 2015. *Greening the Global Economy.* The MIT Press, Cambridge, MA.

Ricke, Katharine, Drouet, Laurent, Caldeira, Ken and Massimo Tavoni. 2018. "Country-level social cost of carbon" in *Nature Climate Change* 8: 895–900. https://doi.org/10.1038/s41558-018-0282

Riehl, Brianne, Guangyu Wang, Sarah Eshpeter, Helen Zhang, John L. Innes, Nuyun Li, Jinliang Li and John O. Niles. (2017, "Lessons Learned in Mandatory Carbon Market Development", in *International Review of Environmental and Resource Economics:* 10: 3-4, 227-268. http://dx.doi.org/10.1561/101.000000878.

Rotman, David. 2019. "The one number you need to know about climate change." *MIT Technology Review*, April 24. Accessed May 13, 2019, at https://www.technologyreview.com/s/613328/the-one-number-you-need-to-know-about-climate-change/

Sovacool, Benjamin J., and Bjorn-Ola Linner. 2016. *The Political Economy of Climate Change Adaptation.* Palgrave-McMillan (St Martin's Press), New York.

Stern, Nicholas. 2006. *The Stern Review Report on the Economics of Climate Change.* London: HM Treasury. Accessed January 5, 2021 at https://webarchive.nationalarchives.gov.uk/20100407172811/http://www.hm-treasury.gov.uk/stern_review_report.htm

Transparency International Bangladesh. 2013. *An Assessment of Climate Finance Governance/ Bangladesh*. Transparency International, Dhaka. Accessed January 5, 2021 at https://www .ti-bangladesh.org/beta3/images/max_file/cfg_pub_Assesment_CFG_11-13.pdf

United Nations Development Programme. 2010. "Staying on track: Tackling corruption risks in climate change." Accessed January 11, 2021, at https://theredddesk.org/sites/default/files/ resources/pdf/2011/staying_on_track_corruption_risk_in_cc-_undp_-_november_2010-_ low_res.pdf

United Nations Environment Program Global Environmental Alert Service (UNEP GEAS). 2013. "The impact of corruption on climate change: threatening emissions trading mechanisms?" Accessed May 14, 2019, at https://na.unep.net/geas/archive/pdfs/GEAS_Mar2013_EnvCorruption.pdf

Verkuijl, Cleo, and Harro van Asselt. 2019. "Paris Rules?" *Environmental Policy and Law* 49(1):11–19.

Weikmans, Romain and J. Timmons Roberts. 2019. "The international climate finance accounting muddle: is there hope on the horizon?" in *Climate and Development* 11:2: 97–111. https://doi. org/10.1080/17565529.2017.1410087

World Bank Group. 2014. "Fossil fuel energy consumption (% of total)." IEA Statistics, 2014. Accessed January 11, 2021, at https://data.worldbank.org/indicator/eg.use.comm.fo.zs?name_ desc=false

World Bank Group. 2015. "Renewable energy consumption (% of total final energy consumption)." Accessed January 11, 2021, at https://data.worldbank.org/indicator/EG.FEC.RNEW.ZS

World Bank. 2017. *Report of the High-Level Commission on Carbon Prices—Executive Summary*. The World Bank, Washington, DC.

World Bank Group. 2020a. "Carbon pricing dashboard." Accessed January 11, 2021, at https:// carbonpricingdashboard.worldbank.org/map_data

World Bank Group. 2020b. "State and trends of carbon pricing." Accessed January 11, 2021, at https://openknowledge.worldbank.org/bitstream/handle/10986/33809/9781464815867. pdf?sequence=4&isAllowed=y

Case Study 7.1

FIDUCIARY RESPONSIBILITY, CLIMATE CHANGE, AND THE UNIVERSITY MOVEMENT TO DIVEST FROM FOSSIL FUEL HOLDINGS

LILLIAN FRAME[1]

In 2014, the Board of Trustees at American University faced a quandary. Students had voted to divest the school from fossil fuels, a committee dedicated to socially responsible investing had recommended it, and now it was in front of the Board to decide (American University Student Government [AUSG] 2020). Yet while individual Board members and administrators expressed support for divesting funds in the university's endowment away from fossil fuel companies, the Board declined to vote. In 2020, the issue came up again, with another student vote in favor of divestment, and the Board was again asked to make a decision. This time, the Board of Trustees decided to fully divest American University from fossil fuels, making it one of four universities to achieve "fossil free" status (Fossil Free 2020b). Both times, student coalitions encouraged—and pressured—the Board, raising public awareness and trying to encourage the student body to mobilize around divestment. Still, there is no direct evidence linking the Board's decision (or even its holding the vote) to the student pressure.

Was the student movement decisive? Long-time observers argue that student and faculty pressure for change played an important role, but also that investment choices for "green investment" and the lower cost of divestment made the Board of Trustees more readily able to divest while still adhering to its fiduciary responsibilities (i.e., its obligation to steward the university's resources responsibly, earning good returns on the university's endowment while trying to shield investments from excessive risk). This case study addresses some of the causes for the change in the Board's position from 2014 to 2020 and the roles of different actors in helping to bring about that change. The central tension, in both cases, was between environmental sustainability and fiduciary responsibility. In the sections that follow, this case study offers more background on the role of universities and students in social change, and then considers the "fossil free" argument and that of fiduciary responsibility. It then presents a role-playing scenario with questions for consideration in the classroom.

University students have played active roles in many famous protests around the world, like the US civil rights movement in the 1960s, the US strike of four million students in response to the Kent State National Guard shooting in 1970, the worldwide Divestment Campaign against South Africa in the 1980s, the Chinese government's Tiananmen Square massacre of civilians in 1989, the Arab Spring in 2011 (centered in several countries, including Egypt and Tunisia), and the US-based Black Lives Matter movement, which began in 2013 but escalated greatly after police slayings of African-Americans in 2020, to address police brutality (Cavanaugh 2019). Few of these movements, however, were as globally united as the contemporary climate movement, which benefits from a pressing issue of international concern and instantaneous communication through electronic platforms like Facebook and Instagram not available before the 1990s. While the broader climate movement has addressed a range of issues, the student movement—involving college students and younger ones, like renowned Swedish high-school activist Greta Thunberg—has focused internationally on generational justice (see Chapter 1) and also, more narrowly in the United States, on divestment of fossil fuels from the investment fund portfolios of institutions like universities.

The current divestment movement was based in part on perhaps the most famous such movement, the agreement by many nations to divest from South Africa in the 1980s when generations of rulers, up to the government of President Frederik Willem de Klerk in the early 1990s (deputy president after the historic 1994 election of the first black president, Nelson Mandela), openly and legally discriminated against the black citizen majority through the apartheid system. From the 1960s to the 1980s, students encouraged their schools and corporations to divest from South African companies and holdings (Fossil Free 2017). By 1990, divestiture reached one billion dollars,[2] with 200 US companies as well as and European and UN bodies divesting (Fossil Free 2017). In the climate change movement, divestiture refers to entities removing their investments in fossil fuel companies.

To consider these broad issues in a narrower and concrete context, this case study focuses on one mid-sized private university in the United States and its efforts to divest from fossil fuels as well as build economic strength. American University is a tuition-dependent university of about 7,000 undergraduate students and 4,000 graduate students (American University 2020). The university's endowment in 2020 (pre-COVID) was approximately 750 million dollars, and in 2014, it was approximately 425 million dollars (AUSG 2020). In 2014, some 4.1 percent of its investments were in fossil fuel stocks (AUSG 2020). The nationally ranked university has improved steadily in academic rankings over the last decade, with an acceptance rate of about 32 percent

(US News & World Report n.d.) and is known for its politically active student body (Princeton Review 2020) and for undertaking efforts to gain its carbon-neutral status. We discuss the arguments for and against divestment at American University, but first the meaning of divestiture, and the panorama of student actions at a range of US universities.

The Origin and Meaning of "Fossil Free" Divestment

The US Fossil Free Schools movement is a student-led campaign aimed at full financial divestment by universities from fossil fuels, which started in 2012 and has student membership of at least 50 universities and colleges nationwide. It has grown in popularity over the last eight years by helping encourage dozens of US universities to divest. As discussed extensively in Chapter 4 of this book, fossil fuel companies are linked to accelerating climate change, and these companies do little to stop it or even acknowledge their own contributions to the climate crisis (Hrynkow 2015, 157). The movement's umbrella group, 350.org, directly links divestment in fossil fuels to its roots in the South Africa anti-apartheid movement of 40 years ago on its webpage for the fossil free movement (gofossilfree.org). The nonprofit 350.org was founded in 2008 and led by climate activist and author Bill McKibben.[3] Named after the notion that 350 parts per million is the clean concentration of CO_2 in the atmosphere (by 2020, the number was well over 400), the group has been funded, at least in part, by the Rockefeller Brothers Fund (see https://www.rbf.org/grantees/350org).

"Fossil free" has several levels of compliance, according to 350.org, which has established standards for what makes an institution "fossil free" (Fossil Free 2020c). In general terms, achieving "fossil free" status means that universities meet the fossil-free standard; stop investing in fossil fuel companies; divest their money from any fossil fuel funds, companies, bonds, and trusts; and reject offers of funding from fossil fuel companies for university research and programming (Fossil Free 2020d). Different levels of divestment, according to Fossil Free, are as follows: fossil free, full, partial, coal and tar sands, and coal only.

To be designated "fossil free," a university must have no investments in any fossil fuel companies of any type and have committed to never investing in them in the future (Fossil Free 2020c). As mentioned, as of 2020, only four universities had reached this status[4] (Fossil Free 2020b). Full status means that the university has made a binding commitment to divest but has not divested yet (Fossil Free 2020c). This is the most populous group, with 28 universities having made the commitment[5] (Fossil Free 2020b). Partial status is given to universities that are partially divested and have made a binding commitment to divest from some companies in all asset classes, or to divest from all companies but only across some asset classes (Fossil Free 2020c). Five universities were considered partially divested as of 2020[6] (Fossil Free 2020b). Coal and tar sands status means that universities have made a binding commitment to divest from coal and tar sand companies only, and coal only status means there is a binding commitment made to divest from coal companies only (Fossil Free 2020c). As of 2020, only two universities were classified as coal and tar sands,[7] and six were classified as coal only[8] (Fossil Free 2020b).

According to the standards set by Fossil Free, universities have five years to honor pledges to divest across the asset classes they promised, or they lose their status. There are over 5,000 universities in the United States, but only 45 have made any commitment to fossil free divestment. It is important to note here that divestment has also been an issue worldwide, and that internationally, the demands may differ. Returning to universities, in the following section we describe some campus movements towards divestment, before specifying the case of American University, which divested in 2020, after considering divestiture (but without divesting) in 2014.

Tactics Used by the Movement

Tools used by advocates of the movement are different for every school. Divestment-related actions are executed through campus divestment organizations or clubs, but sometimes also receive information from outside environmental groups. Tactics used depend on the size of the club, the friendliness of university administration, and the willingness of club members. The most prevalent tactics include hosting rallies, posting flyers, building art exhibits, using social media, and advocating for needed Board of Trustees votes, which determine investment practices at private and public universities. Some campus chapters, however, take it to the next level.

George Washington University (GW) in Washington, DC, is an example of a very active chapter. On GW's campus, there is a Regulatory Studies Center (RSC), which is funded by ExxonMobil[9] and the Koch Brothers[10] (Somayajula 2020). At an anniversary event celebrating the RSC, student members of Sunrise, the campus' "fossil free" movement chapter, disrupted the event, took the stage, and gave a speech (Schonfeld 2020c). Actions like these are meant to show universities exactly how serious students are and tend to garner public attention.

Harvard and Yale students also escalated their activism beyond standard practices. During half-time of a football game on November 23, 2019, between the two universities at the Yale stadium, hundreds of students from Fossil Free groups at both schools stormed the field to call for divestment. While approximately 150 students intended to be arrested that day, over 500 students unaffiliated with either club ran onto the field with them, delaying the game for about an hour (Pontecorvo 2019). About 50 students were arrested in the ensuing chaos (Heaphy 2020). This action, said student Nora Heaphy,[11] was designed to impact the schools in the ways that matter the most: alumni opinion, public image, and donations (Heaphy 2020).

Such public actions were meant to pressure university administrations and to increase public attention to the issue. Disruptive actions, which can sometimes include civil disobedience and potentially being arrested, cannot rely on social media to plan and announce those actions. They are largely organized in secret, relying on word of mouth and secure messaging sites. But media attention is still necessary, which is another important tactic. Disruptive actions are meant to impact public opinion on divestment, raise awareness, and show student commitment to the cause.

Some Fossil Free groups rely on vertical organization, with student leaders making decisions and speaking on behalf of the group, whereas others are more horizontally organized, meaning that there is less "top-down" central leadership. Whatever the organizing principle, the decision to use it is put to vote (Elias 2020). Before action by Fossil Free at American University, meetings are held to decide tactics, anticipate possible repercussions, and air concerns. After the discussion, a vote is taken over whether to undertake a particular action. Student leaders state that use of art, social media, new leadership structures, and democratic decision-making all reflect the core of the movement. It is ultimately a "people power movement" that relies on students, grassroots organizing, and collective action (Elias 2020).

University Reactions to Fossil Free Chapters

Universities are usually fairly tolerant of local Fossil Free chapters. They invite students to address Board of Trustees meetings, as in the case of Yale; ask them to sit on committees to consider divestment, as in the case of GW; and may be sympathetic to the ideas attached to the divestiture movement (Heaphy 2020). However, at some universities, an administration's appearance of support for the divestiture cause may be more focused on public relations than on execution. Still, as one student leader put it, "an adversarial relationship, within certain bounds of reason, is ultimately good for us" (Somayajula 2020). This adversarial relationship is useful, because it draws clear lines between university administrations and pro-divestiture movements and stops the university from exploiting student groups.

At Yale, the students who had been "invited" into conversation about divestment for years believed that it was ultimately a stalling tactic by the university to make students feel as though the university was actually trying to divest (Heaphy 2020). The Yale Fossil Free group then began to escalate actions, having sit-ins at the university's investment office, where students were arrested on multiple occasions (Stannard 2019). Following the sit-ins and the football game protest, the Chief Investment Officer at Yale met with students for the first time, and the Faculty Senate addressed divestment for the first time (Heaphy 2020). Yale has not divested yet, but the goal may be coming within closer reach.

The GW administration has reacted adversely to the actions of their Fossil Free club. GW President Thomas LeBlanc, when asked what he thought about pro-divestment results of a student referendum and whether the university would be divesting, said: "What if the majority of the students agreed to shoot all the black people here? Do I say, Ah, well, the majority voted?' No." (Schonfeld 2020a)." The university president comparing divestment from fossil fuels to hate crimes and murder obviously did not go over well and resulted in many student groups calling for his resignation (Schonfeld 2020a).[12]

In an action before President LeBlanc made these remarks, 100 students planned to march on LeBlanc's campus residence and deliver a letter requesting divestment (Freire 2020). The majority of students intended to stay in the street, while a few, including Victoria Freire,[13] were designated to deliver the letter at the president's doorstep. This action was considered low-risk and was meant to be merely symbolic (Freire 2020). However, upon arrival at the residence, the students were met by a university police presence. Freire said she started to approach the front door and explained to the police officer that no one was expecting her, but that she was a student delivering a letter (Freire 2020). Students chanted in the street while Freire and a few others waited to be allowed near the door (Freire 2020). As captured on video, GW police pushed the students on the stairs away from the door, and Freire was pushed by a campus police officer before the officer appeared to shove Freire down the flight of stairs (Schonfeld 2020b). Freire said she obtained a concussion from the assault, demonstrating that such protesting can carry risks (Freire 2020).

The Fossil Free Argument

Universities are just one piece of this puzzle, as debates about divestment have also played out in pension funds and other institutional trusts and organizations, albeit with perhaps less intensity than on some of the aforementioned college campuses. According to the Forum for Sustainable and Responsible Investment (2020), there exists a 536 billion dollar difference in divestment between 2016 (when divestment was 144 billion dollars) and 2018 (when divestment reached 680 billion dollars), and divestment is likely to have grown in the years since. The membership organization, which represents financial advisors and socially responsible investment instruments, cites a 350.org statement that, worldwide, institutions and individuals representing some 12 trillion dollars in assets have pledged commitments to divestiture (Forum for Sustainable and Responsible Investment 2020).

A common question is whether divestment matters in the grand scheme of things. Universities are such a small part of these companies' investors, so the companies are unlikely to be significantly affected by a single university's decision to divest. The hope is that if enough universities and other funds divest, this will ultimately cause the companies to lose money and, perhaps, force them to "strand oil" (leave hard-to-reach oil reserves in the ground, especially in inaccessible areas where such reserves are particularly expensive to extract or in protected natural lands like the US Arctic National Wildlife Refuge).

The first university fossil fuel divestment campaign began in 2011 at Swarthmore College in Pennsylvania, which has still not divested (Fossil Free 2020a). That divestment campaign grew into an entire movement, which has spread across universities (Fossil Free 2020a). Since the creation of Fossil Free, it is estimated that institutions have divested 14 trillion dollars, with educational institutions accounting for 15 percent of that divestment, amounting to just over US 2.1 trillion dollars (Fossil Free 2020).

The monetary repercussions to companies aside, there are other reasons to divest. Pranay Somayajula,[14], a member of Sunrise GW,[15] says that universities are "supposed to be places that are dedicated to advancing the best interests and knowledge of humanity and society," and that espousing those values while remaining invested in fossil fuels is "a contradiction" (Somayajula 2020). Universities are educating the future world leaders, so they should be part of ensuring that the future exists. Such a responsibility should not be taken lightly, activists argue, and universities should listen to students when they vote on issues directly pertaining to them.

The Fiduciary Argument

The role of the Board of Trustees is to make all major decisions for the school. This includes approving budgets, creating goals, investing decisions, and more. The Board wants to best protect their investments, which will then protect the school's endowment.

Green investments are relatively untested, but they are rising in viability and popularity. Eric McCarthy, former general counsel for Proterra,[16] says that there has been an "uptick in very focused investing, specifically when it comes to environmental and sustainability issues" (McCarthy 2020). He also notes that there has been an increase in green energy companies, which will make it possible for universities to maintain a diverse portfolio while still being socially conscious.

It is also possible to be eco-friendly without divesting. Some examples are becoming a carbon-neutral campus, implementing recycling programs, purchasing electric transportation, and limiting energy waste. American University has become carbon neutral and sources its electricity from renewable sources (Sine 2014). For many schools that are unable to divest currently, taking other steps is seen as a way to make campus as "green" as possible while remaining invested.

One of the principles of investing is that it is best to maintain a diverse portfolio. This ensures that if one investment fails, other investments will remain steady. So full divestment from fossil fuels and having that money reinvested in green funds is relatively risky. Many universities argue that it makes sense to divest partially from fossil fuels and invest that money into green funds, but not to go 100 percent. Green funds, though showing solid rates of return, have not existed for very long, meaning they may be a greater investment risk. Fossil fuels have yielded strong returns and been good investments for decades, and while green alternatives are being created, fossil fuels were still reliable investments in 2014. However, as per the AUSG (2020, 13–17), fossil fuel firms have fallen out of the Standard & Poor's 500 stock index, which had seven fossil fuel firms in the Top 10 in 1980 but zero by 2020. Fossil fuel firms had negative growth even before the COVID-19 pandemic dropped the actual price of oil below zero (as usage was so low that it was accumulating in storage containers all over the world).

This ties into a broader fear, that of changing societal norms. Somayajula (2020) says that "refusal to divest is reflective of a broader tendency of many universities towards . . . wanting to preserve the status quo." In other words, some universities may seek to remain apolitical, and not have their campuses become testing grounds for every new social movement. Part of this has to do with "sunk costs" made by most universities, which might include, for example, oil-powered heating systems and usage of low-mileage vehicles in their institutional fleets. Some institutions, however, feature commitments to social justice as part of their basic tenets, which might imply that they would be more open to student activism on issues like climate change and environmental sustainability.

What Changed at American University from 2014 to 2020?

In 2013, 79 percent of American University students voted to divest the school from fossil fuels in a referendum (AUSG 2020). American University then convened an Advisory Committee on Socially Responsible Investing

(ACSRI) to evaluate divestment (AUSG 2020). The ACSRI released a report in 2014 with their recommendation, stating that "[g]iven AU's many climate efforts and its guiding principles, a divestment policy appears to clearly align with the university's core purpose" (ACSRI 2014, 4). The report outlined a roadmap to divestment, which would ensure that the university remained financially protected while also divesting over the course of several years (ACSRI 2014). Divestment included reallocating all fossil fuel holdings, screening investments, and continuing to monitor green investments, then creating a "Green Investment Fund" with the specific goal of investing in green technology and, finally, engaging students in more decisions as well as fossil fuel companies to encourage them to become more environmentally friendly (ACSRI 2014).

The Board declined to act on the report, and refused to bring the issue to a vote, because it perceived divestment as too risky given the narrowing of diverse investments and the substantial costs of requiring its endowment managers to de-select fossil fuel funds in co-mingled holdings (Wapner 2020).[17] Indeed, a trustee's memo cited estimates that divestment would cost 1.1 million dollars per year in additional management fees, which, when compounded, would add up to millions in lost investment income over the coming decades (AUSG 2020, 10).

Six years later, in 2020, another student vote occurred, with 93 percent of students voting in favor of divestment. With the advent of new technology to make it easier to pull funds, along with the creation of better green funds, the decision for the Board became much easier (Wapner 2020). Following the recommendations of the student government, student body, and ACSRI, American University fully divested the 12.9 million dollars remaining in fossil fuels in February 2020 (University Communications 2020).

Where Does the Movement Go from Here?

Activists argue that students often enter college feeling somewhat strongly about climate change, and then join Fossil Free to learn more (Heaphy 2020). By the time they leave, students often are "committed and trained climate justice organizers," who can use the skills they have learned in their everyday life or in any other movements they may join (Heaphy 2020).

For those still in school when their university divests, this may be an important consideration. After American University unexpectedly announced their transition to fossil free status, the Fossil Free club's previously laid plans for the next year crumbled (Elias 2020). Elias describes this as "a wrench, but a very good wrench," and remarks that her entire college career has been focused on divestment, so she's not fully sure what she'll do in her senior year (Elias 2020). Fossil Free may join with other campus climate organizations, forming a coalition that will continue to bring climate justice to the forefront and hold the university accountable (Elias 2020).

Some students want their universities to go even further, adopting some version of the Green New Deal in terms of prioritizing spending to improve environmental sustainability (Freire 2020). Somayajula (2020) wants GW to be at the forefront of climate activism, and thinks that they can accomplish that, but that divestment is only one step. Opening a climate research center, cutting ties with the RSC, and continuing to hold the university accountable are a few ways to do this (Somayajula 2020).

Others think that divestment from fossil fuels is just the beginning. At Yale, students' divestment campaign goes well beyond fossil fuel companies. Heaphy (2020) said the movement also includes demands to divest from prisons and student loan companies. Other ideas include divestment from the US military-industrial complex (i.e., the defense industry), as well as divestment from products made in countries with poor human rights records.

Is It Financially Responsible to Divest?

Even though it has historically been seen as irresponsible to divest, more studies are coming out demonstrating the opposite. For example, a 2018 study comparing returns on one representative portfolio to another similar portfolio (but without fossil fuel investments) found little difference in returns (Trinks et. el. 2018). A 2019 study did even better, finding that low-carbon stock portfolios earned higher returns between 2010 and 2018 than the overall market returns (Halcoussis & Lowenberg 2019). And another recent study assessed that replacing fossil fuel energy and utility stocks with clean-energy stocks in an investment portfolio improved the rate of return (Henriques & Sadorsky 2018).

Investing always comes with a level of uncertainty and an assumption of risk. Some risks have been seen as too great, and that may have stopped universities from investing in green funds. However, a study conducted in 2019 analyzing recent years of university investment showed that "divestment from fossil fuels presents a viable investment strategy," and that "there was not a significant abnormal risk-adjusted return for fossil-free portfolios" (Ryan & Marsicano 2019, 13–14). Aside from just the investing risk results, the study also looked into the effect these investments had on university endowments. It proved that there was "no discernible impact of fossil fuel divestment [to] the natural log of endowment assets," and that there was no "consistent, average impact of divestment on endowment assets and no conclusive evidence of negative effects to private university endowments (Ryan & Marsicano 2019, 40–46)."

EXERCISE

SCENARIO FOR CLASSES

Students should split into two groups. One will be Board of Trustees members, and the other will be Fossil-Free student advocates, invited to give their position before the Board. The instructor will serve as moderator (Board chair, taking a neutral position). Once the class has been separated, each side should prepare questions and select a representative of their group to lead the debate. Students may also wish to split into a group presenting their position inside the Board meeting (respectfully and based on facts rather than emotional appeals) and students waiting outside the Board meeting, whose views might be sought by local news media in attendance outside the boardroom. The student representative outside the Board meeting should also be respectful and composed but might be a protestor, making emotional appeals as well as more fact-based arguments.

In this hypothetical scenario, 10 percent of the university's endowment (roughly sixty million dollars[18]) is invested in fossil fuel companies like ExxonMobil, Shell, BP, and Chevron. The Fossil Free student club is pushing for a divestment commitment. The next Board of Trustees meeting (after the one occurring today) is in two months. Last week, there was a student election where a referendum to divest was approved. With a response rate of 70 percent, about 90 percent of students voted to divest.

The class should take five to ten minutes to consider the preparation questions. The Fossil Free students should then present their reasoning for why the school should divest. The Board of Trustees should then take a few minutes to discuss, and then announce their decision, including their reasoning. Finally, the whole class should debrief, talking through the discussion questions. Both sides should research their institution's financial conditions, environmental sustainability programs, and history of student activism and administrative concern for climate and the environment. Students should feel free to consider the culture of their own university, as well as successful means of protest at their university. These questions are meant to leave room for improvisation.

BOARD PREPARATION QUESTIONS

1. Are there other ways for your university to help the environment without divesting?
2. Will you be able to divest in five years?
3. Where would you reinvest funds?
4. How will you ensure student voices, and those of other constituent groups, are heard on this matter?

Continued

5. Will you ultimately divest?
6. What are the costs and benefits of divesting versus not divesting?
7. Can the endowment be used to express university values?
8. What are the Board of Trustees' fiduciary responsibilities, and how can you ensure those are met?

FOSSIL-FREE PREPARATION QUESTIONS

1. Are you asking for full or partial divestment?
2. What tactics will you use?
3. How willing are you to alienate certain constituencies (like administrators), and what would be the cost of this alienation?
4. If the Board does not vote to divest, what will be your plan moving forward?
5. How will you work with the Board of Trustees, and not just against them?
6. How will you convince the general student body to support you?
7. How important is divestiture as part of a broader climate change student movement?
8. How does this student movement fit into the actions of other groups?

DISCUSSION QUESTIONS

1. How did the tactics used impact the Board's decision?
2. Did the student vote impact the Board's decision?
3. Is your school divested/does it have a divestment group? How effective is it?
4. Did this exercise change your opinions on divestment?
5. What are the pros and cons of university divestment from fossil fuels?
6. Can endowments be used to address climate change or mostly to draw attention to the issue?

CASE STUDY 7.1 NOTES

1. Frame was a sophomore at American University in 2020, studying in the School of Public Affairs, where she majored in Political Science. She has worked with the student climate justice movement.
2. At this time, the South African GDP was approximately 59 billion dollars.
3. Fossil Free is an offshoot of 350.org. It began as a movement on college campuses but now extends to religious institutions, companies, and more. 350.org focuses less on divestment and more on overarching climate justice issues.
4. Goddard College, Salem State University, University of California, and American University (Fossil Free 2020b). This list is accurate as of May 21, 2020.
5. Antioch University, Brevard College, Brown University, California Institute of the Arts, Chico State University, College of the Atlantic, Georgetown University, Hampshire College, Lewis & Clark College, Middlebury College, Naropa University, Northland College, Oregon

State University, Peralta Community College, Pitzer College, Rhode Island School of Design, Seattle University, Penn State, Sterling College, Syracuse University, The New School, Unity College, University of Dayton, University of Hawaii, University of Maryland, Warren Wilson College, and Whitman College (Fossil Free 2020b). This list is accurate as of May 21, 2020.

6. Humboldt State University, Prescott College, University of Puget Sound, Western Oregon University, and Yale University (Fossil Free 2020b). This list is accurate as of May 21, 2020.
7. Boston University and San Francisco State University (Fossil Free 2020b). This list is accurate as of May 21, 2020.
8. Brandeis University, Columbia University, Johns Hopkins University, Stanford University, University of Maine System, and University of Washington (Fossil Free 2020b). This list is accurate as of May 21, 2020.
9. Fourth on the list of the most destructive companies (fossilfreeindexes.com).
10. The Koch Brothers and their foundation have donated over 150 million dollars to climate change denial programs since 1986 (Greenpeace USA n.d.).
11. Heaphy was a junior at Yale University in 2020, where she majored in ecology and evolutionary biology.
12. LeBlanc has since apologized, stating that he was trying to prove that the majority should not always rule.
13. Freire was a sophomore at George Washington University in 2020, where she majored in International Affairs.
14. Somayajula was a junior at George Washington University in 2020, where he double majored in Political Science and International Affairs. He ran communications for Sunrise GW.
15. A hub of the larger national organization, Sunrise. Sunrise GW absorbed Fossil Free GW.
16. A company that makes electric buses.
17. Co-mingled holdings refer to mutual funds and other financial instruments combining investments in a range of stocks.
18. This is based on the median university endowment, which is 65 million dollars (Kerr 2019).

8

Fixing Eden

Ratcheting Up Ambition, Deep Decarbonization, and "Plan B"

The small Caribbean island of Dominica suffered some of the worst recorded storm damage ever in 2017, when Hurricane Maria wreaked havoc on the island's 73,000 people, generating damage equivalent to 226 percent of a year's gross domestic product (World Bank Group 2018, 5). In other words, well over two years of economic production for the entire island nation was obliterated in a couple of days. Luckily, more than half the island's public budget is based on "citizenship by investment" (Skerrit 2018), a sale of citizenship (passports to citizens from nations with restrictive travel regulations like China and Iran), which is a function of the legal flexibility the country's passport provides more than the state of tourism infrastructure.[1] Nevertheless, tourism and light industry were decimated, and more importantly, more than 90 percent of the island's housing was damaged, with over a third severely damaged.

How could the government come up with millions of dollars overnight to rebuild bridges and restore infrastructure? Luckily, in 2007, the government had joined an innovative institution for sharing risk (like insurance companies do) among all the nations of the Caribbean: the Caribbean Catastrophe Risk Insurance Facility (CCRIF). In fact, Dominica received that institution's first payout ever, for earthquake relief. Funded by international donors, the CCRIF by 2018 had paid out 19 million dollars for 38 catastrophic events in over a dozen member nations over its first 11 years.

The CCRIF has been criticized on the basis that at least one member nation, Jamaica, pays a six million dollar per year premium but has never received a payout (Artemis 2017). Authorities at the World Bank, which helped design the CCRIF, acknowledge that joining an organization and paying annual "dues" or premiums does put pressure on politicians to get something back. Recognizing this, the CCRIF in 2019 announced that it is launching parametric[2] insurance, which it already had for catastrophic events like hurricanes and earthquakes, for extreme rainfall amounts during non-hurricane events. Overall, the institution is recognized as a pioneering effort to spread risk among small, fragile nations that would not be able to respond at scale to the magnitude of damage if not for this risk-pooling system, which is also being attempted in Africa and among the small island states of the South Pacific (Broberg 2019, Vincent et al. 2018).

Overall, some 93 percent of natural disasters are climate related, according to Palmer (2013, 857), and while floods and cyclones (hurricanes) are sudden disasters that can be readily measured, other slow-onset problems, like drought, are less readily addressed until after they have reached critical stages. To one analyst (Palmer 2013, 858), only four percent of disaster financing is used for prevention and preparedness. This raises broader questions regarding adaptation.

The Challenges of Acquiring the Adaptation Mindset

Recall that adaptation is the routinization of climate change effects—that is, their acceptance and accommodation into peoples' daily lives. The Intergovernmental Panel on Climate Change (IPCC; 2007) defines adaptation as "adjustment in natural or human systems in response to actual or expected climatic stimuli or their effects, which moderates harm or exploits beneficial opportunities" (9) This contrasts with mitigation, which the IPCC (2007) writes is "an anthropogenic intervention to reduce the anthropogenic forcing of the climate system; it includes strategies to reduce greenhouse gas sources and emissions and enhance greenhouse gas sinks" (9). While means have still not been found to readily measure adaptation, mitigation has been boiled down to the second part of the definition—reducing greenhouse-gas (GHG) emissions—so that it can be measured. Indeed, we can relatively easily model the effectiveness of climate change mitigation (diminishment) based directly on emissions reductions, which can be readily calculated.

As mentioned, there is no such ready-made unit of adaptation to be measured. And as stated well by disaster management specialist Adam McAllister (2019), there is also a need for more private sector involvement, in addition to public sector leadership:

> There is a tendency to treat climate change adaptation as basic linear algebra. Rather, we should consider it as calculus; we can approach certainty, but never quite arrive. We cannot wait for a perfect climate model. Instead, through bold policy that accepts uncertainty in our models, we must create the conditions for the private sector to adapt and thrive.

The point is well taken that certainty about the future may not be as well-defined in future business models, but some private sector interests are accepting this uncertainty. For example, the Kenya-based Eastern Africa

Grain Council (EAGC) is promoting adaptation among its members by striving to anticipate and preempt climate threats. "The anticipated changes in crop supply are already forcing us in the grain industry to adapt. Grain is the staple for all these countries and we are looking at how to improve the varieties of grain and also how to restructure trade so there is a more predictable supply process," said Janet Ngombalu (2019), EAGC regional programs coordinator. This means improving storage over longer terms, in the event that increasing weather fluctuations diminish harvests; adjusting to climate-related threats to maize and other grains, such as the fall armyworm and aflatoxin; strengthening regional transportation and trade networks, including importation in times of scarcity; and the provision of access to credit and insurance markets for producer members of the trade association. Ngombalu said that her association is also undertaking extensive public outreach campaigns to help grain producers understand how best to protect their crops and investments in times of greater uncertainty, and is partnering with meteorological services to help farmers better understand the implications of climate changes on their production.

Of course, the public sector is also a fundamental actor, if not the fundamental actor, especially in climate-vulnerable nations like those in the Caribbean. St. Lucia Finance Ministry Economist Cecil Charles said recently that adaptation planning "is not directly in the budget as of yet," although it is being accorded priority in the long-term planning of the hurricane-ravaged island nation.[3] Dominica Finance Ministry Permanent Secretary Rosamund Edwards (2019) stated that a literal "rainy day fund" to help the country plan for weather disasters had been allocated in the national budget right before Hurricane Maria wiped it out. Moreover, after Hurricane Maria, Dominica's prime minister, Roosevelt Skerrit, declared that his nation of 73,000 people was to be the first climate-resilient nation in the world. "I come to you straight from the front line of the war on climate change," Skerrit told the General Assembly days after Hurricane Maria, adding that he made the journey from Roseau to New York "because these are the moments for which the United Nations exists! . . . The stars have fallen, Eden is broken. The nation of Dominica has come to declare an international humanitarian emergency" (United Nations News Agency 2017).

The Debate over Measuring and Delivering Adaptation Finance

The mainstreaming of adaptation, its integration into broad facets of everyday life, is increasing within large organizations. But in the Dominica of the "fallen stars," this was viewed as urgent. In 2018, the government launched a cross-government task force, the Climate Resilient Execution Agency for Dominica

(CREAD). CREAD's former director Pepukaya Bardouille (2019), said that resilience, the ability to recover from disaster, is highly related to climate adaptation, which is the day-to-day capacity to respond. While the precise parameters of planning for resilience as the highest priority are still being finalized, Bardouille emphasized the importance of considering climate resilience within a broader context of "overall resilience," and that a major portion of the task will be assessing risk and finding ways to instill all projects with resilience components.

In its theoretical definition, resilience conceptualizes the ability of individuals or groups to take proactive adaptation actions over time (McDaniels et al. 2008). In practice, "resilience is the new adaptation, which is sustainable development," said Bardouille (2019), referring to the context of Dominica, adding that "these are buzzwords." In other words, despite government statements that Dominica would become the world's first resilient country, this meant largely that climate considerations would be mainstreamed into all planning and government functions, and that Dominica would have to consider climate risk more prominently as a matter of course in government planning. Integrating adaptation and resilience does not seem to mean creating new concepts, but rather applying existing strategies of mainstreaming and risk assessment more rigorously and more broadly.

THE SCARCITY OF INTERNATIONAL FUNDING FOR ADAPTATION

Worldwide, only about four percent of the estimated more than 500 billion dollars in public and private climate finance in 2017 was destined to adaptation (Climate Policy Initiative 2018, 7). It turns out mitigation is a public good (everyone benefits from the reduction in emissions), and while adaptation may be the result of a "public bad" (mostly extreme weather events exacerbated or caused by climate change), analysts tend to view it as a private good (Sovacool & Linner 2016, 14–15). Indeed, adaptation spending by wealthy nations on their own preservation may merely reiterate and deepen the North-South divide. The famous Dutch dikes do not benefit the drowning Pacific islands of the Maldives or Tuvalu, whereas the emissions-reducing effect of substituting solar energy for all coal-powered plants would mitigate climate change markedly over time and, hence, presumably benefit everyone. Even within nations, adaptation through the construction of water embankments in the southern delta of Bangladesh does not help climate-caused droughts in the northern Himalayan foothills.

Despite the "lumpiness" (inequality and non-transferability) of adaptation as a non-public economic good, some institutions are starting to spend more, as if it was, to paraphrase Prevost, "the new way to undertake sustainable development." Indeed, the largest development institution, the World Bank, pledged in the fall of 2018 to spend 50 billion dollars over the next five years on adaptation, which would seemingly double the 2017 total.[4]

In a separate study, Eisenstadt, Olawole, and Toman (2020) have indeed found that World Bank climate lending over the last eight years has achieved parity between adaptation lending and mitigation lending, and that the World Bank's direct climate lending in 2018 constituted about 14 percent of the institution's portfolio.[5] Of course, the economic downturn in 2020 due to the effects of the COVID-19 pandemic will likely make funds for long-term and discretionary public expenditures, such as those for adaptation-related projects, even more scarce for the next several years. And private sources of funding have not yet taken a commanding role in adaptation, as business models in which adaptation spending can be profitable are still being formulated.

The World Bank, like the government of Dominica, has also started mainstreaming climate considerations into all of its projects, claiming that by 2019 nearly 30 percent of its financial flows directly addressed climate issues (even if as components of other projects).[6] Nonetheless, even at the World Bank, perhaps the world's largest development lender, means of measuring adaptation, while undertaken, have not yielded precise measures of how to assess when development projects achieve adaptation goals, which tend to be diffuse and long-term. It may be that, regarding ways of measuring adaptation, "if you build it, they will come";[7] if means are found to precisely measure units of adaptation, the ability to assess progress will translate into concrete objectives and more precise budgets.

WHAT HAPPENED TO THE UN PLEDGES FOR ADAPTATION?

Part of the challenge of getting donors to make adaptation a core part of the development agenda is that the United Nations Framework Convention on Climate Change (UNFCCC) has rejected efforts to operationalize demands by the developing world to require that the UNFCCC fund adaptation in the form of "loss and damages" to vulnerable nations from industrial nation polluters (recall the "carbon majors" and their national origins as discussed in Chapter 4). After the Warsaw Mechanism was created by the UNFCCC in 2013 to adjudicate loss and damage, the issue was largely dropped from discussion, as diplomats and observers seemingly held their breath for several years, waiting to see if the pledged 100 billion dollars per year by 2020 started to flow. (It mostly has not.)

Adaptation finance was addressed broadly and vaguely in the 2015 Paris Agreement, but as anyone who has studied laws and rules knows, sometimes the best way to "kill" a bill is to pass something broad and sweeping but then fail to further codify and specify mechanisms for implementation. Even before the United States started undermining implementation of the Paris Agreement with President Trump's withdrawal from the agreement in 2016 (completed before his term ended but reversed by climate champion Biden during his first week in office), the United States (and other industrial powers) had worried that recognizing loss and damage would empower major lawsuits by developing nations, which could cost industrial nations billions of dollars.

When Adaptation Fails: Legal, Economic, and Political Adjustments for Climate Migrants

Meanwhile, climate migration is finally being recognized as an important empirical phenomenon. The International Organization for Migration (2018) estimates that some 200 million climate migrants will likely have been displaced by 2050, with estimates ranging from a low of 25 million to a high of one billion, depending on the rate of climate change. As usual, poor and vulnerable nations in Global South "hot spots" will be disproportionately affected, with World Bank analysts estimating in *Groundswell* that some 86 million people will be displaced from sub-Saharan East Africa (the hot spot from Eritrea to Mozambique), some 40 million will be displaced from South Asia (from Afghanistan to Bangladesh, dominated by India), and 17 million will be forced to resettle from Latin America (with a focus on Mexico and Central America) (Kumari Rigaud et al. 2018).

Managed migration can be part of a "sensible climate change adaptation strategy if managed carefully and supported by good development policies and targeted investments" (Kumari Rigaud et al. 2018, 182). The *Groundswell* authors insist that mitigation is the first means of diminishing and rationalizing migration, and that migration needs to be embedded in development planning and understood through investment to contextualize climate migration (Kumari Rigaud et al. 2018, 183–189). A recent study of the most climate-vulnerable people suggests that migration decisions are strongly tied to whether survey respondents had suffered strong economic impacts from extreme weather events over the last five years (Wright et al. 2020).

THE LEGAL DEBATE OVER MIGRATION AS "LOSS AND DAMAGE"

Like "loss and damage," with which it is sometimes—but not always—factored in to cost calculations, climate migration has been a controversial topic given the contentiousness of adjudicating responsibility for financing and receiving displaced populations. The UN has issued several recommendations, including one to ensure adequate resettlement opportunities for those who are temporarily and permanently displaced, and another to ensure that the least-developed countries are provided with technical assistance and financial resources to cope with climate-related displacement (Kälin 2010, Rolnik 2009). Still, states have full sovereignty to decide whether they choose to address the issue, or how to avoid it.

A range of proposals has been presented, including one that says climate migrants should be provided permanent residency abroad, with recipient countries accommodating these migrants proportionate to receiving nations' historic national emissions (Gerrard in Holthaus 2016). Another, in collaboration with the UN Development Programme, proposed to negotiate the creation of a "new protected status applying specifically to environmental migrants" to protect the rights of those forcibly displaced by slow-onset process impacts (Wentz & Burger, 2015). Additionally, it proposed establishing a climate change displacement coordination facility as a step toward developing a legal framework to negotiate future resettlements.[8]

These legal debates about compensation and resettlement are transpiring with insufficient information from those who are the most affected and the potential program recipients: migrants themselves. The urgent need of systematically collecting valid and reliable data to measure climate migration, as this project proposes, was underscored at the UN Climate Change Conference in Bonn in May 2016 (UN Newsroom 2016). As the UN commences extensive debates to establish a system for cost-sharing loss and damage across member nations, no systematic surveys attempting to understand comprehensive cross-regional attitudes across a range of levels of vulnerability have been reported anywhere. Gaining a deeper understanding of the link between displacements and vulnerability is necessary for creating institutional mechanisms to empower vulnerable populations. Vulnerable nations continue to advocate for considering climate displacements as "loss and damage," acknowledging that there are climate impacts to which nations cannot adapt (Karsapan 2015).

DOES "MIGRATION WITH DIGNITY" JUSTIFY CLIMATE MIGRATION?

In drowning island countries, whether and how climate migrants gain legal protection is . . . well . . . academic. The people of Tuvalu do not have lots of time to ponder the thought exercise of how to wedge "climate migrants" into international human rights law. Their homes are flooding every February, and they have much more practical problems to solve. Some 2,000 Tuvaluans (close to 20 percent of the population) already live in New Zealand (Curtain and Dornan 2019, 9); and Fiji has offered to resettle Tuvaluans displaced by climate change (Roy 2019). But Siliga Kofe, chief of Tuvalu's largest island (2015), argued vehemently that his people do not want to relocate, and that the international community should live up to the pre-Paris pledge of restricting climate change to the targeted maximum of 1.5°C over this century (rather than the 2°C goal accepted at Paris). "How many more people must be drowned or have their livelihoods washed into the oceans before we act?" asked Kofe (Eisenstadt 2017). The leadership of Tuvalu and other islands, organized in the UNFCCC as the Alliance of Small Island States, consider themselves to be the conscience of the UN, and as such, they fear that migrating, rather than staying put to take a stand for their nation's continuance, would be tantamount to "giving up" the fight against climate change.

Others see large-scale migration as inevitable, and claim that the best approach is "migration with dignity." For example, Kiribati, another Pacific atoll nation, with a much larger and burgeoning population of about 110,000, "has developed the most concrete migration strategy at an international level" (McNamara et al. 2018, 113). Kiribati's government has commenced negotiating international seasonal labor programs with larger nations in the region, modeled after seasonal employment programs already in place in Australia and New Zealand. The nation is seeking donors to help invest in education and English training for Kiribati migrants to ensure their economic competitiveness; and started buying land in Fiji in 2014 for resettlement (McNamara et al. 2018, 114). And closer to home (slightly) for US-based readers, the 350-plus residents of Newtok, Alaska, have voted several times to permanently relocate after losing acres of land to erosion and thawing permafrost. While the Newtok villagers have not completed their move, in 2018 they received 15 million dollars to relocate the entire village to a site several miles away, acquired in a 2003 land trade with the US Fish and Wildlife Service. Climate migration is only now starting to receive publicity, but has become quite commonplace, especially in the Global South.

CLIMATE MIGRATION AND THE NORTH-SOUTH DIVIDE

Recall the quote from Barbados Prime Minister Mia Mottley (2019) in Chapter 1, which to us epitomizes the difference between the climate victims of the South and the voyeurs of the North: "It is only when moral and ethical leadership is given, both at the national level and at the international levels, that we will summon the courage to fight these [climate] battles. Until such time, it [addressing the issue] is a form of idle entertainment for those who choose to watch." Recent research has shown also that vulnerable communities "on the ground" suffer extreme weather events—cyclone after cyclone—without really knowing who to blame or, about half the time, that the intensifying weather patterns are even due to climate change (Eisenstadt et al. 2020).

Climate migration seems to be the problem of the most vulnerable nations, which are many of the same nations of the Global South referred to, during the Cold War era, as the Third World.[9] If a decade of voluntary efforts by the UNFCCC to channel new resources toward adaptation has yielded only the beginnings of a shift in funding to the most vulnerable nations (Eisenstadt, Olawole & Toman 2020), then perhaps means of incentivizing nations to pay more attention to climate change must be found. In the following section, we address efforts to incentivize compliance with stepped-up climate action across the full array of issues.

Fossil Fuel Subsidies and the Need for Creative Destruction of that Lobbying System

Subsidies of fossil fuels by national governments—for exploration and production, in the form of not charging fossil fuel producers for emissions, and in the form of consumption subsidies to lower costs to consumers—are undermining pledges by developed nations to help finance climate change mitigation by developing nations. In fact, even as the rich countries have pledged—but not collected—some 100 billion dollars to mitigate climate change costs to the disproportionately affected poor, these same affluent OECD countries spend over five times that amount, on an annual basis, to promote oil and coal extraction, refinement, and delivery. And even more importantly, by increasing profits of fossil fuel producers, some 40 governments that subsidize fossil fuels are incentivizing the increase in emissions.

Perhaps former Mexican President Felipe Calderón said it best, at a side event of the UNFCCC 2015 Paris meeting (Eisenstadt 2015): "The main reason for preserving fossil fuel subsidies is politics. There is a clear relationship between fossil fuel subsidies and votes for political incumbents." Citing an International Monetary Fund (IMF) report, Calderón said that very few of these fossil fuel subsidies went to supporting the poor, the claimed beneficiaries of such actions. The IMF study makes the most ambitious projection, that total energy subsidies (including those to fossil fuels) cost 5.3 trillion dollars in 2015, equivalent to 6.5 percent of worldwide gross domestic product (or over 30 percent of national budgets, Calderón pointed out, which could be much better used for education and health care). This projection is based mostly on the fact that most countries charge energy taxes way below levels needed to also reflect environmental damage from energy consumption.

Pollin (2015) argues that governments defend these subsidies on several grounds. They claim that by subsidizing fuel they are lowering energy costs to low-income households and promoting economic growth and jobs and, in fossil fuel–producing countries, they are promoting leading domestic industries. However, "policymakers could achieve all of these basic objectives through alternative measures that would be much more productive, starting with committing their economies to a large-scale clean energy investment program that would deliver affordable energy and expand domestic job opportunities" (56). As demonstrated in Chapter 4 for the United States, probably the most blatant case outside the Middle East, fossil fuel interests have captured national governments, leading to suboptimal environmental policies (Eisenstadt et al. 2019) and even the subsidy of these bloated "fat cat" relic industries, rather than looking to the future and supporting renewable energy.

Some of the indigenous communities in the Ecuadorian Amazon featured in Case Study 3.1 (Chapter 3) led a movement, seized upon by President Raul Correa in 2008, to "leave the oil in the soil," at Yasuni National Park, one of the world's most biodiverse protected areas, where the Andean highlands roll down to the lush Amazon Basin (Eisenstadt & West 2019, 174–76, 184–86; Martin 2015, 119–144). Correa shrewdly predicated the protection of Yasuni, which he referred to as a patrimony of all humanity, on raising funds internationally to offset profits forgone by not drilling, and in 2016, after nearly a decade of fundraising, regional drilling commenced, but just outside protected parklands. Since the Yasuni movement began, US universities (like American University; see Case Study 7.1 in Chapter 7) and European governments (like Ireland and Norway) have banned oil investments from their endowments and pension funds. Recall from Chapter 5 that the 2018 IPCC special report argued that overall emissions must be cut by 45 percent by 2030, with fossil fuel extraction falling by nearly a trillion dollars by 2050, and with coal extraction ending entirely. How might this be achieved, given that the world's GHG emissions have still not yet peaked?

Some argue that reforms on the fossil fuel supply side (i.e., extraction and production) are needed, because those on the demand side (fossil fuel usage) are not achieving our objectives. To this end, several countries have either banned or partially banned oil and gas exploration and extraction as of 2018: Belize and New Zealand had banned offshore exploration, Canada and the United States Arctic oil and gas exploration (the Trump administration had sought to allow drilling but the Biden administration insisted on maintaining the ban), Denmark onshore oil and gas drilling, Costa Rica all oil exploration and extraction, and France exploration and future extraction (Erickson et al. 1038).

While the economic impact of the COVID-19 pandemic on the energy sector was still unfolding as this book was being finished in 2021, it seems that clean energy growth and decarbonization cannot happen without some supply side reductions in the production of oil and gas and the elimination of coal. In the following sections, we consider what acceleration of a clean energy growth agenda might look like, but accompanied by more rational urban planning, construction, and transportation. Yet before addressing substantive changes to reduce the role of fossil fuels in the economy, we offer an innovative means of inducing change through procedures. If Chapter 7 addressed economists' means of slowing climate change, here we focus on the views of urban planners and political scientists, starting with an innovative idea to move the international relations debate.

Thinking Outside the Gas Tank: The Fossil Fuel Non-Proliferation Treaty

If the Montreal Protocol discussed in Chapter 3 represents the world's biggest unqualified success to date in regulating environmental pollution through an international treaty, the Nuclear Non-Proliferation Treaty (NPT), implemented in 1970 and now signed by 191 countries, represents our best effort to limit the arms race, which nearly spiraled out of control during the end of the Cold War.[10] While nonstate terrorists are probably viewed as the greatest threat to global nuclear (and non-nuclear) security since the September 11, 2001, non-nuclear terrorist attacks on New York City's World Trade Center and the Pentagon, nation-state signatories of the NPT continue monitoring each other, seeking to prevent the spread of world-endangering nuclear capabilities to nations like Iran. That is a nifty little history lesson, smart and gentle readers like you may be thinking, but what does this have to do with reducing the extraction of fossil fuels?

Similar to the NTP, a possible Fossil Fuel Non-Proliferation Treaty would be a multilateral effort in which national governments strive to control fossil fuel extraction from the supply side. As per Newell and Simms (2019):

> In the case of the NPT, it is in the interest of the world's most powerful countries—who already possess nuclear weapons—to ensure that others do not acquire similar capabilities. In the case of climate change, however, the majority of countries in the world already have at least some reserves of fossil fuels, even though these are unevenly distributed. The challenge is then how to negotiate their non-use in a fair and credible way at the speed and scale required to avoid a collective threat to all states and their citizens. In this sense multilateral responses may be attractive to powerful countries wanting to ensure other states do not free-ride on commitments they are now making to leave fossil fuels in the ground. (3)

As Newell and Simms point out, the NPT has stronger incentives, as even one additional nuclear power in a strategic corner of the earth (like Iran) poses a much more immediate and urgent risk than one additional oil driller (whether Mexico, Kuwait, or Norway), regardless of that driller's location. But a supply side prohibition against fossil fuel extraction, with the community of nations enforcing the ban, may be worth attempting. Essentially, governments would have to agree to force companies to strand oil (discussed in Chapter 6).

The proposers (Newell & Simms 2019, 4–6) state that an extraction-based accounting system is needed to track countries' production levels and emissions, and that an existing agency like the UN Energy Program could regularly update member-states on the gap between the "stranding" of fossil fuel assets and the impact of this on an agreed-upon international emissions budget (another contentious issue). Furthermore, a "disarmament plan" would be implemented, which in this case would mean coordinating a "managed and accelerated decline of existing fossil fuel infrastructures" (Newell &Simms 2019, 5). Nations with the greatest ability to pay and which are most responsible for the present level of GHG emissions would be asked to lead.

The Fossil Fuel Non-Proliferation Treaty may help us more rapidly transition out of fossil fuels. In addition to the nations mentioned earlier with moratoria on oil and gas exploration, France, Germany, Ireland, Scotland, Uruguay, and Wales have moratoria in place against hydraulic fracturing ("fracking"), and China, Spain, and the United Kingdom have set imminent timetables for phasing out coal and other fossil fuels. A "Powering Past Coal Alliance" was launched in 2017, with more than 25 nations pledging to build no more coal-fired power stations. And internationally, precedent does exist for banning mining in Antarctica, and some 500 organizations signed the 2017 Lofoten Declaration calling for an end to fossil fuel development and a more active ramping down of existing production. Even if a non-proliferation treaty does not emerge, might it be possible to sue oil companies for their roles in climate change? Such a suit might be along the lines of an inter-generational lawsuit filed by scientist James Hansen and dozens of others based on the generational inequality posed by climate change, as it hurts the young disproportionately, as per the logic mentioned in Chapter 1, or the suit against ExxonMobil discussed in Chapter 4. But this one would just be more basic: Can fossil fuel companies be held accountable for contributing to climate change?

Can the Courts be Used to Hold Fossil Fuel Companies Responsible for Climate Change?

Saúl Luciano Lliuya believes that fossil fuel companies can be held accountable, and residents of his high mountain community of Huaraz in the Peruvian Andes are suing the German utility giant Rheinisch-Westfälisches Elektrizitätswerk (RWE) for 20,000 dollars to see. With the help of the nongovernmental organization (NGO) Germanwatch, Lliuya's case received standing in Germany's national courts in 2017, in what may be an international precedent. The evidence is somewhat indirect. RWE (which according to the Carbon Majors Database has contributed about 0.30 percent of international emissions – CDP 2017) is being sued for contributing to glacial melt damage to the small alpine city of 120,000 people. Whatever the outcome of the case, it will not be the first to assess "blame" for climate change, although it would, if culpability is established, seemingly be the first to tie the activities of a private company emitter to historical emissions to climate impacts. Case Study 8.1 at the conclusion of this chapter offers many of the facts of the case for your consideration.

Another landmark case, *Urgenda Foundation v. Kingdom of The Netherlands*, may be the first victorious case assessing responsibility for climate change, but in this 2015 lawsuit, the defendant was a national government. Urgenda, a Dutch environmental NGO, sued the government for relaxing GHG emission goals set by that country's prior administration. Urgenda won in District Court in 2016, and the Hague Court of Appeals upheld the ruling in 2018.

The court ruled that, in failing to execute its pledge to stay on track toward reducing GHG emissions by at least 25 percent by the end of 2020, the Dutch government was failing to uphold the European Convention on Human Rights, which protected the human rights of its citizens. In other parts of the world, too, governments are being held responsible for implementing climate policies. Also in 2015, *Leghari v. Republic of Pakistan* used a similar argument as the *Urgenda* case to hold the government accountable for the implementation of the Climate Change Policy of 2012 and the Framework for Implementation of Climate Change Policy (2014–2030). In the United States, a 2007 Supreme Court ruling in *Massachusetts v. Environmental Protection Agency* confirmed that the US EPA must regulate GHG emissions under the Clean Air Act in cases where GHG emissions endanger public health or welfare.

Countries suffering some of the worst abuses from gasoline-related GHG emissions may include Nigeria, where petroleum "burn-off" flaring has presented great public health risks for decades. "Courts in Nigeria are beginning to recognize the interdependence of civil and political rights (as guaranteed in the constitution) and socioeconomic rights, such as the right to a healthy environment," said Oluwatoyin Adejonwo-Osho (2019), a Nigerian environmental and human rights lawyer. She cited the 2005 case of *Jonah Gbemre v. Shell Oil*, in which a Nigerian federal court ruled that oil companies must stop flaring gas in the Niger Delta as a strong start toward enforcement of such socioeconomic rights, which might include climate rights in the future. "Awareness of global issues and a better understanding of climate change and other environmental challenges have pressured the courts to start ruling in favor of these socioeconomic rights," Adejonwo-Osho said.

Worldwide, the number of national climate change laws at the disposal of legal advocates has increased drastically, from some 60 in 1997 to over 1,500 laws and policies by 2018 across the 197 Paris Agreement signatories (Nachmany 2018). Over 1,200 climate cases have been processed in US courts between 1986 and 2020, according to the Climate Change Litigation of the World Database (Setzer and Byrnes 2020), with only some 374 originating in nations other than the US. . As Laurel Wolf notes in Case Study 8.1, many cases are being brought by companies against governments, but that a few seek to hold companies accountable. There are several landmark climate litigation cases that try to strategically link rights and climate change by urging mitigation, adaptation, and/or compensation. According to Nachmany (2017), some two-thirds of climate rulings reinforce existing climate regulations, although strategic litigation to hold corporations and governments accountable is on the increase (Schäfer et al. 2018, Nachmany 2017). Judicial activism will clearly be an important strategy to force governments and private companies to adhere to climate policies, but this strategy needs to be complemented by the tightening of the policies themselves at national levels, as well as internationally.

Can the United Nations be Climate-Proofed?

Diplomats have roundly lauded the Paris Agreement, even as IPCC scientists have warned that the agreement's low ambitions, consisting of "bottom-up," voluntary Nationally Determined Contributions to reduce emissions rather than emissions reductions mandated from a "top-down" UN carbon budget, are inadequate. Indeed, the voluntary pledges, consisting of whatever nations "want to give," will not hold climate change to a 2.0°C by the end of the century, and the window may soon close on holding climate to a 1.5°C increase. Still, some academic analysts have argued that the Paris Agreement just needs a bit of tinkering around the edges rather than major reform (see, e.g., Hohne et al. 2017 and Pauw et al. 2019). We argue, on the contrary, with Keohane and Victor (2016) that a major institutional overhaul, or at least the creation of a "climate club" to negotiate most of the international community's climate change implementation and management, is sorely needed. How might the Paris Agreement be ratcheted up? Keohane and Victor have constructed a hierarchy of six potential strategies towards what they call "deep cooperation" (4) In descending order of rigor/ambition, these are:

1. *Universal agreements* with emissions targets that bind member-states and with timetables for completion. The Kyoto Protocol was an example, although many nations opted out or "defected" from participation. The Paris Agreement does not have a "carbon budget" (i.e., specified emissions reductions with a concrete timetable assigned by a central authority) or timetable. Rather than a "top-down" assignment of targets, it relies on "bottom-up," voluntary pledges of emissions-reducing actions. Kyoto was extremely ambitious but ultimately applied to nations governing only to about one third of global GHG emissions. A Fossil Fuels Non-Proliferation Treaty would qualify as an ambitious universal agreement.
2. Small group *climate clubs* that, rather than needing the consensus approval of some 195 UN members, require only that a few of the largest emitters monitor each other and pledge sanctions in case members fail to cooperate. While these have never been fully tried in practice, it seems the "peer pressure" implicit (or explicit) in these could ratchet up ambition. China and the United States alone emit some 45 percent of GHGs, and those two nations plus the European Union, India, Japan, and Russia generate over two-thirds of global emissions. If groups of nations could coordinate binding targets and timetables, they could generate emissions reductions nearly as large as those of the entire UN, and without all the necessary consensus-forming and coordination.
3. Incentives to generate *coordinated research on new technologies* to fight climate change might offer intermediate-ambition opportunities for nations to cooperate. As discussed extensively in Chapter 5, technology (and the

new economies of scale it may bring) could be a big part of the answer. Solar power, battery storage, as well as carbon capture and storage are all cutting-edge technologies, and the "so-called" disruptive technologies may also have a role to play. However, it does remain unclear how these technologies would be best harnessed by nations rather than by firms.

4. The "*pledge and review*" approach, codified in the Paris Agreement, represents a modestly ambitious approach, in which countries pledge emissions reductions, which are then tracked by a central authority. As seen in the Paris example, voluntary pledges have thus far been insufficiently ambitious to hold warming by century's end to even 2°C, and the "rule book" of conditions to review each country's execution of pledges is only being written in 2019.
5. *Multilateral national actions* (i.e., those between two or more countries) represent a reduced-ambition version of the climate clubs. Here, nations might undertake actions in their national interests (e.g., reducing toxic air pollution from coal-fired power plants) and get credited with the assumption of climate change leadership. The 2014 China-US bilateral emissions pledge by each nation has this character. However, political conditions changed in the United States after 2016, and it was unclear until President Biden's election that the will remained to enforce this bilateral agreement.
6. Lowest common denominator *universal agreements* such as those codifying the post-Paris process of UNFCCC annual meetings, do not by themselves promise emissions reductions, but they do keep parties in dialogue. As Keohane and Victor (2016) point out: "Such agreements for minimal coordination lead to few actions beyond what countries would have done anyway" (4). The question at this point may be how much better these are than no cooperation at all.

Certainly, the world needs a fortified universal agreement. We need a "Paris Agreement on Steroids" with carrots and sticks. The carrots would have to be generated from funds dedicated to research on pathways and technological innovations leading to net-zero carbon emissions worldwide by 2050, funding for adaptation and resilience for the most vulnerable nations, and incentives and tools from the UN itself and other international institutions, such as the World Bank and International Energy Agency, for firms, private citizens, and nations to measure and monitor our own carbon footprints. But how will we reach that point?

THE PARIS AGREEMENT'S POWERLESS "PLEDGE AND REVIEW" STATUS

The "pledge and review" Paris system is really only a "lowest common denominator" approach until the rule book of what is to be pledged and reviewed is implemented, hopefully within a few years of the set-before-COVID original

2020 target date. The Paris system is falling short, as Kyoto did. It may be time to accept that a real universal agreement is beyond the scope of possibility given existing resources and wills, and to move down the hierarchy Keohane and Victor present to the second-most powerful option, climate clubs. Nordhaus (2017), lamenting the failure of quantitative targets for emissions reductions like those of the Kyoto Protocol, argued that "they have shown excessive price volatility, lose precious governmental revenues, and have not lived up to their promise of equalizing prices in different regions" (119).

Indeed, this claim that the carbon markets created by Kyoto—and now Paris—have been inadequate is consistent with our demonstration in Chapter 7 of the broad range of carbon prices offered around the world in carbon taxes and in emissions trading systems The answer for Nordhaus (2017, 118)—and an increasing number of other social scientists—is to have a small number of countries join together in requiring a singular, set price for carbon traded among and within the group, and to have penalties charged for all countries that do not participate, such as a tariff on all imports from nonmember countries, of—say—two percent, which is not fatal to commerce, but is costly.

CLIMATE CLUBS AS MORE EFFECTIVE INTERNATIONAL ORGANIZATIONS?

A climate club, starting, for example, from adding a few more members to the US-China bilateral agreement in 2014, would have several advantages. First, it would abnegate both the limits of national sovereignty and the ineffectiveness of international law. Member countries would hold each other accountable for enforcing agreements, but there would be no unwieldy overarching body like the UN, which is often unable to agree on policies. Second, it would incentivize participation rather than free-riding. The implementation of a penalty on nations that fail to join or welch on the terms will keep nations from joining the group without contributing meaningful emissions reductions (which seems, even at this early stage, to be an emerging problem with the Paris Agreement, as it was with the Kyoto Protocol). Third, a focus on the main emitters will make negotiating and implementing agreements much easier. The top two emitting countries produce 45 percent of emissions, and if they were able to agree on a plan—without the noise of some 195 other countries with voices and votes in the UNFCCC—this would streamline bureaucratic processes markedly.

The economists Paroussos et al. (2019) modeled incentives for participation in climate clubs into climate policy models, assuming that participation in climate clubs could, in addition to preventing cheating through a system of mutual accountability, also give members two important positive benefits: modeling exemptions by climate club members to tariffs, and access to preferential finance terms and technology to low-carbon sectors (for renewables and carbon capture and storage). With positive results, "we find that the key benefits that could structure the club are enhanced technological diffusion and the provision of low-cost climate finance, which reduce investment costs and also enable developing countries to take full advantage of technological diffusion" (Paroussos et al. 2019, 542).

Even with climate clubs, the international will may not presently exist to establish a carbon cost regime with mandatory emissions reductions and real enforcement mechanisms, even in the face of increasingly dire IPCC reports. The UN has been bandying about such ideas for decades now, but foot-dragging governments like the United States may need more California, Montana, and New Mexico wildfires along with Louisiana, Florida, Texas, South Carolina, New Jersey, and Puerto Rico hurricanes before we finally and unequivocally agree that anthropocentric climate change is behind many of these climate shifts. But Millennials and Gen Zers like you are becoming politically active, beginning to vote, and advocating for international and domestic "silver buckshot" policies (remember the Bill McKibben quote from Chapter 1: "There are no silver bullets; only silver buckshot"). You may be able to punish elected climate deniers by prioritizing this issue and moving policymakers toward implementation of the changes we need. In the following sections, we discuss what such policies may look like in terms of deep decarbonization.

The Move to Deep Decarbonization in 2050

Deep decarbonization is the process of removing fossil fuels from the global economy, and it "is a daunting task," according to Bernstein and Hoffman (2018), "as carbon lock-in arises from overlapping technical, political, and social and economic dynamics that generate continuing and taken-for-granted use of fossil energy" (248). For example, if a new and extremely expensive-to-build coal plant is coming online in China this year with an expected use for 40 years, how can that plant contribute to China's efforts to eliminate coal by 2050? Indeed, the 2018 IPCC report offered a dire warning, although Sims Gallagher (2018) does argue that without Paris, things would be a lot worse:

> Putting all the commitments together, scientists have estimated that emissions will be limited to close to current levels with the Paris Agreement, thus avoiding a near doubling of emissions in the absence of the agreement. That's real progress. Countries need to focus on implementing this agreement so that the planned emissions-reductions can be realized. Temperatures would then be limited to approximately 2.5–3.0°C from what they otherwise would have been: 4–5°C. That gets us halfway there, which isn't bad. In fact, it is an excellent start.

Indeed, Sims Gallagher points out that Regional Greenhouse Gas Initiative, a subnational ETS in the northeastern United States, has lowered power sector GHG emissions 36 percent between 2005 and 2015. The United States, more generally, has reduced its GHG emissions by 11 percent between 2005 and 2016, and Europe has lowered its GHG emissions 23 percent between 1990 and 2016. China's emissions seem to have stopped their dramatic increase of the last few decades, with little change over the last three years, ahead of the national schedule to peak emissions in 2030.

MOUNTING SCIENTIFIC PRESSURES TO DECARBONIZE SOON

Still, as per our discussion in Chapter 2, the IPCC took the more pessimistic (or realistic) view that unless emissions are cut by 45 percent below 2010 levels worldwide by 2030, we will surpass the 1.5°C barrier and start experiencing more significant effects of climate change even before 2050. Bernstein and Hoffman (2018) note that some US policymakers have accepted climate change, but view it fatalistically:

> One startling example of this attitude can be seen in July's US Department of Transportation's draft environmental impact statement on future vehicle emission standards, which acknowledged the scientific consensus that without significant action, global temperatures by 2100 will be 4°C higher than preindustrial levels. Instead of framing the accepted science as a call to radical action, it used the stark numbers to justify no policy change, reasoning that the inevitable temperature rise was so great that increasing fuel efficiency standards would make little difference . . .

Contrary to the Trump administration's "ostrich with its head in the sand" approach, much of the world (including the Biden administration) is woke to the fact that we are on a likely course for 3°C of global warming by 2100. One means of addressing this is to embark on deep decarbonization, starting with the burning of fossil fuels in electricity and heating, which cause about a quarter of the world's GHG emissions.

A Deep Decarbonization Pathways Project has been established, consisting of researchers in 16 nations accounting for 70 percent of world emissions. The researchers seek to diminish emissions at a minimal cost, while at the same time continuing to expand coverage on national utility grids to cover the 1.1 billion people (some 15 percent of the world's population) still without access to electricity (Patel 2017). China and India, both among the world's biggest coal consumers, are planning transitions to non–fossil fuel sources, but India is finding that as interest rates for business borrowing drop, more private solar power suppliers are emerging and there is actually a surplus of electricity on the grid. However, "this challenges utilities, which also incur huge losses from a weak grid infrastructure, politicians syphoning power for their constituents, and millions of people not paying for electricity" (Patel 2017, 632).

LEADERS AND LAGGARDS AND THEIR APPROACHES TO DECARBONIZATION

Researchers in Sweden, Germany, Austria, and Switzerland—all consistent climate change policy leaders—as well as in Australia have proposed a "carbon law" that would halve anthropogenic carbon dioxide (CO_2) emissions every decade (Rockström et al. 2017, 1269). Rather than trifling with justice and equity issues (or, perhaps, failing to do so), the authors argue that in these desperate times all sectors (recall the discussion from Chapter 4) and countries must be required to decarbonize, and that carbon capture and storage will have to make up the shortfalls. The strategy "pushes renewables and other zero-emissions technologies up the creation and dissemination trajectory, while simultaneously pulling fossil-based value propositions from the market (Rockström et al. 2017, 1269)."

Even in the short term, implementation of this radical decarbonization formula would require a carbon tax, ETS, feed-in tariffs, and quotas to promote renewables that would need to roll out immediately, with a necessary ratcheting up of each nation's Paris pledges, elimination of all fossil fuel subsidies, and a moratorium on all new investment in coal-based energy generation, with richer, coal-heavy nations leading the departure from coal and other countries, such as India and Indonesia, following (Rockström et al. 2017, 1270). By 2020, the 49 countries committed to carbon neutrality by 2050 would have needed to increase in number to over 100, and given the food industry's possibilities for biological carbon removal, a worldwide plan to sustain healthier and lower-meat diets would have had to have been developed, along with food waste being reduced. Cancellation of the 2020 UNFCCC annual meeting due to COVID left many of these 2020 goals up in the air. But efforts to rachet up mitigation ambitions do continue, and diplomats, nonprofits, and, increasingly, the private sector (through firms like Beyond Beef and USA Solar) continue seeking to hold governments responsible.

In the medium term (between 2030 and 2040), Rockström et al. (2017, 1271) envision carbon pricing worldwide, starting at 50 dollars per metric ton presently and then ramping up to over 400 dollars per ton by 2050, phase-outs by 2030 of fossil fuel–based combustion engine cars (following those already in place in Germany, The Netherlands, and Norway), as well as increased research and development spending on battery-life extension, energy storage, carbon capture and storage, energy efficient industrial production, smart power grid development, the design of alternative aircraft propulsion systems, and the rise of sustainable urban growth (with energy-conserving buildings and transportation). By 2030, the authors estimate that leading decarbonizers like Denmark, Norway, and Sweden should be able to achieve net-zero emissions, and that combustion engines will be replaced by synthesized fuels, biomethane, and hydrogen alternatives (Rockström et al. 2017, 1271). Bioenergy will be combined with carbon capture and storage, direct air carbon capture and storage will be absorbing emissions not reduced through other means, and financial mechanisms will be established to foster needed investments in decarbonization.

Rockström et al. (2017) conclude by arguing that "climate stabilization must be placed on par with economic development, human rights, democracy, and peace," and that therefore "the design and implementation of the carbon roadmap should therefore take center stage at the UN Security Council" (1271)._However, as noted by Scott and Ku (2018, 235), "climate change has for some time now been regarded as a threat multiplier and possible contributor to destabilizing already fragile states, but has not itself been labelled a threat to peace and stability," making it difficult for the UN Security Council to find a point of entry. The authors argue (234–236) that climate change may present multiple security threats, ranging from the disappearance of entire island nations like the Maldives, Kiribati, and Tuvali; exacerbating food and water supply threats and internal migration, such as in South Sudan and the Lake Chad region (and many other parts of the world in time); and in obviating development and economic growth, as extreme weather events have shown time and time again in countries like Honduras and Bangladesh and as slow-onset droughts have shown in Ethiopia and Somalia, to name just a few.

Scott and Ku (2018, 239–243) further envisioned that the UN Security Council could, if the five permanent members—China, France, Russia, the United Kingdom, and the United States—agreed, sanction violators of carbon emissions agreements (adding sorely needed enforcement "teeth" to a less voluntary and more mandatory renegotiation of the Paris Agreement), send peacekeepers to help contain climate-related losses due to extreme weather or sea level rise, pass international climate regulations, or try climate "criminals" in new courts. The Security Council would have to "take" this authority, but precedent does exist, such as in its enforcement, under universal jurisdiction, of the 1949 Geneva Conventions on human rights. If climate stabilization is on par with human rights, then it should be an issue the Security Council could incorporate into its domain. And if decarbonization fails and the world has to turn to "Plan B" (geoengineering), then international oversight will be even more crucial.

"Plan B": Geoengineering, AKA Aerosol Injection and Solar Radiation Management

Geoengineering encompasses a range of techniques for adapting the environment to trap or repel GHG emissions. In Chapter 5, we discussed carbon capture and storage as possible solutions proposed by some advocates. Here, we address sulfate aerosol injection, the process by which sulfate aerosols are shot into the stratosphere to reflect sunlight away from Earth as a sort of protective shield, before it enters the atmosphere where its heat energy would otherwise remain.

This process has occurred in nature, most notably in 1991 eruption of Mount Pinatubo in the Philippines, which sent 20 million tons of sulfur dioxide into the upper atmosphere (the stratosphere) and lowered temperatures worldwide by about 0.5°C between 1991 and 1993 (Montgomery 2019). The volcanic gas had negative effects as well, including reducing global precipitation by 10 percent (and causing drought), increasing acidic precipitation, increasing stratospheric temperature, and changing the wavelengths of light that passes through the aerosol (and perhaps other secondary effects). Given this, rinsing our planet in acid rain and changing global precipitation patterns to keep the temperature from rising does not seem at first thought to be the best decision. Also, if we were to apply the sulfate geoengineering Band-Aid , massive carbon emissions would be more likely to continue. And if we ever had to peel off the Band-Aid, the climate would whiplash quickly back to an even warmer climate as the sun's energy was recaptured.

ETHICAL AND POLITICAL PROS AND CONS OF GEOENGINEERING

In part because of the desperation it conveys, geoengineering was "taboo—too dangerous and too uncertain" (Stilgoe. 2013) to publicly consider until about 2010. But as the severity and permanence of climate change have become more widely accepted, the cons of geoengineering have been bracketed in favor of considering it as a low-cost option, perhaps as cheap as 25 billion dollars in total, as compared to the 1.2 trillion dollars per year (about 1.5 percent of the world's economic output, estimated at some 80 trillion dollars in recent years) for a "suite" of decarbonization policies.

The moral hazard inherent in geoengineering is also considerable. This is the risk that once policymakers resort to sulfate aerosol injection, people will not then support the transition to a lower carbon economy. But while many relegate geoengineering to "Plan B," or "the morning after pill" for climate change, Jamieson (2014) writes, consistent with the "silver buckshot" argument we make in this book, that "the real problem we face is not choosing the right class of responses to climate change but rather in selecting and implementing a portfolio of specific strategies for responding to a problem that is urgent, multidimensional and extended in space and time" (219—220). While we are more sanguine about drinking the geoengineering Kool-Aid before depleting other options, it does bear consideration. At the very least, it carries the prospect of a technological fix, which may be perfectible even if it presently offers only a possible solution, but at a very high cost in terms of impacts.

As mentioned, adding sulfate to the stratosphere is not terribly expensive. For as little as a few hundred million dollars, one could invest in a regional delivery system that would have an impact short term (at which point one would have to reinvest money to keep it going, of course). Some wealthy citizens (or subjects if one is in a kingdom) have openly thought of taking geoengineering into their own hands, as the world governments fight over their competing interests and therefore cannot seem to find an effective path forward. Both Sir Richard Branson and Bill Gates have funded research investigating effective ways of geoengineering, including sulfate aerosol injection.

For Sir Richard, who owns Virgin Atlantic Airways among other aeronautic interests and thus has significant interest in continuing an air travel business, it might be as simple as putting SO_4 aerosol canisters on his airplane fleet. They fly through the stratosphere on long flights anyway. This raises the interesting problem around who is permitted to conduct geoengineering? Nation-states independent of each other? A UN committee of some sort? Private individuals with independent means and a sense of (hopefully well-guided) civic philanthropy? Could one country use the technology against another as some kind of climate change warfare, perhaps by releasing climate-cooling aerosols upwind of an enemy's agricultural zone during the growing season to destroy the crops? Many questions beyond cost and effectiveness of aerosol geoengineering arise quickly when we consider the practice. We are not saying it is not worth exploring, but that it is much more complex than it seems at first glance.

Part of what bothers people, aside from questions about who has the authority to manage the atmosphere, is an ethical aspect of geoengineering the climate. The problem can be summarized this way: We are engaging, with full knowledge and intent, in dramatically altering the Earth's atmosphere to the detriment of almost all living systems (with the exception of the deep ocean). We know this and have the power to stop it, since we are the ones lighting the fossil fuels on fire. It is not as if we are helpless. It seems odd then, that upon examining a range of options for fixing the problem, most of which involve reducing the behavior that started the problem in the first place, we are contemplating even more artificial atmospheric manipulations. Those manipulations have uncertain outcomes, problematic side effects, and dubious respect for national sovereignty. Plus, even if atmospheric geoengineering were adopted as a solution to climate change, it would do nothing with regards to CO_2 increases in the atmosphere and associated problems such as ocean acidification that risk collapsing food webs. There are myriad other ethical issues as well.

Svoboda (2016, 57) identifies several such ethical issues. First, he argues, injecting sulfur dioxide would reduce precipitation in Latin America, Africa, and Southeast Asia, the most vulnerable nations in the world and the ones least responsible for climate change. Second, he references the "termination problem," the fact that the aerosol would "wear out" and need to be re-injected, and that if "this replenishment should cease for any reason (e.g., terrorism, war or malfunction), the result would be extremely rapid global warming due to the anthropogenic greenhouse gases still in the atmosphere" (59) (Figure 8.1). Additionally, by altering agricultural productivity and the availability of freshwater in many areas, the process "picks winners." Finally, Svoboda says that if all impacted people are not consulted, this generates another issue of equity in deliberation and decision-making (60), and really, how could everyone (or even close to everyone) be consulted? For these reasons, the idea of fixing our altered atmosphere by altering it even more sounds fishy to people.

These ethical dilemmas are formidable, and Preston (2016) warns that without an agreement on when to stop solar radiation, the "termination problem" would not be solved and that "the task of maintaining climate would become the art form of a full-time technical elite . . . [that] departs from the rhetoric currently offered" (103). Indeed, geoengineering would be a short-term intervention to stabilize climate so that more conventional decarbonization and other approaches (e.g., carbon capture and storage) could solve the problem more comprehensively and holistically.

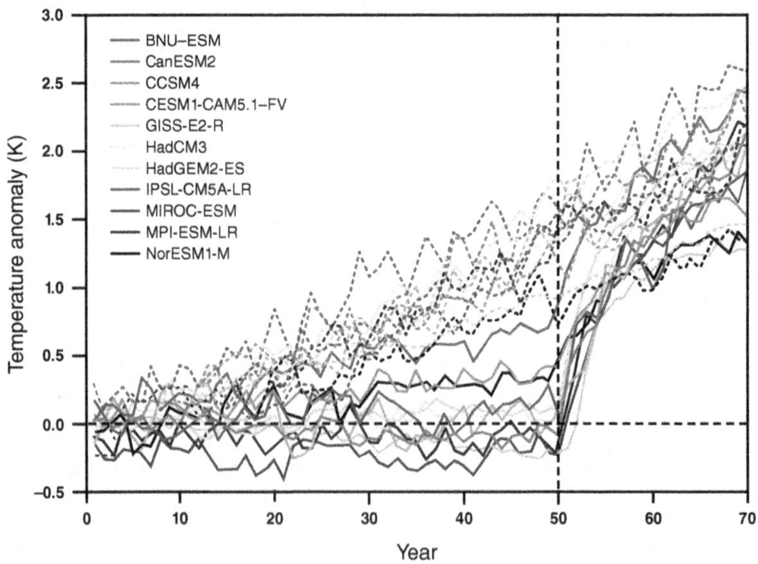

Figure 8.1 Eleven computer model outputs showing the temperature anomalies that would occur if a sulfate aerosol injection program were terminated 50 years from its start (using temperature during the year of initiation set at 0.0). Sources/notes: National Research Council of the National Academies 2015.

ECOLOGICAL PROS AND CONS OF GEOENGINEERING

In addition to the thorny questions around who should be making decisions about intentionally changing the Earth's climate and whether it is ethical to do so, there are serious side effects of injecting sulfate aerosols into the stratosphere. Although some side effects are known and, to a certain extent, predictable, the massive complexity of the atmosphere and our lack of knowledge about some aspects of its response to aerosols (e.g., effects on troposphere/stratosphere interactions), make it likely that unknowns will crop up. As US Secretary of Defense Donald Rumsfeld famously said about a different complex system: "There are known knowns. These are things we know that we know. There are known unknowns. That is to say, there are things that we know we don't know. But there are also unknown unknowns. There are things we don't know we don't know" (Rumsfeld 2002)." Mr. Rumsfeld could have been discussing atmospheric geoengineering and its ramifications.

All that having been said, let us look at some of the issues just in terms of climate science and atmospheric chemistry. Atmospheric geoengineering generally means modifying the way the atmosphere reflects incoming radiation so that it does not contribute to heat energy building up in the troposphere. This can be accomplished by increasing the "albedo" (reflectivity) of clouds or by increasing sulfate aerosols, which scatter incoming radiation as it passes through the stratosphere. We will focus on the aerosols, because those are receiving the most attention and have proven to be effective at cooling the planet.

As mentioned, changes in amounts and patterns of precipitation would result from injecting enough SO_4 aerosol to curtail warming. Mount Pinatubo injected 20 million tons of SO_4, after which the planet cooled by 0.5°C over three years and global precipitation decreased 10 percent, but neither of these effects were uniformly distributed. Precipitation is affected because the aerosols reflect the sun's rays back to space, so there is less energy to evaporate water at the earth's surface (about one meter of seawater evaporates each year normally) (Figure 8.2). To have any effective cooling in a climate management program, one million tons of SO_4 would have to be delivered by aircraft every year (National Research Council of the National Academies [NRCNA] 2015). As observed with the Mount Pinatubo eruption, however, the cooling effects of aerosol injections would not be uniform, nor would be the changes in precipitation.

While models have projected that the warming effects of doubling CO_2 can be cut in half with sulfur aerosol geoengineering (Irvine et al. 2019), scientists disagree on how much reduction would actually be achieved, and where that reduction would occur. Some models show that less than 0.5 percent of the terrestrial non-ice surface would experience changes if aerosols

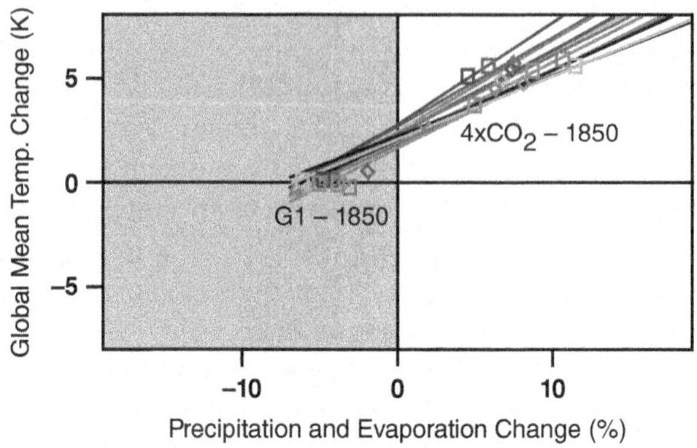

Figure 8.2 Worldwide mean precipitation and temperature modeled from the Geoengineering Model Intercomparison Project (which models the effects of solar dimming on temperature and precipitation). Here we assess G1 versus the preindustrial state. Each symbol represents an individual model run. The G1 scenario adds enough sulfate to maintain temperatures as they were in 1850 (preindustrial). The 4xCO_2 scenario models unmodified (by sulfate aerosols) atmosphere when atmospheric CO_2 in 1850 is quadrupled. Source: National Research Council of the National Academies 2015.

were used to reduce future warming by 50 percent (Irvine et al. 2019). As has been shown over and over again with climate change effects, however, regional differences can be dramatic. For example, models show that areas of the Arctic would still experience an approximate 3°C warming in a scenario that quadrupoles CO_2 even if there were stratospheric aerosol injections to keep overall warming close to 0°C (NRCNA 2015). Additionally, India's and China's precipitation and temperature futures have been modeled out to 2050 under a solar reduction plan via sulfate aerosol. These model results show both precipitation and warming changes that are very different, even though the countries are in the same region of the planet (Figure 8.3). Both countries experience cooling, but China also experiences lower amounts of precipitation.

With these scenarios, doubling CO_2 is modeled to increase temperatures 5°C relative to 1850. If enough aerosol is released into the stratosphere to compensate, precipitation decreases six to eight percent. The goal of the model was to reduce surface temperature, and the precipitation change is a side effect. The radiation balance could be adjusted to allow 1 or 2 K (or °C) of warming, and precipitation would be less effected. For any aerosol solar radiation management (SRM) program, stratosphere concentrations would have to be maintained of course (NRCNA 2015, 52).

There are other risks and side effects aside from changing precipitation or uneven cooling associated with stratospheric aerosol enhancement that are

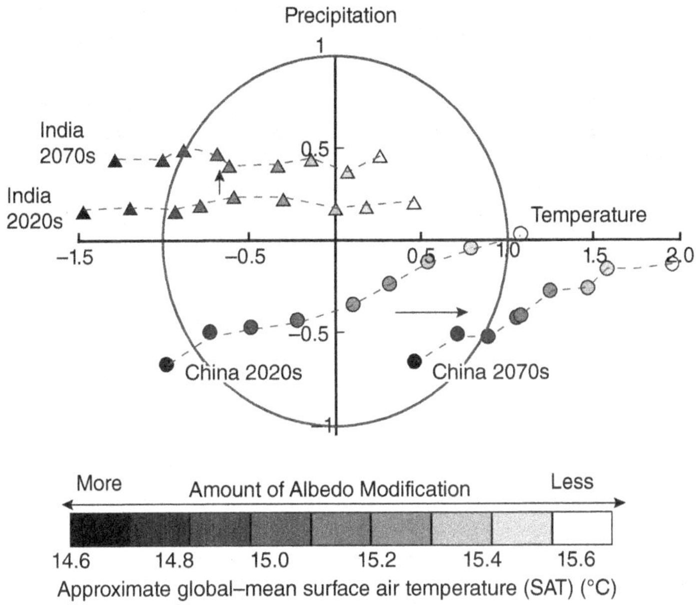

Figure 8.3 Modeled precipitation and temperature changes following aerosol injections to increase albedo (reflectivity). Axes are anomalies in interannual variability normalized for regional temperature and precipitation. Red and blue are lower and greater albedo modification, respectively. China and India are shown for 2020 and 2070 minus 1990 amounts of precipitation and warming. Both countries experience cooling, but China also experiences lower amounts of precipitation. Sources/notes: Ricke et al. 2010.

less well understood. To borrow from Secretary Rumsfeld again, these are among the "known unknowns":

1. *Ozone will be lost:* Ozone protects the planet's surface from cancer causing ultraviolet C radiation, and it is created in the upper atmosphere by interactions between O_2 and sunlight. Since SRM restricts sunlight, it will impact ozone formation.
2. *Stratospheric temperature will rise:* The radiation scattered by the sulfate aerosols will add heat to the stratosphere even as the troposphere cools. The effects of this are unknown, but it is reasonable to expect some changes to global air circulation patterns and in the rate of chemical reactions in the stratosphere.
3. *Ocean circulation patterns will change:* Uneven cooling around the planet (and warming at the poles) would create different patterns of cooler and warmer oceans, altering current circulation patterns. The effects are, again, unknown, but it is likely that current areas of nutrient-rich upwelling will

shift or disappear, and that the speed of ocean basin circulation (called gyres) will be impacted.

4. *Changes will occur in the characteristics of sunlight:* Sulfate aerosols scatter the wavelengths of sunlight in the visible spectrum (380–700 nm), which means less of that light will reach the troposphere. This may have an impact on photosynthesis, depending on how much radiation is scattered.
5. *Changes will occur in stratospheric chemistry:* The chemical reactions that will take place as the sulfate is constantly injected into the stratosphere are not well understood, although some of this sulfate will certainly become incorporated with water to make sulfuric acid rain.
6. *Oceans will acidify:* Artificially reducing of tropospheric warming to compensate for increasing CO_2 will not stop the negative effects of the rising CO_2 itself.

As just mentioned in the list, a major side effect of rising CO_2 is the acidification of the oceans, which have already dropped more than 0.1 pH unit since 1800 (ocean pH has only changed ±0.1 pH unit over the past million years or so). This increase in acidity makes it more difficult for large groups of plankton that use calcium carbonate in their shells (called "tests") to grow. These plankton form the base of large marine food webs, which will shrink (and produce less fish for humans). These are some of the major concerns regarding geoengineering through stratospheric sulfate aerosol that need more study (the known unknowns). As we examine these concerns, we may become aware of some of the unknown unknowns, of which there are undoubtedly a few.

IS SRM PART OF THE CLIMATE SOLUTIONS TOOLBOX?

What if we must undertake Solar Radiation Management (SRM), or at least do more research to sufficiently consider the possibilities? Nicholson et al. (2018, 323) advocate for creating a governing structure to regulate SRM management, encourage and legitimize valuable research through engagement with the public, and ensure that SRM is only one part of a more comprehensive policy agenda. To achieve these objectives, they propose developing a transparent mechanism for managing research, a global forum to inform the public regarding the option, and a means of considering SRM in the Paris Agreement "report back" to Paris member nations scheduled every five years starting in 2023 (Nicholson et al. 2018, 324–326).

Geoengineering is still early in its development, and there are legitimate concerns that this newest technology will be viewed as a "silver bullet," leading us to grow complacent in allowing technology to save us at the last minute rather than striving to address the climate threat through other, more disciplined means. Keary (2016, 24), in arguing against modeling technological change into climate models, endorses a universal carbon tax, which even with its political difficulties "would be several orders of magnitude less" difficult to implement. Implementation of geo-engineering would offer greater challenges because of the special interests that would intercede, the uncertainty of effectiveness and side effects introduced by geo-engineering, and the questions about if and when humanity could ever move beyond dependency on the new, technological elite who would start with disproportionate power which would only grow over time. Let us retreat now from the precipice of a dystopian science fiction novel and move back to more concrete and less risky alternatives.

In the final and concluding chapter, we will consider what individuals can and should do to help stem the advance of climate change. While we each have a role to play, we also caution against making this about what individuals can do. Governments, corporations, interest groups, and social movements are best equipped to diminish the climate crisis. Our main obligation is to pressure for change.

Exercise 8.1 Introduction to the Broken Cities Game on Urban Climate

One of the major failings of the climate movement has been to rely on dry science and reports to communicate the urgency of climate impacts. Many of the best-remembered movements in American history came to a tipping point when the public engaged with the facts through storytelling or other non-scientific means, like with Rachel Carson and the "Silent Spring." Likewise, there are small ways that we can help our minds better understand and conceive of climate change in everyday life, including playing games.

The "Broken Cities" game, developed by designers Pablo Suarez and Janot Mendler de Suarez (https://www.helsinki.fi/sites/default/files/atoms/files/uep_broken_cities.pdf), is a board game that attempts to deal with the challenges of sustainable development facing cities around the world. Cities consume large amounts of the world's energy, contain most of the world's people, and will continue grow. As such, they are a nexus for climate-related decisions, dealing with important trade-offs concerning development and sustainability. The "Broken Cities" game helps us conceptualize of these trade-offs.

While most of us may believe that all technologies should be adopted right away in order to reduce GHG emissions, "the day-to-day decision space for most planners working with climate change is constrained by

Continued

insufficient resources, political limitations, and legal/regulatory impediments" (Juhola et al. 2013). As cities try to plan for the increasing effects of climate change and attempt to reduce their global impact, they face many challenges in weighing how best to accomplish the most good for the least cost.

Using games to help us better understand these challenges has great potential for better communicating the complexity of the climate crisis. Since the 1970s, researchers and urban planners have used games to understand these highly interconnected problems (Juhola et al 2013), and studies have shown increased learning rates among participants (Poplin 2012).

The game play for "Broken Cities" involves four individual players (or four teams) on a single board. Each "City" has its own area with natural features, and players using money tokens to build up their cities. The goal of the game is to reach a certain income level, although there will be complications along the way as other player cities develop and rising emissions wreak climate havoc upon the entire board. The game encourages a large amount of improvisation, and much of the information is left out intentionally. This allows players (and even your whole class!) to bring as much or as little complexity as you are comfortable with into the game. Can your "world" develop sustainably, or will the pursuit of wealth drive everyone to disaster?

PREPARATION AND GAME FACILITATION

Game pieces can be downloaded from https://www.helsinki.fi/sites/default/files/atoms/files/uep_broken_cities.pdf

Setup

1. Practice facilitation ahead of time with a "safe" audience (family, friends, etc.) to discover what can go wrong before playing with an audience where it's not OK to make mistakes.
2. Check materials: Are all the pieces prepared? Do you have makeshift money for players?
3. Arrange tables, chairs, and materials ahead of time.
4. Groups: Consider gender balance.
5. Anticipate potential disruptors (e.g., people with authority, preexisting tensions within group, problems with sensitive topics, and timing of session).
6. Optional: It may be useful to have a board fully set up in advance to show how each team should set up their board and how the parts work if time is a problem.

Introduction and game rules

- Introduce facilitators, and try to maintain a warm, light atmosphere.
- Show all players the demonstration board if you choose to set one up. Familiarize them with the pieces and their functions.
- The game is deliberately simpler than reality. Tell players: "Please do not question the rules."
- Explain rules in the right sequence, and with a clear narrative, display the materials. Tell players that like in the real world, you may not have enough time to think. Just decide and do your best. Be prepared for time pressure.
- Establish an atmosphere of serious but fun gameplay. Confusion is part of the fun of discovery. Tell players: "Confusion is natural. Don't worry; as you play the game, it will become clearer".
- Respond to questions on game rules only. Do not give away strategy.

Facilitation

- Explain objectives clearly and briefly; start the game as soon as possible.
- Manage your time well: Let people learn, but if someone is too slow, then keep going.
- Highlight instances of gameplay where key learning and dialogue opportunities emerge.
- Have a plan on how to shorten the game if time runs out (shorter rounds, eliminate complexity, etc.).
- If something unexpected happens, be flexible; try to think of alternative solutions! The show must go on. Keep the "end game" (objective) in focus. Keep the game flowing, using humor to deal with surprises.

Debrief

- Winners: Ensure swift and correct math to calculate the winner, and award prizes.
- Ask key questions about gameplay experience: What lessons were learned about the theme of the game?
- Share key aspects of particular theme not covered in gameplay, and encourage questions. Remind players that the game is not really the world (if time allows, ask for important differences between gameplay and reality).
- Give players opportunity to discuss how they would modify the game to enhance the learning and dialogue experience.

MODIFICATIONS

A full set of instructions is included in the link (https://www.helsinki.fi/sites/default/files/atoms/files/uep_broken_cities.pdf), and a set of proposed

Continued

alterations for making the game more difficult (if you find it too easy to make it to the end of the game) is presented here.

Instead of having the atmospheric effects constantly proceeding along the effects trail, anything below the 56-point mark for emissions would allow the atmospheric effects marker to move backward instead of forward. Board alterations would be needed to change the points in the triangle of spaces moved for atmospheric effect to be negative if below 56. Options include:

- *Create levels of climate effects:* On the atmospheric effect "trail" of clouds, create three levels—the first is the first few clouds, containing just a couple of red clouds; the second level (after a fair amount of warming has occurred) has a healthy mix of red and white clouds; and then the third level (the most warming) has even more red clouds.
- *Create mixed levels of climate effects:* Similar to creating levels of climate effects, but instead of exclusively red clouds, add in some yellow clouds (with colored construction paper) to create half-effects throughout the trail. In this version, the Climate Change Events card would do an estimated half the usual amount of damage (e.g., "lose all your properties that are next to water" becomes "lose half your properties next to water" or "downgrade all your properties next to water," depending on what the group chooses).
- *Red cloud effects:* Every time you pass a red cloud on the atmospheric effects trail, an effect happens.

NOTES

1. According to authorities, aspiring citizens of Dominica pay tens of thousands of dollars and submit to an extensive background check, or—and here the tourism infrastructure does come into play—they invest considerably more in island developments, some of which are tourist hotel "time shares" in which the new citizens invest for several years.
2. This term refers to payoffs based not on "automatic payments" (subtracting a deductible) for damages, but rather payment of a fixed sum at a threshold of severity in catastrophic weather (e.g., landfall of a Category 4 hurricane, or seismic activity reaching a pre-prescribed Richter ccale measure).
3. Statement at Understanding Risk Caribbean Conference in Barbados, May 29, 2019, sponsored by the World Bank, the Caribbean Disaster Emergency Management Agency, the European Union, the Global Facility for Disaster Reduction and Recovery, and Barbados (national government).
4. Humble analysts estimating climate finance expenditures insist that their numbers are extremely rough, require many assumptions, and should be taken with a pellet (not just a grain) of salt. Ever since the Copenhagen 2009 "100 billion dollars for climate change by 2020" pledge was made (and not kept) by leading UNFCCC rich-country members, critics have assailed the lack of transparency, or even accounting mechanisms, to calculate nations' contributions to climate finance under the Paris Agreement.
5. These totals relied on coding of project descriptions into dozens of categories, including adaptation, mitigation, and disaster relief (closely related to adaptation) coordinated by Knowledge Management Analyst Harsh Anuj.

6. We must reiterate endnote 4.
7. This line is from the acclaimed 1989 Kevin Costner movie *Field of Dreams*, in which Costner builds a baseball field on his Iowa cornfield to draw magical baseball teams to play there.
8. Australia vehemently opposed the creation of this facility before and during the COP 21. See Milman 2015 and Karsapan 2015.
9. The dominant geopolitical strategy of international relations realists at the time was that the West (or First World, led by the United States) and the East (or Second World, led by the then-Soviet Union) competed for what President Kennedy called "the hearts and minds" of the rest of the world (or Third World). Struggles between East and West were waged in places like Vietnam, Korea, Cuba, Chile, and Egypt, whereas the emerging front lines of climate change seem to be small island nations in the Caribbean and Pacific, Central America, the Sahel region of Africa, and South Asia. But these vulnerable nations are also part of the Third World.
10. Most agree the October 1962 Cuban Missile Crisis, a white-knuckle stand-off between US President John F. Kennedy and Soviet Union Premier Nikita Khrushchev, may have been the world's most dangerous moment since World War II—if not ever. It was resolved, after a 13-day impasse including entire days without direct communications between the parties, by Soviet removal of their nuclear armed missiles from Cuba in exchange for US removal of nuclear armed missiles from Turkey.

REFERENCES

Adejonwo-Osho, Oluwatoyin interview. October 8, 2019, African Climate Risks Conference Addis Ababa Ethiopia. Nigerian environmental and human rights lawyer.

Artemis. 2017. "Jamaica questions CCRIF model after floods fail to trigger policy." Accessed June 25, 2019, at https://www.artemis.bm/news/jamaica-questions-ccrif-model-after-floods-fail-to-trigger-policy/

Bernstein, Steven, and Matthew Hoffman. 2018. "What the new IPCC report can't say." *MIT Press*, October 17. Accessed August 5, 2019, at https://mitpress.mit.edu/blog/what-new-ipcc-report-cant-say

Broberg, Morten. 2019. "Parametric loss and damage insurance schemes as a means to enhance climate change resilience in developing countries." *Climate Policy*. DOI: 10.1080/14693062.2019.1641461

CDP. 2017. "New report shows just 100 companies are source of over 70% of emissions." Accessed January 7, 2021 at https://www.cdp.net/en/articles/media/new-report-shows-just-100-companies-are-source-of-over-70-of-emissions

Climate Policy Initiative. 2018. "Global climate finance: An updated view 2018." Accessed July 12, 2019, at https://climatepolicyinitiative.org/publication/global-climate-finance-an-updated-view-2018/

Curtain, Richard and Matthew Dornan. 2019. "A pressure release valve? Migration and climate change in Kiribati, Nauru and Tuvalu." Working paper of Development Policy Centre of Crawford School of Public Policy, The Australian National University. Accessed January 6, 2021 at https://devpolicy.org/publications/reports/Migration-climate%20change-Kiribati-Nauru-Tuvalu.pdf

Edwards, Rosamund interview. June 1, 2019. Roseau, Dominica. Finance Ministry Permanent Secretary, Government of Dominica.

Eisenstadt, Todd A. 2017. "COP-23: Can More Transparency, New Technology Save Small Island States?" *New Security Beat* blog published November 16. Accessed January 6, 2021 at https://www.newsecuritybeat.org/2017/11/cop-23-transparency-technology-save-small-island-states/.

Eisenstadt, Todd A. 2015. "Giving With One Hand and Taking With the Other: How Fossil Fuel Subsidies Undermine "Climate Finance" Fund Donations Pledged in Run-up to Paris UN Negotiation" December 8 *Huffington Post* blog. Accessed January 6, 2021 at https://www.huffpost.com/entry/giving-with-one-hand-and-_b_8751530?utm_hp_ref=climate-change

Eisenstadt, Todd A., Daniel Fiorino, and Daniela Stevens. 2019. "National Environmental Policies as Shelter from the Storm: Specifying the Relationship Between Extreme Weather Vulnerability and National Environmental Performance," in *Journal of Environmental Studies and Sciences* 9:1 (March: 96-107). DOI: 10.1007/s13412-018-0523-4

Eisenstadt, Todd A., Tawfique Haque, Michael A. Toman, and Matthew Wright. 2020. "Bureaucracy's "Black Box." In "Bangladesh: Undifferentiated Blame Attribution by the Climate Vulnerable Among Levels of Government," typescript.

Eisenstadt, Todd A., Ifeoluwa Olawole, and Michael Toman. 2020. "Measuring International Funding for Climate Change Adaptation in Economic Development Programs," typescript.

Eisenstadt, Todd A., and Karleen Jones West. 2019. *Who Speaks for Nature? Indigenous Movements, Public Opinion, and the Petro-State in Ecuador*. Oxford University Press, New York.

Erickson, Peter, Michael Lazarus, and Georgia Piggot. 2018. "Limiting fossil fuel production as the next big step in climate policy." *Nature Climate Change* 8:1037-1043.

Höhne, Niklas, Takeshi Kuramochi, Carsten Warnecke, Frauke Röser, Hanna Fekete, Markus Hagemann, Thomas Day, Ritika Tewari, Marie Kurdziel, Sebastian Sterl & Sofia Gonzales. 2017. "The Paris Agreement: resolving the inconsistency between global goals and national contributions,"in *Climate Policy*, 17:1, 16-32, DOI: 10.1080/14693062.2016.1218320

Holthaus, Eric. 2016. "The rising tide." Sabin Center for Environmental Studies, Columbia Law School. Accessed July 26, 2016, at http://www.law.columbia.edu/magazine/613512/the-rising-tide?platform=hootsuite#.Vx559VDQ9Cs.twitter

Intergovernmental Panel on Climate Change (IPCC). 2007. *Climate Change 2007: The Physical Science Basis. Contribution of Working Group I to the Fourth Assessment Report of the Intergovernmental Panel on Climate Change*, edited by S. Solomon, D. Qin, M. Manning, Z. Chen, M. Marquis, K. B. Averyt, M. Tignor, and H. L. Miller. New York: Cambridge University Press.

Intergovernmental Panel on Climate Change (IPCC). 2018. IPCC, 2018: Global Warming of 1.5°C. An IPCC Special Report on the impacts of global warming of 1.5°C above pre-industrial levels and related global greenhouse gas emission pathways, in the context of strengthening the global response to the threat of climate change, sustainable development, and efforts to eradicate poverty [Masson-Delmotte, V., P. Zhai, H.-O. Pörtner, D. Roberts, J. Skea, P.R. Shukla, A. Pirani, W. Moufouma-Okia, C. Péan, R. Pidcock, S. Connors, J.B.R. Matthews, Y. Chen, X. Zhou, M.I. Gomis, E. Lonnoy, T. Maycock, M. Tignor, and T. Waterfield (eds.)].

International Organization for Migration. 2018. "Migration and climate change." Accessed August 4, 2019, at https://www.iom.int/migration-and-climate-change-0

Irvine, Peter,Kerry Emanuel, Jie He, Larry W. Horowitz, Gabriel Vecchi & David Keith. 2019. "Halving warming with idealized solar geoengineering moderates key climate hazards." *Nature Climate Change* 9:295–299.

Jamieson, Dale. 2014. *Reason in a Dark Time: Why the Struggle Against Climate Change Failed—and What it Means for Our Future*. Oxford University Press, New York.

Juhola, Sirkku, Patrick Driscoll, Janot Mendler de Suarez, and Pablo Suarez 2013. "Social strategy games in communicating trade-offs between mitigation and adaptation in cities." *Urban Climate* 4:102–116. DOI: 10.1016/j.uclim.2013.04.003.

Kälin, Walter. 2010. "Report of the Representative of the Secretary-General on the human rights of internally displaced persons." UN Doc. A/HRC/13/21, January 5. Accessed January 11, 2021, at https://www.brookings.edu/research/report-of-the-representative-of-the-secretary-general-on-the-human-rights-of-internally-displaced-persons-2/

Karsapan, Omer. 2015. "Refugees: Displaced from the Paris climate change agreement?" Brookings Institution, December 7. Accessed July 26, 2016, at http://www.brookings.edu/blogs/future-development/posts/2015/12/07-refugees-climate-change-cop21-karasapan

Keary, Michael. 2016. "The new prometheans: Technological optimism in climate change mitigation modelling." *Environmental Values* 25:7–28.

Keohane, Robert O., and David G. Victor. 2016. "Cooperation and discord in global climate policy." *Nature Climate Change*. DOI: 10.1038/NCLIMATE2937

Kofe, Siliga, chief of Funafuti Island, Tuvalu, interview at 2017 Bonn meeting of the United Nations Framework Convention on Climate Change, November 15, 2017.

Kumari Rigaud, Kanta, de Sherbinin, Alex; Jones, Bryan; Bergmann, Jonas; Clement, Viviane; Ober, Kayly; Schewe, Jacob; Adamo, Susana; McCusker, Brent; Heuser, Silke; Midgley, Amelia. 2018. *Groundswell: Preparing for Internal Climate Migration*. Washington, DC: World Bank, Washington, DC. https://openknowledge.worldbank.org/handle/10986/29461.

Martin, Pamela. 2015. "Leaving oil under the Amazon: The Yasuni-ITT Initiative as a postpetroleum model?" In *Ending the Fossil Fuel Era,* edited by Thomas Princen, Jack P. Manno, and Pamela L. Martin. Massachusetts Institute of Technology Press, Cambridge, MA. pp. 119–144.

McAllister, Adam interview. May 29, 2019. Bridgetown, Barbados. CEO of McAllister & Craig Disaster Management.

McDaniels, Timothy, Stephanie Chang, Darren Cole, Joseph Mikawoz, and Holly Longstaff. 2008. "Fostering resilience to extreme events within infrastructure systems: Characterizing decision contexts for mitigation and adaptation." *Global Environmental Change* 18(2): 310–318.

McMickle, Marvin. 2014. "Web Essay: The passion of individuals," in *Rochester Democrat and Chronicle*, June 4. Accessed January 8, 2021, at https://www.democratandchronicle.com/story/opinion/guest-column/2014/06/04/web-essay-passion-individuals/9925707/

McNamara, Karen E., Bronen, Robin, Fernando, Nishara, and Klepp, Silja. 2018. "The complex decision-making of climate induced relocation: adaptation and loss and damage." *Climate Policy* 18(1):111–117.

Milman, Oliver. 2015. "UN drops plan to help move climate-change affected people." *The Guardian*, October 7. Accessed August 18, 2016, at https://www.theguardian.com/environment/2015/oct/07/un-drops-plan-to-create-group-to-relocate-climate-change-affected-people

Montgomery, Carla. 2019. *Environmental Geology.* Mcgraw-Hill New York, New York. ISBN: 9780078022951

Mottley, Mia. 2019. Speech to inaugurate the "Understanding Risk (UR) Caribbean Conference," May 27, Bridgetown, Barbados. Accessed July 1, 2019, at https://www.youtube.com/watch?v=-ZGp9qohcu8

Nachmany, Michal. 2017. *Global Trends in Climate Change Legislation and Litigation.* Grantham Research Institute on Climate Change and the Environment, London. pp. 1–34.

Nachmany, Michal. 2018. *Global Trends in Climate Change Legislation and Litigation: 2018 Snapshot.* Grantham Research Institute on Climate Change and the Environment and Centre for Climate Change Economics and Policy, London. pp. 1–8.

National Research Council of the National Academies (NRCNA). 2015. *Climate Intervention: Reflecting Sunlight to Cool Earth.* National Academies Press, Washington, DC.

Newell, Peter, and Andrew Simms. 2019. "Towards a fossil fuel non-proliferation treaty." *Climate Policy.* DOI: 10.1080/14693062.2019.1636759

Ngombalu, Janet interview October 8, 2019 at African Climate Risks Conference, Addis Ababa, Ethiopia.

Nicholson, Simon, Sikina Jinnah, and Alexander Gillespie. 2018. "Solar radiation management: A proposal for immediate polycentric governance." *Climate Policy* 18(3):322–334.

Nordhaus, William. 2017. "Climate clubs and carbon pricing." In *Global Carbon Pricing: The Path to Climate Cooperation,* edited by Peter Cramton, David JC MacKay, Axel Ockenfels and Steven Stoft. The MIT Press, Cambridge, MA. pp. 109–124.

Palmer, Lisa. 2013. "Providing aid before climate disaster strike." *Climate Change* 3:857–858.

Paroussos, Leonidas, Antoine Mandel, Kostas Fragkiadakis, Panagiotis Fragkos, Jochen Hinkel and Zoi Vrontisi, 2019. "Climate clubs and the macro-economic benefits of international cooperation on climate policy," in *Nature Climate Change, Nature Publishing Group*, 9(7), pp. 542–546. DOI: 10.1038/s41558-019-0501-1. Accessed January 7, 2021, at https://ideas.repec.org/p/hal/journl/halshs-02334592.html

Patel, Prachi. 2017. "Deep decarbonization faces deep challenges." *Energy Quarterly* 42(9): 632–633.

Pauw, W. P., P. Catsro, J. Pickering, and Shikha Bhasin. 2019. "Conditional Nationally Determined Contributions in the Paris Agreement: Foothold for equity or Achillies heel?" Climatic Policy, 143: 503. https://doi.org/10.1080/14693062.2019.1635874

Preston, Christopher J. 2016. "Climate engineering and the cessation requirement: The ethics of a life-cycle." *Environmental Values* 25: 91–107.

Prevost, Careen interview. June 5, 2019. Roseau, Dominica. Executive Director of Climate Resilient Execution Agency for Dominica.

Pollin, Robert. 2015. *Greening the Global Economy*. The MIT Press, Cambridge, MA.

Poplin, Alenka. 2012. "Playful public participation in urban planning: A case study for online serious games." *Computers, Environment and Urban Systems* 36(3): 195–206. DOI: 10.1016/j.compenvurbsys.2011.10.003

Ricke, Katharine L., Granger M. Morgan, and Myles R. Allen. 2010. "Regional climate response to solar-radiation management." *Nature Geoscience* 3(8): 537–541.

Rockström, Johan, Owen Gaffney, Joeri Rogelj, Malte Meinshausen, Nebojsa Nakicenovic and Hans Joachim Schellnhuber. 2017. "A roadmap for rapid decarbonization: Emissions inevitably approach zero with a 'carbon law." *Science* 355(6331):1269–1271.

Rolnik, Raquel. 2009. "Report of the Special Rapporteur on adequate housing as a component of the right to an adequate standard of living, and on the right to non-discrimination in this context." *UN Doc.* A/62/25, August 6. Accessed January 11, 2021 at https://www.ohchr.org/en/issues/housing/pages/housingindex.aspx

Roy, Eleanor Ainge. 2019. "'One day we'll disappear': Tuvalu's sinking islands," in *The Guardian*, May 16. Accessed. Accessed January 6, 2021 at https://www.theguardian.com/global-development/2019/may/16/one-day-disappear-tuvalu-sinking-islands-rising-seas-climate-change

Schäfer, Laura, Vera Künzel and Christoph Bals. 2018. "The significance of climate litigation for the political debate on loss & damage." *Germanwatch*. Accessed December 7, 2020, at www.germanwatch.org/en/15104.

Scott, Shirley V., and Charlotte Ku. 2018. *Climate Change and the UN Security Council*. Edward Elgar Publishing Limited, Cheltenham, UK.

Setzer, Joana and Rebecca Byrnes. 2020. Global trends in climate change litigation: 2020 snapshot," policy paper. London: Grantham Research Institute on Climate Change and the Environment, London School of Economics.0020Accessed January 7, 2021, at https://www.lse.ac.uk/granthaminstitute/wp-content/uploads/2020/07/Global-trends-in-climate-change-litigation_2020-snapshot.pdf

Shermer, Michael. 2005. "Rumsfeld's Wisdom: Where the known meets the unknown is where science begins," in *Scientific American* online (September 1). Accessed January 7, 2021, at https://www.scientificamerican.com/article/rumsfelds-wisdom/

Sims Gallagher, Kelly. 2018. "Virtuous climate cycles needed." *MIT Press*, October 17. Accessed August 5, 2019, at https://mitpress.mit.edu/blog/virtuous-climate-cycles-needed

Skerrit, Roosevelt. 2018. *The National Budget of the Commonwealth of Dominica, Fiscal Year 2018–2019*. Government Printery, Roseau, Dominica.

Sovacool, Benjamin J., and Bjorn-Ola Linner. 2016. *The Political Economy of Climate Change Adaptation*. Palgrave-McMillan (St Martin's Press), New York.

Stilgoe, Jack. 2013. "Why has geoengineering been legitimised by the IPCC?" in *The Guardian*, September 27. Accessed January 12, 2021 at https://www.theguardian.com/science/political-science/2013/sep/27/science-policy1

Svoboda, Toby. 2016. "Aerosol geoengineering deployment and fairness." *Environmental Values* 25: 51–68.

UN News Agency. 2017. "'To deny climate change is to deny a truth we have just lived' says Prime Minister of storm-hit Dominica." Accessed June 25, 2019, at https://news.un.org/en/story/2017/09/566742-deny-climate-change-deny-truth-we-have-just-lived-says-prime-minister-storm-hit

UN Newsroom, 2016. "New data brings new answers on climate migration experts set out latest advances in Bonn." Accessed July 26, 2016, at http://newsroom.unfccc.int/unfccc-newsroom/human-mobility-and-the-paris-agreement/

Vincent, Katharine, Stéphanie Besson, Tracy Cull, and Carola Menzel. 2018. "Sovereign insurance to incentivize the shift from disaster response to adaptation to climate change—African Risk Capacity's Extreme Climate Facility." *Climate and Development* 10(5): 385–388.

Wentz, Jessica, and Michael Burger. 2015. "Designing a climate change displacement coordination facility: Key issues for COP 21." Sabin Center for Environmental Studies, Columbia Law School. Accessed July 28, 2016, at https://web.law.columbia.edu/sites/default/files/microsites/climate-change/unfccc_climate_change_displacement_coordination_facility.pdf

World Bank Group. 2018. *Disaster Risk Management in the Caribbean: The World Bank's Approaches and Instruments for Recovery and Resilience.* The World Bank, Washington, DC.

Wright, Matthew, Todd A. Eisenstadt, Tawfique Haque, and Michael Toman. 2020. "Economic vulnerability as a primary cause of climate migration: Evidence from Bangladesh." Typescript.

Case Study 8.1

CAN A GERMAN UTILITY COMPANY BE SUED FOR FLOODING DAMAGES IN THE PERUVIAN ANDES? ASSESSING THE EVIDENCE OF RWE VERSUS HUARAZ

LAUREL WOLF[1]

Just before Christmas 1941, the residents of Huaraz, a small provincial capital and agricultural hub nestled in the Andes of Peru, experienced a catastrophe. Lake Palcacocha, which is filled by runoff from glaciers, sits high above the town at the top of a valley carved deeply into the mountains by centuries of erosion. The lake is held back by a natural dam called a moraine, and on the morning of December 13th, a huge chunk of glacier broke off and tumbled into the lake, causing a wave of ice, boulders, water, and mud to breach the dam and careen down the valley to the town below. Within 15 minutes, the mudslide reached the city, completely destroyed almost a third of the buildings, and killed around 5,000 people (Carey 2010).

The city of Huaraz rebuilt, sometimes on top of the rubble (Jarvis 2019), and the Peruvian government reinforced the dam with additional stabilizing features and drainage pipes to keep the water level low (Somos-Valenzuela et al. 2016, 2543). That security made way for economic prosperity, and as a result, the city of Huaraz has grown to its current size of 120,000. But since that time, anthropogenic climate change has shifted the calculations that keep the city of Huaraz safe from another disaster like that in 1941. Almost 50,000 residents now live directly in the path of a future flood, and the lake has already grown fourfold since 2003 thanks to climate change (Darby 2016).

Saúl Luciano Lliuya is one of those who has built his home and life in the shadow of the 1941 flood. The house, built by his father, is in the "Nueva Florida" neighborhood, where residents still find huge boulders deposited by the 1941 flood if they dig into the ground. "If I said, 'It's a flood zone, I won't fix the house,' that would look crazy," Lliuya mused. "But it's also crazy to do it knowing the danger, no? From both sides, it's crazy" (Jarvis 2019). Lliuya, a father of two, is a farmer as well as a mountain guide in the tourist season, professions passed down to him from his father, just like the house he lives in.

Lliuya is acutely aware of the dangers of a potential flood, particularly in the Andes, where climate change is making extreme seasonal fluctuations 1.5 times more likely (Christidis et al. 2018). A friend connected him to the German NGO Germanwatch, who suggested that he bring his case against the German energy producer Rheinisch-Westfälisches Elektrizitätswerk (RWE). The company doesn't operate in Peru, but they are mentioned as a "carbon major" in a recent study outlining the top historical contributors to climate change. RWE, according to this study, has emitted 0.47 percent of total global greenhouse-gas (GHG) emissions (Heede 2013).

Lliuya and his lawyers believe the companies that have contributed the most to the problem should help those who have contributed the least—and who bear the worst risks. They are asking for 0.47 percent of the total cost of a project to increase Lake Palcacocha's drainage pipes and further reinforce the dam. There are many other companies that have contributed much more to global climate change than RWE, however, and the company has taken great strides to become a clean energy leader in the past two decades. In addition, it argues that the effects of climate change are far too complex for a farmer from across the world to require them to pay for localized effects in his community.

As of this writing, Lliuya's case sits in the German courts, waiting to proceed slowly through the evidentiary phase while court-appointed scientists collect evidence at Lake Palcacocha. A judge threw out the first case that Lliuya and his lawyers brought with the support of Germanwatch, but their team is confident that their appeal will fall on sympathetic ears at the higher court in Hamm. The long legal process is certainly not helping Lliuya's quest to prevent the dam from being breached by an avalanche or rockslide, however, as the waters continue to rise from melt every day. All of these delays make it seem like the case may have been for nothing—but if he succeeds, Lliuya's could set a forceful precedent that requires polluting companies all over the world to help rebuild after climate change damages.

Origins of a David vs. Goliath Climate Litigation Story on Two Continents

Lliuya had never been outside Peru before he decided to fight for his community through the court system in Germany. A farmer and father of two, as mentioned, he leads people from all over the world as a mountain trekking guide. Outside the tourist season, he and his wife farm corn, potatoes, quinoa, and mint and raise a few cows for subsistence and to supplement his income after the tourists have left for the year (Jarvis 2019). In both his mountain trekking and farming, Lliuya has noticed significant changes to the glacier, the lake, and the land where he has made his living over the years.

Huaraz, a provincial capital and economic base for the region with a population of around 120,000, is the second-largest city in the Peruvian Andes and has been settled since pre-Columbian times. It sits at the base of a deep valley carved by glaciers over millions of years and is more than 10,000 feet above sea level (Divide, Colorado, the highest elevation city in the United States, is about 1,000 feet lower). At the top of the valley, perched just below the glaciers, is Lake Palcacocha.

The Cordillera Blanca (literally "white rope"), as the Andes are known in the Huaraz area and along the high north-south ridge traversing Peru, forms a majority of the world's tropical glaciers (Leavell & Protocarrero 2003) (Figure 8.4). Studies have also shown that climate change will result in higher temperature increases at higher altitudes, and since Peru is situated in the

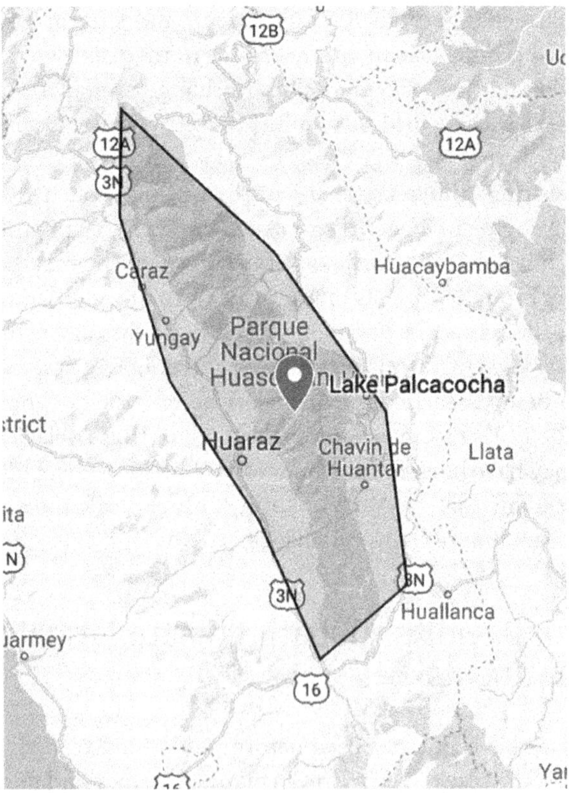

Figure 8.4 Approximate area of the Cordillera Blanca. Sources/notes: Google Maps, 2020. Google and the Google logo are registered trademarks of Google LLC, used with permission. Annotations by Laurel Wolf 2019.

tropics and exposed to even more radiation, the Cordillera Blanca likely will experience maximum temperature increases (Bradley et al. 2006).

Tropical glaciers are immensely important—they store water up during the wet season and release it slowly during the dry season, ensuring continuous supply in times with otherwise sparse rainfall. More than 50 percent of Peru's agriculture, mostly corn and potatoes, is on land fed by glaciers (US DOE), which are the primary source of water for the estimated six million people in the Andean highlands who still rely on small-scale agriculture for their subsistence (Heikkinen 2017).

For farmers and others who rely on this type of land in Peru and elsewhere in South America (including Colombia, Ecuador, Bolivia, Chile, and Argentina), a glacial dam breach would mean not only physical devastation through the loss of life and property destroyed by flooding, but also a complete lack of water for agriculture if the waters from the lake were not available year-round. "There's an annual input and return from those glaciers, like a bank account," says Bryan Mark, a glaciologist at the Ohio State University, "If you are withdrawing more than you are putting in, eventually your account is going to go dry" (Konkel 2008). Andean agriculture is already threatened by climate change, and irrigation needs will grow while the glaciers diminish. Therefore, a catastrophic event like a dam collapsing or overflowing would cause widespread water scarcity in the area.

The mountains do not experience large seasonal temperature fluctuations, but they do have wet and dry seasons, with the wet season being when the majority of precipitation is stored in the glaciers as ice or snow. However, a changing climate brings changing weather patterns to the area. In March 2017, Peru's Andes experienced one of its wettest ever months, which resulted in flooding and landslides affecting 1.7 million people, 177 deaths, and an estimated total damage of $3.1 billion (Christidis et al. 2018).

As the glaciers melt at a higher pace than ever before, the many high-altitude lakes, like Lake Palcacocha (at almost 15,000 feet), are at an elevated risk of bursting the natural dams (referred to as moraines) that hold them back. Higher lake levels could result in a catastrophic flood if falling rocks or avalanches from these now-unstable glaciers suddenly increase the lake's volume. This type of event is called a Glacial Lake Outburst Flood (Jarvis 2019). Although these floods are rare, when they do occur they send ice, water, and rock careening down the steep valleys to the villages, towns, and cities settled at the base of the mountains. In their paper modeling the results of a large avalanche into Lake Palcacocha if the lake's level is not lowered, Somos-Valenzuela and his team (2016 demonstrated the flow volume and area impacted downstream from the lake in Huaraz (Figure 8.5).

The rapid melt of the glaciers and increase in extreme weather is a problem not just for the safety of the inhabitants below but also a cultural and spiritual crisis, as the Quechua culture of the region believes that mountain

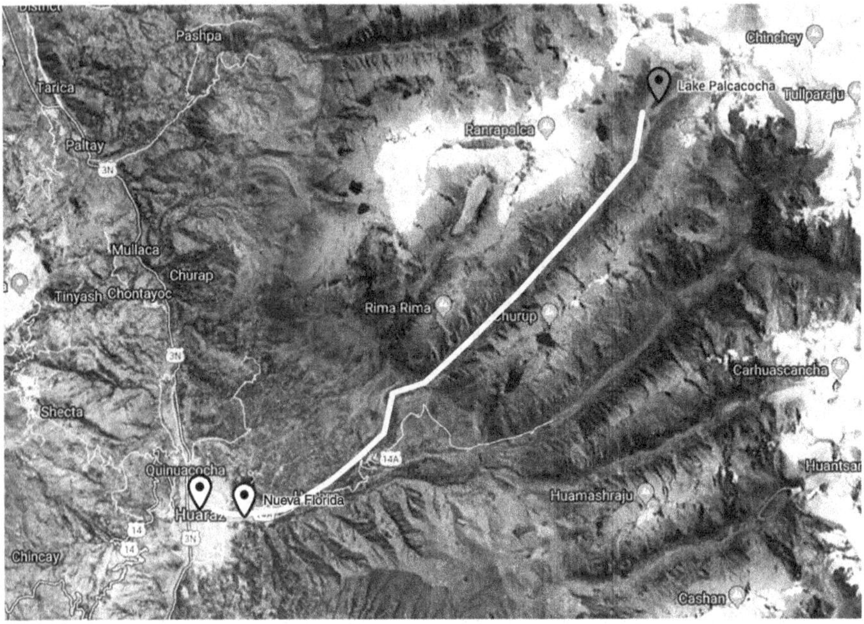

Figure 8.5 Blue marker is Lake Palcacocha, with Huaraz and the Nueva Florida neighborhood where Saúl Lliuya lives marked in the lower right-hand corner. The red line indicates the valley where a flood would flow. Sources/notes: Google maps and TerraMetrics, 2020. Google and the Google logo are registered trademarks of Google LLC, used with permission. Annotations by Laurel Wolf 2020.)

ecosystems are *apu,* ancient ancestors who control water, health, and community well-being (Bacigalupo 2018). In many communities like Lliuya's up and down the Andean range, across half a dozen nations, these glaciers have been relied upon for water and agriculture for centuries.

Indeed, many indigenous cultures in Latin America, including Lliuya's Quechua culture, revere the earth a living being, with its own inherent rights. *Sumak kawsay* is a Quechua concept that can vaguely be translated as "harmonic life," a fluid and rational concept that among Ecuadorian Andean indigenous communities (in the same cordillera) establishes "a constantly adapting bond among humanity, nature, and spirituality, and a responsibility for maintaining this bond for future generations (Eisenstadt & West 2019, 152)." In the Peruvian Andes, where water is scarce and precious, water is often used as an offering (Bacigalupo 2018), and annual religious pilgrimages are made to the glaciers. The mountain ecosystem in Peru is sacred and treated as a sacred being. Meanwhile, the rights of this being are eroding because of climatic change created by people living far away. For Quechua people, it is an outside force that has thrown off the balance.

Peru contributes only about 0.2 percent of the world's climate emissions (World Resources Institute [WRI] 2021), so it can feel to many of the farmers and others who rely on the mountains for their livelihoods as if their home is being destroyed by forces far beyond their control. "The people who caused these effects aren't doing anything about it," Lliuya says. "Someone has got to" (Nugent 2018). Lliuya, who lives a very low-emissions lifestyle, especially compared to consumers in nations like the United States and China, who rely much more heavily on automobiles and air conditioners, wanted to identify those responsible for the risk that he and his neighbors live with every day (Jarvis 2019).

Forming of an Alliance to Stem the Browning of the Cordillera Blanca

Lliuya was not the only one concerned about the effects of climate change on Lake Palcacocha and the ramifications it might have in the community. A friend in Huaraz connected him with the German NGO Germanwatch, which has helped advise him throughout the process of pursuing his lawsuit. "Some people have called me crazy," he said. "But the mountain is everything to us. If you'd seen what's happening to it, you'd do the same thing" (Nugent 2018). It does seem a bit far-fetched that a Peruvian farmer would sue the German energy giant RWE to pay for some of the negative effects of climate change, but this lawsuit is part of a group of lawsuits that seek to answer the question—who pays for the damage?

For Lliuya, a new strategy was needed to combat climate change: one holding corporations liable for damages caused by rising global temperatures. RWE is Germany's biggest energy provider, and one of the largest single historical emitters of carbon in the world according to the "carbon majors" study (Heede 2013). The data, collected on 83 of the world's largest oil-, gas-, and coal-producing entities (and CO_2 from the largest cement-producing entities) helped Richard Heede to estimate these producers' historical carbon and methane emissions (Heede 2013).

Until 2013, no one had used a scientific methodology to attribute specific portions of climate change to corporate emissions. As argued in discussions of the Carbon Majors in Chapter 3, it is possible to compare not only a company's contributions to total global emissions, but also a country's total emissions. Peru itself contributes about 0.2 percent, while the Top 10 emitting entities (China, the United States, the European Union, India, Russia, Japan, Brazil, Indonesia, Canada, and Mexico) are responsible for over 73 percent of total GHG emissions (WRI 2020). Based on Heede's list (including not just companies but nationalized energy sources like China's coal production), RWE is only number 41 on the list of historical emitters, far below many of the biggest offenders, including Peru's neighbor Brazil, whose company Petrobras is responsible for 0.77 percent of historical emissions (Heede 2013) and whose current president, Jair Bolsonaro, has taken active steps against climate action.

Still, Lliuya and his lawyers believe that RWE should at least help to pay for the damages that he and his community face as a result of the effects of climate change in Peru. Along with Germanwatch and several other international sponsors, the team is trying to prove that a German energy company's contributions to climate change are partially responsible for the rapid melt of the glaciers above his home that have endangered the people who live in the valleys below these glacial lakes, and that those companies (or, in this case, one specific company) should be held liable for the potential damages and provide funds to help prevent and protect against that damage. And it is getting standing (being allowed to proceed, having met a minimum threshold of evidence) in Germany, the largest nation in the European Union, which has been the world's most consistent champion of policies to fight climate change for decades.

RWE does not operate anywhere near Peru or Lake Palcacocha. The company was started as in 1900 and its operations are entirely based in Germany, where it is a public utility (owned by the government) and has boasted substantial profits most years. While the company has produced electricity from coal, gas, and nuclear in the past, it is hoping to achieve carbon neutrality by 2040 (Schunck 2019). Clearly, while the company has contributed significantly to climate change, they are also adapting to both European and German laws to reduce emissions and provide renewable energy.

The case raises two important questions: Who pays for climate change, and who *should* pay for climate change? At the UN and in other international fora, the onus has been on nation-states, and the debate has been whether rich industrial nations, those responsible for historic emissions, should pay disproportionately for damages done over the last 150 years to the world's climate. This case, however, introduces corporate actors as possible responsible parties. The resolution of this case could have implications for how we deal with the climate crisis for decades to come, rebuilding and saving what we can while relocating or adapting what we cannot. Although Lliuya is asking for a pittance, 20,000 dollars in damages that the town of Huaraz has incurred in reinforcing the dam, the amount is symbolic, and important in the signal that it sends to companies around the world. "What we are saying is this is happening and it's putting us at risk. The damage caused is irreversible; for that no amount of money is enough" (Collyns 2018).

In nations around the world, tort or nuisance laws address compensations for harms from another's actions or lack of actions, investigating who will qualify as a victim, what is counted as a misdeed, and how to "make it right" through compensation, if that is a possibility. For instance, if Farmer 1 is using toxic chemicals that then run downhill onto Farmer 2's property and kill some of her cows, what is the best way to compensate Farmer 2 for her loss? In order to recuperate her loss, Farmer 2 would have to prove that her loss was connected to a specific harm (called standing), that Farmer 1 breached her responsibility to avoid behaviors that could foreseeably cause harm to others (called duty of care), that Farmer 1's actions caused Farmer 2 harm, and that the Farmer 2 suffered real losses as a result (sometimes these losses can be projected into the future as well). It's not always obvious in the courts that these connections exist, and proving them, especially with a relatively young science like climate science, is very difficult.

Paired with German environmental lawyer Roda Verheyen, Lliuya set out to prove that the energy giant had created risk in his hometown and that, as a result, they should pay to help him mitigate that risk. Christopher Bals, the Policy Director of Germanwatch, asserted:

> Those who harm others have to be held accountable for their actions. Thus, Germanwatch expects RWE to develop and implement a new business model so as to avoid causing further damages. In addition, the company has to pay its fair share of financing measures to protect those at risk. If the "polluter pays" principle is not implemented in a market economy, profits are privatized while risks are socialized. (Germanwatch 2015).

Acknowledging that RWE was only an emissions contributor—and a small one compared to the "worst" carbon majors, ChevronTexaco and ExxonMobil, responsible for 3.52 and 3.21 percent, respectively, of worldwide emissions (Heede 2013)—the complainants asked the German national courts, under Section 1004 of the German Civil Code,[2] to force that country's largest energy provider to pay around 20,000 dollars (0.47 percent of the project's total cost) to help shore up the dam and prevent catastrophe. The District Court of Essen dismissed the case in 2016, saying that the climate activists' argument was unfounded and that no link could be established between RWE's CO_2-emitting power plants and the endangerment of Lliuya's house in Peru. Lliuya and his lawyers appealed that decision to the Higher Regional Court of Essen in 2017 (Germanwatch 2017).

RWE, for its part, maintains that the damages of climate change are too complex and interlinked to be attributed to one company alone. In its initial defense, RWE pointed out that even the Intergovernmental Panel on Climate Change (the UN's scientific research "arm" reports on climate change state clearly that local temperatures cannot be extrapolated from overall global warming averages. Hence, the effects of global trends felt in Peru cannot be reliably assessed, and tying such effects to emissions from one company, located half a world away, is impossible. RWE successfully made that argument in Essen (Baldrich 2017).

Outside the narrow scope of legal proceedings, there are additional reasons why suing an individual energy company for climate damages may not be the most efficient strategy. RWE has taken great strides to get their emissions under control by modernizing their power plants and investing billions of dollars in renewable energy ("German Court" 2017). A 2018 merger made them Europe's second-largest producer of wind energy, and the company continues to eye other renewable energy investments around the world, becoming Europe's third largest renewables producer by 2020 (Busvine 2020). RWE serves a huge population, and it is working to modernize its existing plants and to rely more on clean energy going forward. Were Lliuya to win his case, this might signal that people around the world could take the company (or other companies, for that matter) to court based on this precedent. And if further claimants came forward seeking larger damages built on the precedent set by Lliuya and Germanwatch, this could damage one of Europe's largest purveyors of clean energy—exactly what is needed to turn away from fossil fuels.

A spokesperson for RWE also claims that trying to receive damages from individual companies through this process reduces the pressure on governments to come up with sensible policies and laws regarding climate change. "We think of climate change as a global social and political issue," he says. "It is national governments and international agreements, such as at the UN climate conference, that have to develop instruments to reduce carbon dioxide" (Nugent 2018). Many have faulted the 2015 Paris Agreement, which requires national contributions to mitigating climate change, as lacking in ambition, because it fails to require mandatory emissions cuts by polluting nations and instead just asks nations to set forth and follow voluntarily presented Nationally Determined Contributions (see Chapter 3).

The Higher Regional Court of Hamm in 2017 decided that the case was admissible, following many of the arguments made by Lliuya and his lawyers. The court said that while RWE's emissions are not responsible for the whole flood risk due to glacial retreat, it is enough that these emissions are partially responsible for the actual, present risk. According to the court's reasoning, there is no basis in the law to argue that partial causation does not exist, and evidence to be used in court should be collected (Germanwatch 2017). Unfortunately, RWE and the complainants could not agree on which experts would deliver evidence on the actual threat of flooding, the amount of historical CO_2 released by RWE, or its share in creating climate change.

The German court was required in July 2018 to find its own experts to investigate, but as there exists no interstate legal assistance agreement between Peru and Germany, German the court had to ask the Peruvian government to allow the gathering of evidence in Huaraz. An "on site" visit by German authorities to Huaraz had been planned in 2020 when the COVID-19 epidemic struck, but the process was deferred until after the lifting of pandemic travel restrictions (Valdrich and Milke 2020).

Beyond the Andes: Climate Litigation in Global Perspective

The case has grown much larger in scope than just a Peruvian farmer and a German energy utility. This is not the first case of using litigation to force action on climate and given the growing desperation of climate activists to achieve change even as governments are slow to act, it will definitely not be the last. There is far-reaching significance if an individual can sue a company that does not operate in that person's country for damages caused by worldwide climate change. The mere existence of a case precedent like this will likely be much more important than the verdict.

The number of national climate change laws has increased significantly in recent years, with more than 1,500 laws and policies by 2018 across 197 country signatories to the Paris Climate Agreement, compared to just 60 such laws in 1997 (Nachmany 2018). Countries have chosen to enact these laws in different ways, some by executive action and some by legislation; these strategies for enacting climate laws vary widely based on the local context and government. It follows that with this increase in environmental lawmaking, there have been significant developments in climate litigation and lawsuits as well. While the majority of climate litigation cases address specific projects, a small but important number link climate effects and human rights (Nachmany 2018, Setzer & Bynes 2019).

The Climate Change Litigation of the World Database lists over 1,000 climate change cases currently in US courts, and over 250 others internationally (Setzer and Bynes 2019). Interestingly, a plurality of those cases are brought by companies (generally against governments to overturn or contest climate legislation). A small portion of climate litigations are seeking to hold governments or companies to account. Several landmark climate litigation cases try to strategically link rights and climate change by urging mitigation, adaptation, and/or compensation.

One such case is the 2007 *Massachusetts v. Environmental Protection Agency* ruling that confirmed that the US EPA must regulate GHG emissions under the Clean Air Act if it was determined by the agency that public health or welfare was endangered by these emissions. For the United States, this helped the EPA establish that climate-related emissions were indeed a danger to human health and welfare. This was an important precedent when fewer than 60 percent of Americans even believed in climate change (Leiserowitz et al. 2012), much less that it had anthropogenic causes.

While *Massachusetts v. EPA* made waves in the United States, a whole slate of new climate cases was kicked off by *Urgenda Foundation v. Kingdom of The Netherlands*. As discussed in the Chapter 8 text, in this 2015 case, a Dutch environmental group sought to hold their government accountable when it relaxed GHG emission goals set by that country's previous government.

The Foundation won in the District Court in 2016, and in 2018, the Hague Court of Appeal upheld the ruling, concluding that in failing to reduce GHG emissions by at least 25 percent by the end of 2020, the Dutch government is neglecting its duty of care under Articles two and eight of the European Convention on Human Rights), which protect the right to private life, family life, home, and correspondence. Invoking a laundry list of environmental principles, the court successfully linked climate mitigation to human rights, perhaps for the first time, rather than just enforcing an existing law.

European courts are not the only ones where the climate-aggrieved seek redress for climate change wrongs. A 2015 case in Lahore, Pakistan (*Leghari v. Republic of Pakistan*) used a similar argument as the *Urgenda* case to hold the government accountable for failing to implement the Climate Change Policy of 2012 and the Framework for Implementation of Climate Change Policy (2014–2030). The complainant cited the constitutional rights of Pakistani citizens, including the rights to life, human dignity, information, and property, saying that since Pakistan was not a major contributor to climate change, it must do all that it could to adapt. An appellate court granted these claims. Subsequently, in early 2019, a coalition of Pakistani women filed a case that builds on the rights violations established in the *Leghari* case, stating

> the Respondents have failed to undertake the implementation or processing of even a single renewable energy power project for the past fourteen months (and counting) in Pakistan. Such failure on part of the Respondent Federation amounts to a clear breach of its stated commitment under the Paris Agreement to encourage and foster the development of renewable energy sources. Furthermore, the Respondents through their inaction and laissez-faire attitude towards development and support of renewable energy projects have abdicated their responsibilities as trustees of the natural resources of the country under the Public Trust Doctrine, thereby violating the rich jurisprudence of this Honourable Court on environmental justice, intergenerational and intragenerational equity and, most importantly, climate justice. (Honourable Lahore High Court 2019)

Cases have also been filed in Belgium, Switzerland, and New Zealand confronting governments for not responding adequately to climate change (UN Environment Programme [UNEP] 2017). Pursuing a slightly different concept, cases in Colombia, Pakistan, and Norway challenge oil or coal extraction on a climate change basis (UNEP 2017).

The concept of a "public trust" is used in several other cases around the world. One well-publicized case in the United States is that of *Juliana v. United States*, in which 21 youth plaintiffs filed a lawsuit in federal district court against the federal government to compel it to reduce carbon emissions

(Howe 2018). The plaintiffs say that a stable climate is critical to their constitutional rights to life, liberty, and property, and that the defendants violated their due process rights by allowing fossil fuel emissions to rise to dangerous levels (UNEP 2017).

Finally, there are a handful of cases like Lliuya's seeking to establish that certain bodies can be held accountable for climate change effects by showing that particular emissions are a proximate cause of particular climate impacts. In the United States, the *American Electric Power Co. v. Connecticut* case in 2010 sought to cap emissions from power plants. In another case, *Kivalina v. Exxon-Mobil* (2008), the US-based plaintiffs sought damages for their injuries from fossil fuel companies. Both cases used the theory of public nuisance under US federal common law, and the federal circuit courts in both cases concluded that since the EPA was now able to regulate emissions under the Clean Air Act, it did not need to also rule in favor of the petitioners (UNEP 2017). In other words, while the power of the EPA to enforce climate regulations was reaffirmed, the complainants lost these cases.

The effects of this wave of lawsuits in other nations on the suit regarding possible spillage from Lake Palcacocha are unknown. While precedents do not carry across nations, there is a sense of international urgency regarding the climate crisis, and a growing sense that litigation may actually be faster than the glacial pace of international negotiations (see Chapters 3 and 7). The case against RWE in Germany falls approximately into this category of cases: those that seek to show certain GHG emissions are linked to certain climate effects. Lliuya is hoping that the German courts will acknowledge that, with the publication of Heede's (2013) "carbon majors" study tracking the total emissions of the world's biggest polluters over the decades, it has become increasingly possible to establish that certain bodies are directly and quantifiably responsible for damages being felt around the world.

From all these cases, a few trends emerge that might give the Peruvian activists hope. The first is that two-thirds of court rulings in climate cases strengthen or preserve existing climate regulations (Nachmany 2017). The second is that strategic litigation against corporations and governments is on the rise (Nachmany 2017, Schäfer et al 2018). If these two trends continue, there will be many avenues for plaintiffs to claim loss and damages in courts all over the world.

Important weaknesses exist in climate litigation. Given the international treaties already established on climate change, it might be inappropriate to go after individual companies when the international consensus may be interpreted as one where developed countries take responsibility for their emissions and make payments to developing ones. And even if governments do not believe this "loss and damages" approach to torts, they seem to have accepted that the agents of action (and hence the subjects of lawsuits) may likely be governments rather than corporations. Another concern is enforcement. Even if these companies are held to be responsible by the court, who will enforce these orders and how? Finally, legal systems around the world differ widely; being able to establish a link between companies and climate damages in Germany does not mean that other courts around the world will also take up the cause.

Germanwatch argues (Schäfer et al. 2018) that litigation brings many more benefits than just winning the single case. First, they offer that continuous litigation will force large companies to actually consider the risks of these suits and both change their practices and begin to set aside funds for paying out insurance and damages, or even for pursuing carbon-capture techniques (Krauss 2019). Second, winning some of these climate cases will signal to other companies, their investors, bond rating agencies, and insurance markets that climate litigation will have an impact on the bottom line across the industry, pressuring these actors to shift their investments elsewhere or at least explicitly take climate risk into account. Finally, these lawsuits apply pressure on governments to act and take climate change damages seriously.

Conclusion

Saúl Luciano Lliuya traveled across the world to Germany to hold those he feels are partially responsible for the dangers his community faces accountable for the climate chaos that has accompanied the rampant use of fossil fuels through industrialization. Climate litigation has already seen some important successes, but it is not yet certain that this is how to hold people accountable for "wrongs" committed halfway across the world that involved emissions from their utility plants, to be sure, but also required that consumers purchase their energy.

While the basic science of anthropogenic climate change is settled, the decisions of who should pay who and how are still being fought every day. As discussed in Chapter 1, these considerations require making ethical and moral assumptions as well as following scientific and national laws. The international community continues to debate whether countries that have contributed the least to climate change (often the very same ones projected to suffer the worst consequences) should be entitled to "loss and damages" from developed, high emitters. But individual lawsuits against private companies leave so many out of the process of achieving climate justice—if one man or town is paid, what about the neighboring town? Or all the world's other vulnerable towns? And if one company is held liable, what about the companies that drove demand for energy and have not done nearly as much as RWE to curb emissions, spark technological innovation, and diminish their footprint? Regardless of the answers to these questions, it would be quite an achievement indeed if a Peruvian small farmer "David" was able to force one of the largest energy company Goliaths in the world to admit wrongdoing.

EXERCISE

SCENARIO FOR CLASSES

It is now 2024 and time for the court to issue a final ruling in the case. After carefully considering the information provided by the case study, students should divide into teams. One team will work on the side of the defendant RWE, one team will work for the claimant Lliuya, and one team will be in charge of judging the case based on the following criteria:

1. Does the claimant establish a sufficient link between the potential damages and the actions of the defendant?
2. If the defendant were to cease the action, could the damage be prevented, even partially?
3. If the defendant is responsible, what portion of the damages should they pay?
4. Is this the most appropriate avenue for resolving this dispute, or do other routes exist to right the perceived wrongdoing?

After hearing the arguments from each side, the class, assembled as a "debrief" group, should vote on a ruling that determines who has won the case (after the group of "judges" makes an initial determination), the damages (if any) to be awarded, and further action required by each party. The instructor should lead the debrief and might ably serve as the judge if the discussion stalls otherwise.

CASE STUDY 8.1 NOTES

1. Laurel Wolf is pursuing a Master of Public Administration focused on mitigating and adapting to climate change. She has worked as a grassroots organizer on climate issues in the United States and executed agroforestry projects in Senegal as a Peace Corps Volunteer.
2. Section 1004 of the German Civil Code (BGB) :"If the ownership is interfered with by means other than removal or retention of possession, the owner may require the disturber to remove the interference. If further interferences are to be feared, the owner may seek a prohibitory injunction."

CASE STUDY 8.1 REFERENCES

Baldrich, Roxana. 2017. "Court documents of the Huaraz case." *Germanwatch*, December 13. Accessed June 23, 2020, at germanwatch.org/en/14198

Baldrich, Roxana. 2019. "The Huaraz case in its fourth year." *Germanwatch,* May 8. Accessed June 23, 2020, at germanwatch.org/en/16451

Bradley, Raymond S., Mathias Vuille, Henry F. Diaz, and Walter Vergara. 2016. "Threats to water supplies in the tropical Andes." *Science* 312(5781):1755–1756. DOI: 10.1126/science.1128087

Bacigalupo, Ana Mariella. 2018. "La Política Subversiva de los Lugares Sentientes': Cambio Climático, Ética Colectiva y Justicia Ambiental en el Norte de Perú" in *Scripta Ethnologica*, 40:9–38.

Baldrich, Roxana and Kalus Milke. 2020. "5 Years of climate litigation against RWE – a milestone for climate justice," *Germanwatch* November 26 press release. Accessed January 8, 2021, at https://germanwatch.org/en/19596

Busvine, Douglas. 2020. "Germany's RWE rakes in $2.4 billion to fund renewable energy expansion," August 18, *Reuters newswire*. Accessed January 8, 2021, at

Carey, M. 2010. *In the Shadow of Melting Glaciers: Climate Change and Andean Society.* Oxford University Press, New York.

Christidis, Nikolaos, Richard A. Betts and P.A. Stott. 2018. "The extremely wet March of 2017 in Peru." *Bulletin of the American Meteorological Society* 100(1): 1–5. DOI: 10.1175/bams-d-18-0110.1.

Collyns, Dan. 2018. "Climate change has turned Peru's glacial lake into a deadly flood timebomb." *The Guardian*, June 29. Accessed June 23, 2020, at www.theguardian.com/environment/2018/jun/29/climate-change-has-turned-perus-glacial-lake-into-a-deadly-flood-timebomb

Darby, Megan. 2016. "Peruvian climate lawsuit against German coal giant dismissed." *Climate Home News,* December 15. Accessed June 23, 2020, at www.climatechangenews.com/2016/12/15/peruvian-climate-lawsuit-against-german-coal-giant-dismissed/

Eisenstadt, Todd, and Karleen Jones West. 2019. *Who Speaks for Nature? Indigenous Movements, Public Opinion, and the Petro-State in Ecuador*. Oxford University Press, New York.

"German court to hear Peruvian farmer's climate change case against RWE." 2017. *The Guardian*, November 30. Accessed June 23, 2020 at www.theguardian.com/environment/2017/nov/30/german-court-to-hear-peruvian-farmers-climate-case-against-rwe

Germanwatch. 2015. "Melting glaciers: Peruvian requests German utility RWE to pay for protective measures." *Germanwatch*, March 16. Accessed June 23, 2020, at germanwatch.org/en/10002

Germanwatch. 2017. "Interesting facts: Background information on the decision of the Higher Regional Court Hamm." *Germanwatch*, December 12. Accessed June 23, 2020, at germanwatch.org/en/14831

Heede, Richard. 2013. "Tracing anthropogenic carbon dioxide and methane emissions to fossil fuel and cement producers, 1854–2010." *Climatic Change* 122(1–2): 229–241. DOI: 10.1007/s10584-013-0986-y

Heikkinen, Anna. 2017. "Climate change in the Peruvian Andes: A case study on small-scale farmers' vulnerability in the Quillcay River Basin." *Iberoamericana—Nordic Journal of Latin American and Caribbean Studies* 46(1): 77–88. DOI: 10.16993/iberoamericana.211

Honourable Lahore High Court, Pakistan. 2019. "*Maria Khan Et Al. v. Federation of Pakistan Et Al.* (2019)." *Climate Change Litigation.* Accessed December 7, 2020, at blogs2.law.columbia.edu/climate-change-litigation/wp-content/uploads/sites/16/non-us-case-documents/2019/20190214_No.-8960-of-2019_application.pdf.

Howe, Amy 2018. "Justices refuse to block climate-change trial." *SCOTUSblog*, November 2. Accessed June 23, 2020, at https://www.scotusblog.com/2018/11/justices-refuse-to-block-climate-change-trial/

Jarvis, Brooke. 2019. "Climate change could destroy his home in Peru. So he sued an energy company in Germany." *New York Times Magazine*, April 9. Accessed June 23, 2020, at www.nytimes.com/interactive/2019/04/09/magazine/climate-change-peru-law.html

Juhola, Sirkku, Patrick Driscoll, Janot Mendler de Suarez, and Pablo Suarez. 2013. "Social strategy games in communicating trade-offs between mitigation and adaptation in cities." *Urban Climate* 4: 102–116. DOI: 10.1016/j.uclim.2013.04.003

Konkel, Lindsey. 2008. "What are tropical glaciers and why do they matter?" *Science Line*, December 15. Accessed June 23, 2020, at scienceline.org/2008/12/ask-konkel-tropical-glaciers-melting-andes-mountains-runoff-groundwater/

Krauss, Clifford. 2019. "Blamed for climate change, oil companies invest in carbon removal." *New York Times,* April 7. Accessed June 23, 2020, at www.nytimes.com/2019/04/07/business/energy-environment/climate-change-carbon-engineering.html

Leavell, Daniel N., and Cesar Jirón Portocarrero. 2003. "Sustainability of Peruvian water resources in light of climate change." XI World Water Congress, International Water Resources Association, Madrid, Spain.

Leiserowitz, A., Maibach, E., Roser-Renouf, C., Feinberg, G., & Howe, P. 2012. "Climate change in the American mind: Americans' global warming beliefs and attitudes in September 2012." Yale University and George Mason University. New Haven, CT: Yale Project on Climate Change Communication. http://environment.yale.edu/climate/files/Climate-Beliefs-September-2012.pdf

Nachmany, Michal. 2017. *Global Trends in Climate Change Legislation and Litigation*. Grantham Research Institute on Climate Change and the Environment, London. pp. 1–34.

Nachmany, Michal. 2018. *Global Trends in Climate Change Legislation and Litigation: 2018 Snapshot*. Grantham Research Institute on Climate Change and the Environment and Centre for Climate Change Economics and Policy, London. pp. 1–8.

Nugent, Ciara. 2018. "Climate change could destroy this Peruvian farmer's home. Now he's suing a European energy company for damages." *Time Magazine*, October 5. Accessed June 23, 2020, at time.com/5415225/rwe-lliuya-climate-change/

Poplin, Alenka. 2012. "Playful public participation in urban planning: A case study for online serious games." *Computers, Environment and Urban Systems* 36(3): 195–206. DOI: 10.1016/j.compenvurbsys.2011.10.003

Rheinisch-Westfälisches Elektrizitätswerk (RWE). 2018. "RWE history." Accessed December 7, 2020, at www.group.rwe/en/the-group/history.

Schäfer, Laura, Vera Künzel and Christoph Bals. 2018. "The significance of climate litigation for the political debate on loss & damage." *Germanwatch*, October 18. Accessed December 7, 2020 at www.germanwatch.org/en/15104

Schunck, Stephanie. 2019. "The new RWE: carbon neutral by 2040 and one of the world´s leading renewable energy companies," September 30 press release. Accessed January 8, 2021, at https://www.group.rwe/en/press/rwe-ag/2019-09-30-the-new-rwe

Setzer, Joanna, and Rebecca Bynes. 2019. *Global Trends in Climate Change Legislation and Litigation: 2019 Snapshot*. Grantham Research Institute on Climate Change and the Environment, London. pp. 1–11.

Somos-Valenzuela, M.A., Chisolm, R.E., Rivas, D.S., Portocarrero, C., and McKinney, D. 2016. "Modeling a glacial lake outburst flood process chain: the case of Lake Palcacocha and Huaraz, Peru," in *Hydrological. Earth Systems. Sciences*, 20:2519-2543, DOI:10.5194/hess-20-2519-2016.

UN Environment Programme (UNEP). 2017. *The Status of Climate Change Litigation: A Global Review*. Accessed June 23, 2020, at columbiaclimatelaw.com/files/2017/05/Burger-Gundlach-2017-05-UN-Envt-CC Litigation.pdf

World Resources Institute (WRI)."Climate Watch: Peru: Greenhouse Gas Emissions and Emissions Targets." 2020.Accessed January 8, 2021, at https://www.climatewatchdata.org/countries/PER

Wright, Matthew, Todd A. Eisenstadt, Tawfique Haque, and Michael Toman. 2020. "Economic vulnerability as a primary cause of climate migration: Evidence from Bangladesh." Typescript.

9

Sowing Our Gardens is Not Enough

What Can We Do as Citizens and Consumers?

Climate scientist Ken Caldeira, who studies coral reefs at Stanford University, finds his job discouraging:

> I see reefs that are over 90 percent dead. That's a real tragedy. I see it and experience it . . . I plod on . . . Maybe I've become better at suppressing my feelings . . . I can still see this is really tragic. Fossil fuel emissions killed 90 percent of this reef. But my dominant strategy is to intellectualize it and say, 'What a crazy species we are.' As a scientist, my only role is to generate useful information" (quoted in Corn 2019, 13).

At American University, two instructors (the authors of this book) were asked what they would do to stop climate change, as students came in after the 2016 election and laid on the floor, apoplectic. After trying to "buck up" the students by brainstorming an activist agenda, the students turned the question on us. We told them that our college protesting days may have been behind us, but that we would write a textbook for the class, which did not really have one. Having written most of this text in the third person, we must now, like Caldeira and our students, figure out how we can contribute in our personal lives. We could not complete the text without addressing this, as climate change is such a pervasive and all-encompassing problem.

It may also be true that gentle readers (including you) may by this time be growing weary of reading about the effects of climate change in faraway places like Bangladesh and Dominica, and instead wish to know what is going on beyond the most vulnerable nations, as most readers of this book are undergraduates in industrial nations. Indeed, as noted in Lazrus (2012): "Climate change is something happening somewhere else, tropic islands or arctic ice, but not here. The result could be a reduction rather than increase in concern among Western audiences" (294). Lazrus was writing in 2012, however, and in the years since then, it has become apparent that climate change effects are appearing in the industrialized nations of the Northern Hemisphere: record-setting droughts in California along with record-setting wildfires, 500-year floods occurring yearly in Maryland, hurricanes intensifying more rapidly than have ever in the Caribbean, crop failures in Central America sending economic/climate refugees to the US southern border in jaw-dropping numbers, and the list goes on. Here in the concluding chapter of this book, it is time to de-exoticize climate change and consider not only what is happening closer to home, but also what we each can do about it.

There is a risk of individualizing responsibility ("if I just flew less, recycled, and rode my bike more, etc., I could contribute to solving climate change"), in that governments and corporations may then feel less pressured to collectively offer the most powerful and—ultimately—the only effective solutions. Nonetheless, if we take individual responsibility, we can keep the issue front and center, get a head start on adjusting to the carbonless lifestyle we will all face within about 50 years, and even make some inroads toward solutions.

Let us be clear. We are not recommending individual actions as a *substitute* for group-led actions by public and private sector institutions, which must now step up under pressure from citizens, which we must start to escalate. We can drive electric cars (and take the subway more), and this will help, at the margins. "Feel good" activism is a small part of the solution, however, and we should not let the fossil fuel industry and its allies convince us otherwise.[1] There are actions we can and should take, as greenhouse-gas (GHG) emitters ourselves, to reduce our carbon footprints and thus empower ourselves, even if only to a small degree, while we continue to push our politicians out in front of us. The main reasons to undertake these actions are, first, to show our elected leaders that we are serious and disciplined about solving this problem and, second, so that we can discipline ourselves in moving toward solutions, and slouch toward the post-carbon lifestyles our children will have to embrace. Being just one person or a local community group contemplating how to make a difference in the face of a huge threat can be daunting. But if we think our efforts will be futile, recall what the cultural anthropologist Margret Mead had to say about change: "Never doubt that a small group of thoughtful, committed citizens can change the world; indeed, it's the only thing that ever has (McMickle 2014)."

"The Grasping Problem" and What We Can (and Must) Do

Ethicists have noted "the difficulties in recognizing our moral duties to people who are either temporally or spatially distant from us" (Kasperbauer 2016, 354). Kasperbauer (2016, 358) argues that this is for psychological reasons as well as moral ones. For example, he cites ingroup bias (the propensity for empathizing with people from one's own group rather than with climate victims around the world), low-impact bias (the adoption of small environmental behaviors that make people feel like they are fulfilling moral obligations), and single-action bias (where we are willing to take a single action to fulfill moral obligations expeditiously and drain our "finite pool of worry"). We want to improve the environment, but more importantly, we want to allow ourselves to stop thinking about it.

Mitigating against these anti-environmentalist psychological traits, possibly, is our tendency to conform. Kasperbauer (2016) finds hope in the fact that people do adopt the habits of those they perceive as being "like them." For example, Kasperbauer notes that people who used above-average amounts of energy in a San Mateo, California, experiment were sent "frowny faces" on their bills, whereas those who used below-average amounts were sent "smiley faces" and congratulated. As consumers were given information about averages in their area, the low performers improved their energy consumption (359).

Some people are just not wired to consider moral harm, and in fact, psychologists argue that different political leanings impact peoples' views of the climate crisis. For example, some people value tradition and authority more highly, whereas others, more apt to act on the moral ambiguities presented by climate change, tend to value fairness and caring. These personality traits vary across cultures and within them, according to Kasperbauer (2016), meaning that "individual psychological limitations lead to the result that we must address climate change at the level of policy, but at the level of policy we find that individual limitations are partly due to incomparable moral norms" (364). It would seem that conformity may be redirected in the direction of pro-climate actions through information campaigns and training, but it would also seem that such campaigns may take time and patience.

In the meantime, voluntary individual acts can have impacts. Dietz et al. (2009) estimate that household reductions in carbon dioxide (CO_2) by driving low-emissions cars and weatherizing houses could reduce household emissions 20 percent. This could reduce overall US emissions by about seven percent. This is a significant number, and it would add up over time. But there is not enough "slack" in voluntary household actions to reduce overall national emissions by 30 or 40 percent within the next few years. Furthermore, many consumers are unlikely to undertake such actions without provocation, for reasons mentioned earlier relating to the sometime moral hazard of individuals. Recall that moral hazard is the individual's lack of incentives to protect themselves against risk if that individual feels personally insulated from the risk, such as by insurance, by the future orientation of the manifestations of the problem, or by the fact that "people on the other side of the world" are suffering the consequences.

Ultimately, the 100 "carbon majors" (the fossil fuel companies discussed in Chapter 3) are responsible for 71 percent of the GHG emissions since 1988 (and for an even larger percentage of the GHGs in the atmosphere). Even so, the United States is responsible for some 26 percent of the cumulative GHGs, and while we are no longer the top per capita emitters, US consumers are the largest consumers on the planet, accounting for over 20 percent of the world's energy use but only five percent of the planet's population. US citizens, then, are in a unique position to address climate change, and given the degree of our excess consumption (relative to that of others), we should be able to make meaningful contributions with less sacrifice. So what exactly can we do? Lessons for the most affluent consumers in the world may come from the most vulnerable.

Scientist-Policymaker Collaboration: Small Signs of Hope from the World's Most Vulnerable Region

Africa may be the most world's most climate-vulnerable region. The Sahara Desert, which expanded dramatically in the 20th century, continues to increase at an escalating pace, and desertification has affected well over 80 percent of the nations of Africa (Africa Group of Negotiators, Experts, Support 2020). At the same time, extreme weather events, such as flooding from intense storms, are projected to increase in many areas (Intergovernmental Panel on Climate Change [IPCC] 2018, 3–41). The supply of potable water is diminishing, while the continent's population is expected to more than double between 2010 and 2050. Food insecurity is also projected to increase. While approximately a quarter of the climate change mitigation pledged by the world's countries in the Paris Agreement process will likely come from reforestation and the much less efficient process of afforestation (planting trees where there were none previously), structural changes such as these to biological and hydrological resources need to be complemented by political processes that help address human needs more thoughtfully and comprehensively.

Lusaka, Zambia provides the example of a large city (over two million people in 2019) with increasing public health hazards as a result of increasingly torrential floods associated with climate change. Underground aquifers flood, with drinking water and sewage mixing on the surface in many of the city's 33 housing areas, 28 of which were unplanned and thus do not have formal sewage structures. In 2017, some 100 people died from cholera as a result of this mixing of sewage and drinking water, causing school closures and a widespread public health threat. Yet an extensive, two-year dialogue between climate scientists and local elected municipal water board leaders in Lusaka has led to the creation of a multi-stakeholder advisory board, along with the construction of a pipeline and bore holes to provide additional sources of water to private industry (and free up extant aquifers as potable water for citizens).

Brenda Mwalukanga (2019), coordinator of the nongovernmental organization (NGO) Lusaka Water Security Initiative, said that local leaders had to be informed of the severity of climate change over time and through the presentation of appropriate information. Rainfall charts and heat maps were inadequate to the task. "If a policymaker has not interacted with this kind of science, you have to engage them over time and get them to trust, familiarize themselves, and get the information into a format they can process," Mwalukanga said. "If you go to a [water board] counselor with rainfall data they may well understand, but their first preference is something they can easily use to engage their constituents."

The Lusaka dialogue between climate scientists, local officials, and NGO service providers was conducted as part of the United Kingdom–funded Future Resilience for African Cities and Lands project, based at the University of Cape Town in South Africa, where climate scientist Chris Jack advocates for much more powerful "distillation" of information relevant to affected constituencies. Jack (2019) said that working in Lusaka and several other cities to help policymakers translate climate trends into actionable policies is increasingly vital. Whereas climate scientists used to present models and maps in hopes the public would use them, a more applied stream of "needs-informed science" has grown more prevalent over the last several years. "When we began to ask the users of information about their needs," Jack recalled, "we started to move away from science pushing data to decisionmakers, with the policy community pushing back their information needs to scientists."

Distillation of information by applied climate science, as well as the media and other partisan sources, may be part of the answer. Another part may be to reconsider how science is disseminated more generally. That is, rather than deductively applying the models patterning weather change from the present to the future, Jack and a few other climate scientists are starting to advocate that people project their ideal futures (within the range of constraints climate change will pose) so that climate scientists can then extrapolate backward from 30 or 40 years out. Decisions about how people see themselves gaining their livelihoods and interacting with their environments in the future can allow citizens to shape current policies to regulate and preserve natural resources. According to Jack (2019):

> We have to move away from climate science dominating the storyline and integrate multiple perspectives at even the local level. Unless peoples' voices have been heard, they will not commit themselves to contributing to a solution. It's not a political correctness thing; it's a claiming of stakes in our common problem.

Given the stalemate and polarization in the presentation of information in places like the United States, a greater sensitivity by an emerging cadre of applied climate scientists and climate-informed planners may be central in translating idealized citizen futures into an emission-reducing present that can be connected to a more sustainable future through distillation and dialogue.

In addition to more rationally and reasonably envisioning our shared future based on a more relevant translation of science into policy, we must also act as individuals and consumers. Leading loci for action in this area, based on the size of impact on emissions and the feasibility of needed sacrifices, would seemingly include changes in patterns of diet, air travel, ground transportation, and other forms of energy consumption. In the following sections, we address these, consider whether and how we might actually use climate change mitigation to lower inequality in development, and then conclude with a call to press our diplomats, business leaders, and politicians to implement more ambitious—and sorely needed—large-scale changes.

Eat More Plants and Fewer Animals

As discussed more extensively in Chapter 6, vegetarian diets are healthier, greatly diminishing risks for type II diabetes, cancer, and coronary episodes. The benefits from a vegetarian diet in reducing your carbon footprint are also important. The interdisciplinary *Drawdown* project, one of the most comprehensive efforts to brainstorm solutions to climate change, estimates that raising livestock accounts for some 15 percent of global emissions (Hawken 2017, 39). Humans could get the 50 grams of protein we need daily from vegetables; whereas in the United States and Canada, adults consume more than 90 grams of protein per day—and to our own detriment, as excess consumption of protein creates illness and "increased morbidity and health-care costs go hand in hand" (Hawken 2017, 39).

Beef is by far the worst culprit of emissions. Even just eliminating beef consumption can make a big difference in your diet. Cows are an extremely inefficient use of calories, absorbing only some 10 percent of those consumed. And they expel, on average, some 600 liters of methane a day (which is 25 times as potent as a GHG as CO_2). Ruminates (mammals such as cows, sheep, and goats that obtain their nutrients by fermenting plant-based food in specialized stomach chambers with microbes) produce six times as much emissions as pork and five times as much as poultry (see Chapter 6). Seafood, usually regarded as healthier for diets than any of the red meats, seems to generate more emissions than pork or poultry but less than cows (depending on whether they are in recirculating water systems, which emit more GHGs, or non-recirculating systems, which emit fewer). Eggs and dairy seem to produce the lowest GHG emissions of animal-based proteins. Vegetarian diets on average emit fewer than half the GHGs of poultry and pork eaters, and fewer than one-tenth those of beef eaters.

A 2016 study factored radical diet change into emissions models, claiming that, in the best case, a complete worldwide transition to plant-based diets by 2050 could reduce emissions by some 70 percent, although such a comprehensive change is highly unlikely (Hawken 2017, 39). Convincing "foodies" who savor Argentine churrasco barbeque beef smothered in chimichurri sauce, or Kansas City BBQ beef rib lovers, may be easier said than done. Economists in France (where demand for animal protein may be more elastic than in Argentina, meaning that consumers may more readily substitute chicken for beef), a large tax on beef was found in survey experiments to be more effective in changing consumption habits than a smaller tax on a wider range of meats (Bonnet et al. 2016). Tremendous progress has also been made in coming up with the Impossible Burger™ and the Beyond Burger™, which taste good without using cow ruminates and reduce methane emissions. But are they sustainable in the sense that they are healthier for people to eat? "Nutritionists commend Beyond Meat and Impossible Foods for creating products with a potentially lower carbon footprint than traditional meat production," writes *Huffington Post* blogger Erica Sweeney (2019), "but that doesn't mean they're more healthful."

The *Drawdown* team claims food waste is also as a major cause of emissions, not to mention a cause of hunger and starvation on a planet where 800 million people (out of over seven billion) face daily hunger. In low-income countries, food waste is caused by spoilage along the supply chain, whether from crops rotting on the vine, due to poor refrigeration, poor transportation infrastructure to get products to market (e.g., bad roads), or waste caused by heat and humidity along the supply chain (Hawken 2017, 42). In higher income nations, food goes to waste for more heinous, "willful" reasons: "Retailers reject food based on bumps, bruises, coloring . . . Other times, they simply order or serve too much" (Haken 2017, 42). The *Drawdown* project estimates that if the emissions caused by food waste were a country, they would closely follow the United States and China as the world's third-largest cause of emissions. Ever traveled to other parts of the world and noticed how large meal portions are in the United States compared to elsewhere? These large portion sizes cannot help with regard to the obesity epidemic in the United States, and they are also bad for the environment, generating needless emissions as we waste food on the same scale that we waste energy and other resources.

Flight-Shaming and "Walking the Walk": How Much Must We Reduce Transportation Emissions?

The climate activist Greta Thunberg spoke to many of us in a recent interview, after her trans-Atlantic boat voyage in August–September 2019 to a New York climate summit (Brangham 2019):

> Many people seem to have this double moral. They say one thing and then do another thing. They say that the climate crisis is very important and yet they do nothing about it. If I want to do something, then I go all in. I walk the walk. Walk the talk . . . I want to practice as I preach.

Many of us have, sometimes rightly, been accused of hypocrisy, as per Thunberg's rhetoric. And indeed, this claim hits academics and other frequent flyers quite hard. Many of us are being encouraged to undertake "flightless conferences" (meaning that if you are not within a train rail radius of the event, you should attend virtually). Perhaps a "silver lining" of the COVID pandemic will be that all our virtual conferencing during 2020 and 2021 has generated climate-improving habits that will last. Before COVID, people from Thunberg's climate change–aware nation of Sweden brought spouses to international conferences only with palpable embarrassment and guilt, which now has a word in the language: *flygskam*, or "flight shaming."[2] But can flying really be that bad for the climate?

According to Anglaret et al. (2019), air traffic constitutes three percent of global emissions, which is three times more than the total emissions of France. (France is near the bottom of the Top 20 CO_2 emitters.) Pre-COVID air traffic was on pace to double by 2030, and at this rate, the aviation sector could, by itself consume a quarter of the 2050 1.5°C carbon budget (Carbon Brief 2016). While only a small percentage of the world's population regularly travels by plane, those who do leave long CO_2 and vapor trails. "If you fly coach from Los Angeles to Paris and back, you've just emitted three tons of CO_2 into the atmosphere, 10 times what an average Kenyan emits in an entire year," wrote Climate Scientist Peter Kalmus (quoted in Rusu 2019). "Flying first class doubles these numbers."[3]

Think tanks like the New Economics Foundation in the United Kingdom have proposed that a frequent flyer levy be added to flight costs, with a rising tax on each flight a passenger takes per year after the first one, which they can take for free. Flight shaming has become a means of applying peer pressure, and the #FlyLess and #NoFly movements are gaining momentum. The authors of this book feel guilty enough to try to attend flightless conferences and purchase carbon credit offsets, and it does seem likely that business travel may be at least partly replaced by more online meetings, especially post-COVID. Still, the best parts of most conferences are the informal hallway gossip sessions rather than the formal presentations. We hope that technological breakthroughs in aviation may save biannual conference travel, but in the event that this does not happen, we are networking relentlessly in the hallways while we still can, and recognizing that the days of in-person conferencing beyond a train-travel radius may be waning. That said, there is also room for human contributions in other areas of transportation.

The Rise of Electric Cars

In 2018, the number of electric vehicles (EVs) rose to 3.2 million worldwide, jumping 50 percent over 2016 numbers. The number of new EV registrations in 2018 was 1.2 million! China is the world leader in electrical car use, with 1.23 million on the road, followed by the United States, with 760,000 and Germany with 93,000 (the entire European Union had 820,000 in 2018) (McCarthy 2018, *T&D World* 2018). Growth of EVs per year is astounding (Figure 9.1), and while we may ascribe China's remarkable surge in growth (Figure 9.2) to an authoritarian policy promoting their use, the growth in the United States is also remarkable.

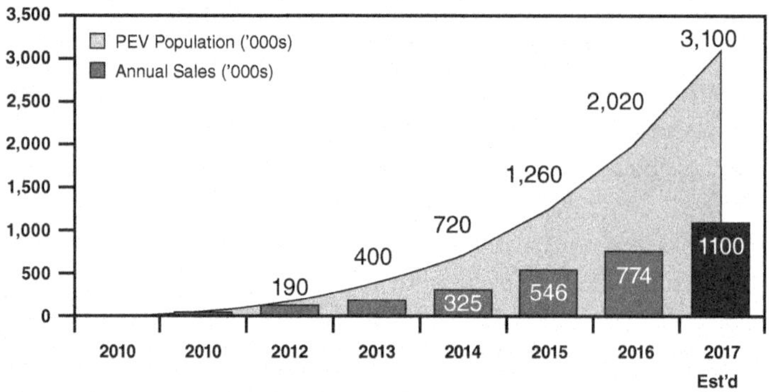

Figure 9.1 Surging growth of plug-in electric vehicles (PEV). (Reproduced with permission from Shahan 2017.)

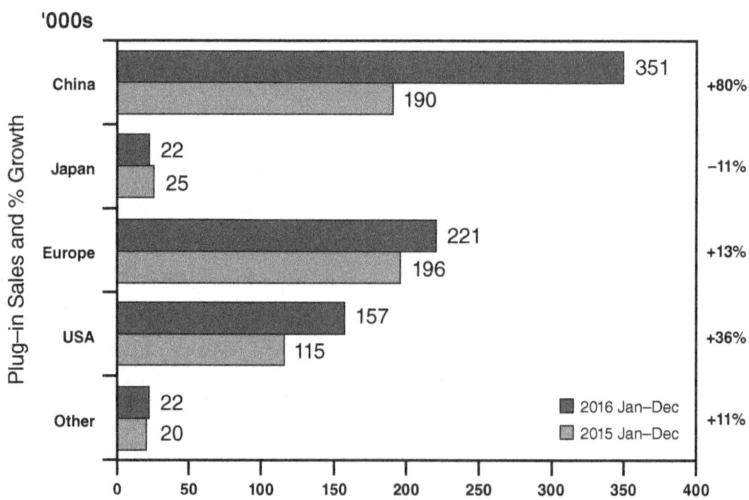

Figure 9.2 Plug-in sales among developed economies mostly increase between 2015 and 2016, with the most remarkable growth occurring in China. (Reproduced with permission from Shahan 2017.)

The fact that EVs increased market share in the United States is particularly revealing, both because fossil fuel interests have actively worked against them (Bade 2019) and because consumers have historically been ambivalent (US consumers have preferred SUVs in recent decades). Perhaps this shows that consumers are coming around to the benefits of EVs. Part of the trend is certainly because the cars are much cheaper to operate than those with internal combustion engines (ICE), and US consumers can obtain tax reductions if they purchase an EV. The average fuel cost for an ICE vehicle is 1,117 dollars per year, while operating an EV is about one-half that, at 485 dollars (Sivak & Schoettle 2018). General maintenance costs are also lower, since there are fewer moving parts to wear out and no exhaust system. The spread of charging stations (discussed in Chapter 6) and the continued development of batteries with ever-increasing storage capacity have alleviated consumers' two dominant concerns: How far can it go, and where will I refuel on trips? Perhaps consumers are also turning to EVs because they realize there are many drawbacks to petroleum-fueled vehicles regardless of CO_2 emissions. The petroleum fuel for ICE vehicles, after all, is (1) expensive, (2) explosive, (3) flammable, (4) carcinogenic, and (5) produces harmful vapors, not to mention that it hamstrings foreign policy for countries that depend on it.

Perhaps consumers also read movie star and politician Arnold Schwarzenegger's 2015 Facebook post regarding a thought experiment related to EVs and traditional combustion vehicles (appearing December 7, 2015). The former *Terminator* star, Humvee driver, and Republican ex-governor of California posted it in response to comments he was receiving regarding his promotion of renewable energy. The post is titled "I don't give a **** if we agree about climate change." Here is an excerpt from his post, which contains his hypothetical:

I see your questions.

> Each and every time I post on my Facebook page or tweet about my crusade for a clean energy future, I see them.
>
> There are always a few of you, asking why we should care about the temperature rising, or questioning the science of climate change.
>
> I want you to know that I hear you. Even those of you who say renewable energy is a conspiracy. Even those who say climate change is a hoax. Even those of you who use four letter words . . . I've heard all of your questions, and now I have (a question) for you.
>
> Let's put climate change aside for a minute. In fact, let's assume you're right . . .
>
> . . . There are two doors. Behind Door Number One is a completely sealed room, with a regular, gasoline-fueled car. Behind Door Number Two is an identical, completely sealed room, with an electric car. Both engines are running full blast.

I want you to pick a door to open, and enter the room and shut the door behind you. You have to stay in the room you choose for one hour. You cannot turn off the engine. You do not get a gas mask.

I'm guessing you chose the Door Number Two, with the electric car, right? Door number One is a fatal choice—who would ever want to breathe those fumes?

This is the choice the world is making right now.

To use one of the four-letter words all of you commenters love, I don't give a damn if you believe in climate change. I couldn't care less if you're concerned about temperatures rising or melting glaciers. It doesn't matter to me which of us is right about the science.

I just hope that you'll join me in opening Door Number Two, to a smarter, cleaner, healthier, more profitable energy future.

Fostering More Sustainable Cities

In 2016, approximately 54 percent of the world's population lived in cities. By 2030, that is projected to reach 60 percent, and by 2050, 70 percent. Globally, the urbanization of the population shows that the big cities are getting bigger. In 2016, there were 436 cities with between one million and five million people, and by 2030, it is projected that there will be 559 (United Nations Habitat 2016). Cities are currently responsible for 40 percent of the world's energy use and 50 percent of all anthropogenic CO_2 emissions. As the cities grow, these percentages will as well. Recognizing that point sources of pollution (including CO_2 pollution) are often easier to mitigate than more diffuse, nonpoint sources, there has been great interest in developing cities that have a lower carbon footprint. Low-energy, sustainable, LEED-certified buildings were discussed in Chapter 6 as an area where improved carbon reductions could be quantified, but what about urban development itself?

There are a few areas that most cities could examine with an eye toward reducing GHG emissions. Capture of landfill methane for power generation in Denton, Texas, has enabled that city to power 1,600 households with the fuel that otherwise would have been released into the atmosphere. Reduction of solid waste production can be achieved by encouraging recycling and composting through education campaigns or by punitive measures, such as charging waste collection fees based on the weight of a household's collected trash. Using this approach allowed Santa Cruz, California, to cut its CO_2 by 11 percent between 1996 and 2005, even as its population increased approximately 10 percent over that period (Tallman 2013).

Both individual households and municipalities can increasingly choose to purchase natural gas and electricity from renewable power vendors. This supports the market for green energy and can make nontransportation energy use approach 100 percent clean. For example, in Maryland, one can purchase either solar or wind power from an energy supplier that would then be delivered via the local grid system. Consumers can choose up to 100 percent of their energy from the alternative sources. Wind power has now become cheaper than even natural gas, so consumers can both save money and drastically reduce their CO_2 footprint.

Also, as cities grow, they are increasingly enabling walkability over promoting vehicle use. Vehicles can cause massive congestion, pollute, interfere with emergency services, and are dangerous (to pedestrians), and they are expensive to maintain, especially in cities. In an effort to encourage residents to ditch the car, cities are promoting walkability, biking, and mass transit. It is projected that encouraging walkability in cities worldwide could reduce driving by 20 to 40 percent, CO_2 emissions by 2.0 gigatons by 2050, and municipal operating costs by a whopping 3.3 trillion dollars by 2050 (Energy Mix 2018).

Can Mitigation Empower the Poor, Women, and the Global South?

Work by Boustan et al. (2017) and Case Study 9.1 in this book show that, even in rich nations like the United States, people relocate from areas stricken by large-scale natural disasters, exacerbating inequality. These impacts are heightened in the most vulnerable nations, largely in the Global South. As argued by Boustan et al. (2017, 18), poverty rates increased by nearly one percentage point in severe disaster areas, worsening the effects of climate change, as those newly poor (whether they migrate into disaster-stricken areas, become poor due to loss, or whether they are impoverished by out-migration of people with resources) are less equipped to address the next severe weather event. They conclude that "areas that do not adapt to natural disaster risk will become poorer over time, as well-to-do residents move away." In climate-vulnerable Bangladesh, it seems that many of the less well-to-do also move away from disaster-prone coastal areas, such as in Barishal, and that their economic futures in the slums of Dhaka and elsewhere are not extremely bright, but also that seasonal and permanent migration are widespread (Eisenstadt et al. 2020).

The continued disenfranchisement of women from community involvement also has had a role in increasing emissions in much of the world. In fact, the *Drawdown* project ranked "educating girls" and "family planning" as sixth and seventh among their 80 proposed "fixes" to climate change, just below "tropical forests" and above "solar farms" (Hawken 2017, 222).[4] A 2010 study cited by *Drawdown* showed that educating girls improves the fertility of the land they till, improves the health and nourishment of entire communities, and gives them higher wages and greater mobility. "It is the most powerful lever available for breaking the cycle of intergenerational poverty, while mitigating emissions by curbing population growth," according to the *Drawdown* team. "Education also shores up resilience in terms of climate change impacts – something the world needs as warming mounts" (81). We must fight the tendency, predominant through time in some countries, that women should stay at home rather than receiving educations. Families with economic limitations have often restricted the education of women, but that must change as girls must be ensured access to affordable, high quality, nonsexist educations. This is a human right as well as a climate mitigation technique, a win-win that is long overdue.

Another important climate-reducing role that can be played more actively by women is family planning. In the 1790s, Thomas Malthus anticipated the coming problem of runaway population growth, and Paul Ehrlich and John Holdren in the 1970s formalized an updated version of Malthus's basic argument, with the statement that "Impact [on Environment] = Population x Affluence x Technology" (Hawken 2017, 78). While much of this book (and climate change planning) has addressed technology and affluence, population continues to be an important issue, as—of course—economic demands drive the economy and larger populations generate higher economic demands.

Family planning programs have reduced birth rates in countries such as Bangladesh, Iran, and Mexico since the 1970s from up to six children to an average of just two per family. However, "the topic continues to be taboo in many countries and institutions, hemmed in by the persistent belief that raising the issue of population, or approaches reduce it, is inherently draconian and an affront to the worth of human life" (Hawken 2017, 79)." Even so, as the *Drawdown* project concludes, family planning honors the dignity of women and children through giving people the knowledge to make good choices. As our planet's women, particularly in nations where they are precluded from receiving the full knowledge needed to make informed choices, they need to be given more education and training than men (not less), and they need to be sensitized to the impacts of the choices.

Empowerment of the poor and women will go far toward mitigating climate change. But another group also needs more tools and, more generally, equitable treatment in adapting to climate change so that inequality does not increase and, with it, economic inefficiency as well as the moral impropriety. We must offer "climate justice" to people from the Global South, usually people of color who have faced discrimination in other realms. These are the citizens, generally, of the tropical areas beset by the worst climate-impacted hurricanes and droughts, who are already seeking to migrate to cooler climates in search of jobs, economic security, and less dangerous places to raise their families. And since 2020, spending on the COVID-19 pandemic has only made inequality worse, crowding public funds worldwide into short-term public health and welfare expenditures and out of long-term planning expenditures on seemingly less urgent issues like climate change mitigation and adaptation.

Rather than rejecting climate migrants (and others) via the torrent of nativist populism that has beset nations from France to Germany to the United Kingdom to the United States to Brazil in recent years, we should welcome these immigrants for a host of reasons, including economic ones. Industrialized nations, such as Germany, Japan, and the United States, face population stagnation (even in the face of overall worldwide population growth) and need more working-age citizens to keep their economies growing. We need to welcome new workers to the US economy rather than disparaging people from other nations, as former US President Donald Trump notoriously did when he told Somali-American Congressmember Ilhan Omar to "Go back (to the) places from which (you) came (Rogers and Fandos 2019)." More people will be seeking to resettle in northern latitudes as the climate warms, and they should be welcomed, as we will need their energy and vibrance and they will need new homes.

Other voiceless groups to be considered include unborn generations of people, as discussed in Chapter 1. Gardiner (2011) argues that the global and intergenerational elements of the climate change conundrum make it all the more difficult to fully grasp. This most striking fact, that "the planet is at serious risk of a shift in global climate comparable in magnitude to an ice age (albeit in the other direction), but occurring over decades rather than millennia" (xi) offers the most profound ethical challenge humanity has ever faced. "What matters most is what we do to protect those vulnerable to our actions and unable to hold us accountable," he writes (xii). While Gardiner would surely include women among the disempowered global poor and among the most vulnerable, he also mentions future generations and "non-human" nature.

John Muir strove valiantly to preserve nature in his debate with Gifford Pinchot over the fate of the Yosemite Valley, back when Teddy Bears represented nature's taming, and climate change was not even in the air (at least conceptually). We must continue to conserve nature, as it is attached to our climate and is perhaps the single best guarantor of our continuance as a planetary ecosystem. The rights of nature have now been enshrined in a few of the world's constitutions,[5] starting with Ecuador's 2008 constitution, which was inspired by the Quechua Andean cosmovision, under which humans are one part of the holistic entity of nature. Indeed, recent scholarship (Eisenstadt & West 2019) has found the Andean cosmovision more conducive to stewardship of nature than Western worldviews, rooted more in the dominion hypothesis, in relation to the role of humans in protecting the earth from anthropogenic climate change. The voices of these groups, and those of other disempowered citizens, urgently need to be brought into the discussion.

Press Politicians, Nation-States, Corporations

More broadly, as Gardiner (2011) also poignantly points out, the logic of collective action, and even the "Tragedy of the Commons" may not fully apply. As game theory, the interactive logical approach embodied by the tragedy of the commons and by intergenerational buck passing, is principally a descriptive enterprise rather than a moral/ethical one. It allows the players in the "game" to be egotistical, or at least self-interested, and that paradigm of thinking has gotten us nowhere (Gardiner 2011, 50). Moreover, the time horizons that predominate in most game theory models may just be too short-term to capture the negative payoffs of continued consumption by individuals, governments, and businesses.

Ten countries do emit 80 percent of the world's emissions, and these countries and their citizens need to be convinced to make do with less. And it is not that they will "overgraze" the commons; it is that they will obliterate the commons, and the planet. The tragedy of the commons contradiction between continuing consumption for individual gain, even at the expense of collective loss, does not fully specify the fact that at a certain point, the entire "game" ends, after which there can be no more individual consumption. What rational human, even a selfish one with lots of grazing animals, could support that outcome?

Recent research has shown that people may be just as likely to "abandon the commons" as they are to "save the commons." A 2014 national survey in Ecuador of 1,700 people, with an oversample in the Amazon region, showed that where the rainforest remains pristine, people concern themselves with protecting it. But where the rainforest has been damaged, such as in Ecuador's northern Amazon region after the reckless 1990s oil spills by Exxon-Texaco and the Ecuadorian parastatal oil company, peoples' attitudes of concern shift from the abstract concept of "saving the environment" to the subordinate logic of "now that the environment is unfit for our livelihoods, what do we do?" They worry instead about migration (to the cities), public health, and economic development (Eisenstadt & West 2019). The shameless 2019 burning season in Brazil's much larger Amazon rainforest, tacitly led by that country's willfully ignorant president, Jair Bolsonaro, offered a vivid display of subordinating the climate and environment to "slash and burn" agriculture.

Extending this message, if we do not actively conserve natural environments like rainforests before they are damaged, people will lose interest in protecting what is left of them later. This finding is highly consistent with the thesis in this book—namely, that if we do not sacrifice a little now (regulating forest burns to preserve carbon "sinks" and, more broadly, taxing GHG emissions) to prevent climate change from reaching dangerous levels, our grandchildren will have to sacrifice a lot more later. And those grandchildren, by virtue of our actions now, will start with fewer resources and a smaller range of options, because of what Gardiner (2011) refers to as an "intergenerational arms race" (185–207).

The COVID-19 pandemic may make it even more difficult for elected officials to summon the will to dedicate the attention and resources to climate change, especially in the United States, where COVID mortality was still by far the worst in the world as President Biden was taking office in late January 2021, and climate denial seemed to be a mainstay belief of many supporters of former President Trump. As pointed out by Bodnar and Grbusic (2020), the average US taxpayer was already paying up to about 1,000 dollars per year on climate-related disaster recovery. In 2017, Congress appropriated 136 billion dollars for this relief, but by the middle of 2020, Congress had appropriated 3.6 trillion dollars (well over 26,000 dollars per taxpayer) on COVID relief. While the COVID pandemic may yet renew faith in the tragedy of the commons by illustrating the cliched but true statement that "we are all in this together," Jamieson and Jacquet (2020) argue that the US COVID response has further dimmed prospects for a strong and comprehensive response to climate change. They argue that rather than cooperating to solve the COVID crisis, the US government, at least in the early months of the pandemic, failed to take responsibility and instead outsourced solutions to the private sector, as they have in climate change, which was denied by presidents Reagan, Bush, and Trump.

The United States under Trump failed to lead and encourage international efforts to solve COVID, according to these critical authors, by choosing instead to disparage international institutions like the World Health Organization, just as they have the IPCC and the United Nations Framework Convention on Climate Change (UNFCCC) during much of the last 40 years. Furthermore, and perhaps most damning, former President Trump's disregard for scientific and technical expertise in the COVID crisis has been compared to his climate change denial, manifested foolishly and lethally in denial of the pandemic for months and failure to provide leadership regarding lockdowns or re-openings. Indeed, US government dysfunction in addressing the COVID-19 pandemic under Trump was dramatic, to be sure, although President Biden responded deftly to both problems. But just as Trump's COVID response seemed to increase US fatalities, his climate change mismanagement (and that of presidents before him with the exceptions of Obama and Clinton) may prove much more costly.

Populist politicians, those who favor short-term gains at the expense of the long-term interests of their country and who lean on nationalistic symbols to reinforce their support, will always deny climate change. We must marginalize them, learn enough about the science to be able to articulately convey our policy options, and move forward, with disciplined sacrifice and purpose. President Ronald Reagan removed the solar collectors that President Jimmy Carter had placed on top of the White House. It is probably time to plant a whole array in the Rose Garden. While we are at it, we must advocate for changing campaign finance laws and better regulating the role of special interests in the legislative process, as we discussed in Chapter 4. And even though citizens must rely on governments to regulate industry, rather than hoping that industry will police itself, board members can push companies toward greener policies, and consumers can boycott companies that fail to go green. We would all do well to remember that there was a world with wilderness and wild beasts before we tamed them as Teddy Bears.

Plant a Garden; Rewild the Planet

To come into closer contact with the earth, the environmental writer Michael Pollan says that it is time to get our hands dirty, but as gardeners. He advocates, in a famous essay, working out the contradictions between our identities as high-flying consumers and emerging climate activists. The metaphor he offers is that of planting a garden. "At least in this one corner of your yard and life," Pollan (2008) instructs us, "you will have begun to heal the split between what you think and what you do, to commingle your identities as consumer and producer and citizen."

While this was certainly therapeutic advice when first offered, it rings incomplete today. We remind our gentle readers that the final passage of Voltaire's French Enlightenment classic *Candide* is also an exhortation that "we must cultivate our garden" (Voltaire 1759, 97). However, in that case, it was an appeal for protagonists to take care of their communities through self-care, without worrying about events beyond themselves. And while it may be possible to work your garden and a community's commons co-equally, this planet's commons has been increasingly ill-tended, and this book must end with an even more ambitious plea, given our collective disregard for the climate in the decade since Pollan wrote his.

Moreover, gardening will not help increase the number of wild spaces. One author was initially surprised, upon visiting Costa Rica's famed Manuel Antonio National Park, renowned for its wildlife viewing, that the entire park is fewer than eight square miles.[6] Guides lead tourists on bird-, reptile-, and insect-viewing tours with binoculars and tripods in a domesticated "wilderness" that could have been straight out of the 2017 Matt Damon movie *Downsizing*. Having hiked days on end in New Mexico's Pecos Wilderness as a teen without encountering other people, the author was initially disappointed at Manuel Antonio's troves of tourists in what amounted to a nature-themed human park. Yet going forward, nature's reduced scope may make such venues all the more precious.

How do we continue to enjoy what is left of nature as a wild force rather than a sterile one? And what does this have to do with climate change?

The journalist George Monbiot popularized the concept of rewilding in 2013. As he wrote in *The Guardian*: "Through rewilding—the mass restoration of ecosystems—I see an opportunity to reverse the destruction of the natural world" (Monbiot 2013). However, later proponents, like the scholar Paul Wapner, have referred to rewilding as an ethical way of coping with climate change and other environmental degradation of the Anthropocene. To Wapner (2020, 114), humans can feel more engaged with the world by accepting and accommodating the unknown (represented by the wilderness we humans must bring back as a state of mind and nature). Specifically:

> Accommodation doesn't equal resignation. It means engagement, but of a certain kind. Befriending [wilderness] involves an attitude shift. Instead of a technocratic view that sees every challenge as a puzzle in need of a solution, accommodation encourages ethical sensitivity and commitment. It invites people to care more about how they treat each other and nature than about how they might solve a given dilemma. (115).

To Wapner, then, the rewilding movement implores people to take responsibility for environmental harms and address them ourselves.

The appeal here might be in recovering the awe that Theodore Roosevelt felt upon encountering the black bear during his Mississippi hunting trip. If we could go back to the moment after Roosevelt was stricken by the majesty of the bear but before he considered killing it (yet spared its life, according to myth, sanitizing the wild beast as a Teddy Bear), we may find the purpose and moral engagement we need for the long struggle ahead. Perhaps we can "rewild" our conscience and our lands, as much as possible, but let us also hedge our bets by continuing to insist upon adhering also to carbon budgets and technocratic fixes. Why must we choose to rewild *or* undertake technocratic solutions? Let's plant nature's carrots, to be sure, but let's also threaten climate destroyers with sticks. The climate-deniers and foot-draggers have held us back for far too long.

Think Big: Plant a Garden, Sure, But Also Empower an International "Climate Squad"

To fix the Amazon, French President Emmanuel Macron offered in September 2019, on behalf of the G7 industrial nations, to contribute a mere 20 million dollars to "save the rainforest," and Brazilian President Bolsonaro called the offer a "lamentable colonialist stance" (Phillips 2019). As these leaders traded barbs, the forest burned. There exists no worldwide institution sufficiently representative, credible, or powerful to stop the slashing and burning. What if the UN Security Council could send "green helmets" to climate crisis areas, as that body has sent "blue helmets" for over half a century to offer peacekeeping and humanitarian relief?

True, this relic of the Allied victory in World War II has scarcely ever looked beyond the narrow national self-interests of its permanent members (the United States, United Kingdom, France, Russia, and the People's Republic of China). And true, the United Nations has been overrun by inefficiency and corruption. And also true, it remains unlikely that aggressive leaders like Bolsonaro would allow "green" helmets to violate their sovereignty, given the contention and cajoling needed before any "blue helmet" mission, and the limits these hamstrung peacekeepers always face.

But what else has worked? Scientific research has brought technological advances, as highlighted in Chapters 5 and 6, but we have not found a substitute for political will and, hence, we must rely on the range of partial solutions that constitute "Plan A," rather than turning to the unstable "Plan B" of geoengineering. The overwhelming scientific evidence of climate change, outlined in Chapter 2, and the complexities of solving the problem mean that we must turn to the realm of politics, as well as continuing to vest some hopes—and research resources—in technology. And while subnational and national climate policies have their place, the international arena is the highest level of aggregation, where political solutions can be applied more sweepingly across nations. The UN-associated IPCC has succeeded in turning attention to the problem, but the UNFCCC has failed to tackle climate change on the order of magnitude depicted by the IPCC.

The Amazon is scorching, and the United Nations has shown flickers of effectiveness (whatever your view of the 2003-initiated Iraq War, consider the centrality of the UN nuclear weapons inspectors—and their bravery—in alerting the world to Iraq's nuclear designs). The five permanent members of the UN Security Council would have to overcome radical geopolitical differences on most matters to support climate actions and bring along the rotating dozen nonpermanent members. We need not look to the Arctic snow melts and Bangladeshi cyclones to know of the increasing rate and ferocity of climate change. US students can look closer to home, at recent extreme weather events in the Bahamas, Brazil, as well as in California, Florida, Louisiana, New Jersey, Puerto Rico, and Texas, to affirm pessimistic scientific predictions and shift public opinion even in the most denial-prone of major emitting nations, this one.

The Paris Agreement, we euphemistically stated in Chapter 7, needs to ratchet up its ambitions. Aside from a few ambitious nations like Norway and France, which are on track for carbon neutrality by 2050 or sooner, the world is running out of time to meet the target of only 1.5°C of warming by the end of the century, or even the more plausible 2°C aspiration. National sovereignty is seemingly going to be eclipsed within 20 or 30 years, as the emissions "ambition gap" grows wider between actual reductions and necessary ones. But against a more dire backdrop, continued calamities could sway leadership in the five permanent members to empower the Security Council to form a "green helmet" climate squad. This may be the moment to push for new climate leadership from China, European Union members (led by Germany and Great Britain), India, the Scandinavian conscience governments (Norway and Sweden). and the United States to use their collective national might to push for more climate action worldwide, bolster the United Nations (or a better institution based on CO_2 emissions rather than World War II–era geopolitics, but born of the UN), and act sooner rather than later, while our range of solution choices is wider, The Amazon rainforest, and the planet, can still rise up from the ashes glowing on its periphery.

Let's implement a more aggressive "Plan A" now and give our grandchildren the serenity to plant gardens. And heck, let's leave them some wild rainforests while we are at it. They will need some wonderment, too.

Exercise 9.1 **Exploring Your Carbon Footprint**

Using the Global Footprint Network's online tool (http://www.footprintcalculator.org/), calculate and analyze your environmental footprint, and then answer the following questions:

1. How many Earths would it take to support the planet's population if everyone used the same amount of resources you do?
2. How many acres (or hectares) of production does your lifestyle need?
3. How many tons of CO_2 do you produce (a year)?
4. Try altering your "lifestyle" and do the exercise again. What did you alter? What was the result (Earths, acres or hectares, CO_2)? Was there much of a difference? Why or why not?
5. Is it even possible to use less than "one Earth's" resources?
6. What does the exercise tell us about the role of individuals in solving climate change?

NOTES

1. The "small individual acts add up" logic is reminiscent of the stream of financial advisor advice that if people just stopped consuming their barista-prepared coffees every morning, they could save enough money to retire years earlier. See Olen 2016.
2. One of the authors thanks colleagues at the 2019 American Political Science Association annual conference in Washington, DC, who flew from Stockholm to Washington (one spouse "just for the weekend") but also spent 20 minutes discussing their *flygskam*. These are kindhearted colleagues who research climate change and surely deserve a break from working constantly about their carbon footprints. Or do they?
3. First class passengers can get up to 45 percent more space, diminishing the number of people the planes can carry, making them less fuel efficient (McCartney 2019).

4. According to the *Drawdown* project, the most important "low-hanging fruit" solutions are (1) improving refrigeration, (2) using more onshore wind turbines, (3) reducing food waste, (4) going on plant-rich diets, (5) tropical forest conservation, (6) educating girls, (7) better family planning, (8) more solar farms, (9) using more silvopasture (mix of "overlapping" grazing and crop lands), and (10) more use of rooftop solar panels (Hawken 2017, 222).
5. On the constitutional rights of nature, see Eisenstadt and West 2019 regarding the Ecuador case, the first constitution protecting nature. For a more comprehensive discussion, including references to New Zealand, Bolivia, and some US city councils, see Kauffman and Martin. 2017.
6. Participant observation by Todd Eisenstadt on July 22, 2017, at Manuel Antonio National Park, Quepos, Costa Rica.

REFERENCES

Africa Group of Negotiators, Experts, Support. 2020. "Desertification and Climate Change in Africa," March. Policy paper accessed January 8, 2021, at https://cgspace.cgiar.org/bitstream/handle/10568/107808/PB1_Desertification.pdf

Anglaret, Xavier, Chris Wymant, and Kévin Jean. 2019. "Researchers, set an example: Fly less." *The Conversation*, February 13. Accessed September 16, 2019, at https://theconversation.com/researchers-set-an-example-fly-less-111046

Bade, Gavin. 2019. "The oil industry vs. the electric car." *Politico*, October 16. Accessed December 9, 2020, at https://www.politico.com/story/2019/09/16/oil-industry-electric-car-1729429

Bodnar, Paul, and Tamara Grbusic. 2020. "Your climate disaster tax bill is growing." *New York Times*, June 23. Accessed June 23, 2020, at https://www.nytimes.com/2020/06/23/opinion/climate-change-financial-disaster.html

Bonnet, Celine, Zohra Bouamra-Mechemache, and Tifenn Corre. 2016 "An environmental tax towards more sustainable food consumption: Empirical evidence of the French meat and marine food consumption." Working Paper TSE-639 of the Toulouse School of Economics, Toulouse, France.

Boustan, Leah Platt, Matthew E. Kahn, Paul W. Rhode & Maria Lucia Yanguas. 2017. "The Effect of Natural Disasters on Economic Activity in US Counties: A Century of Data," National Bureau of Economic Research Working Paper 23410. Accessed January 8, 2021, at https://www.nber.org/system/files/working_papers/w23410/w23410.pdf

Brangham, William. 2019. "Climate activist Greta Thunberg on the power of a movement." *PBS News Hour,* September 13. Accessed January 7, 2021 at https://www.pbs.org/newshour/show/climate-warrior

Carbon Brief. 2016. "Analysis: Aviation could consume a quarter of 1.5°C carbon budget by 2050." Accessed September 16, 2019, at https://www.carbonbrief.org/aviation-consume-quarter-carbon-budget

Corn, David. 2019. "It's the End of the World as They Know It: The distinct burden of being a climate scientist," in *Mother Jones*, July 8. Accessed January 8, 2021, at https://www.motherjones.com/environment/2019/07/weight-of-the-world-climate-change-scientist-grief/

Dietz, T., Gerald T. Gardner, Jonathan Gilligan, Paul C. Stern, and Michael P. Vandenbergh 2009. "Household actions can provide a behavioral wedge to rapidly reduce US carbon emissions." *Proceedings of the National Academy of Sciences* 106(44):18452–18456.

Eisenstadt, Todd A., and Karleen Jones West. 2019. *Who Speaks for Nature: Indigenous Movements, Public Opinion and the Petro-State in Ecuador*. Oxford University Press, New York.

Eisenstadt, Todd A., Tawfique Haque, Jie Lu, Michael Toman, and Matthew Wright. 2020. Forthcoming.

Energy Mix. 2018. "Walkable cities could save 2.92 gigatons of carbon by 2050." *The Energy Mix,* November 5. Accessed January 7, 2021 at https://theenergymix.com/2018/11/05/walkable-cities-would-save-2-92-gigatons-of-carbon-by-2050-2/

Gardiner, Stephen M. 2011. *A Perfect Moral Storm: The Ethical Tragedy of Climate Change.* Oxford University Press, New York.

Hawken, Paul, ed. 2017. *Drawdown: The Most Comprehensive Plan Ever Proposed to Reverse Global Warming*. Penguin Books, New York.

Intergovernmental Panel on Climate Change (IPCC). 2018. IPCC, 2018: Global Warming of 1.5°C. An IPCC Special Report on the impacts of global warming of 1.5°C above pre-industrial levels and related global greenhouse gas emission pathways, in the context of strengthening the global response to the threat of climate change, sustainable development, and efforts to eradicate poverty [Masson-Delmotte, V., P. Zhai, H.-O. Pörtner, D. Roberts, J. Skea, P.R. Shukla, A. Pirani, W. Moufouma-Okia, C. Péan, R. Pidcock, S. Connors, J.B.R. Matthews, Y. Chen, X. Zhou, M.I. Gomis, E. Lonnoy, T. Maycock, M. Tignor, and T. Waterfield (eds.)].

Jack, Chris interview. October 8, 2019. African Climate Risks Conference, Addis Ababa, Ethiopia. University of Cape Town climate scientist and research director of FRACTAL project.

Jamieson, Dale, and Jennifer Jacquet. 2020. "The US response to COVID-19 and climate change endangers the country and the world." *One Earth* 2 (June 19):503–505.

Kauffman, Craig M., and Pamela L. Martin. 2017. "Can rights of nature make development more sustainable? Why some Ecuadorian lawsuits succeed and others fail." *World Development* 92:130–142.

Kasperbauer, T. J. 2016. "The implications of psychological limitations for the ethics of climate change." *Environmental Values* 25:353–370.

Lazrus, Heather. 2012. "Sea change: Island communities and climate change." *Annual Review of Anthropology* 41:285–301.

McCarthy, Naill. 2018. Electric car sales are surging in China." *Forbes,* June 1. Accessed January 7, 2021 at https://www.forbes.com/sites/niallmccarthy/2018/06/01/electric-car-sales-are-surging-in-china-infographic/?sh=2f174560d1f7

McCartney, Scott. 2019. "First Class vs. Coach: A Game of Square Inches," in *The Wall Street Journal*, April 24. Accessed January 8, 2021, at https://www.wsj.com/articles/first-class-vs-coach-a-game-of-square-inches-11556098201

Monbiot, George. 2013. "My manifesto for rewilding the world." *The Guardian,* May 27. Accessed October 5, 2020, at https://www.theguardian.com/commentisfree/2013/may/27/my-manifesto-rewilding-world

Mwalukanga, Brenda interview. October 9, 2019. Africa Climate Risks Conference, Addis Ababa, Ethiopia. Coordinator of Lusaka Water Security Initiative.

Olen, Helaine. 2016 "Buying coffee every day isn't why you're in debt: Debunking the personal finance advice industry's favorite myth." *Slate*, May 26. Accessed August 26, 2019, at https://slate.com/business/2016/05/the-latte-is-a-lie-and-buying-coffee-has-nothing-to-do-with-debt-an-excerpt-from-helaine-olens-pound-foolish.html

Phillips, Tom. 2019. "Jair Bolsonaro demands Macron withdraw 'insults' over Amazon fires." *The Guardian*, August 27. Accessed September 16, 2019, at https://www.theguardian.com/world/2019/aug/27/jair-bolsonaro-demands-emmanuel-macron-withdraw-insults-over-fires

Pollan, Michael. 2008. "Why bother?" *New York Times*, April 20. Accessed September 16, 2019, at https://www.nytimes.com/2008/04/20/magazine/20wwln-lede-t.html

Rogers, Katie and Nicholas Fandos. 2019. "Trump Tells Congresswomen to 'Go Back' to the Countries They Came From," in *New York Times*, July 14. Accessed January 8, 2021, at https://www.nytimes.com/2019/07/14/us/politics/trump-twitter-squad-congress.html

Rusu, Anca. 2019. "Interview Peter Kalmus: Climate Scientist, Activist, and Author," October 28 blog at ethical.net.com. Accessed January 8, 2021, at https://ethical.net/ethical/interview/interview-climate-scientist-peter-kalmus/

Schwarzenegger, Arnold. 2015. "I don't give a **** if we agree about climate change." *Facebook*, December 7. Accessed January 7, 2021, at https://m.facebook.com/nt/screen/?params=%7B%22note_id%22%3A10158687852561760%7D&path=%2Fnotes%2F%7Bnote_id%7D&_rdr

Shahan, Zachary. 2017. "Tesla Model S & Nissan LEAF clocked as world's best-selling electric cars in 2016." *CleanTechnica,* February 4. Accessed December 8, 2019, at https://cleantechnica.com/2017/02/04/tesla-model-s-clocked-as-worlds-best-selling-electric-car-in-2016/

Sivak, Michael and Schoettle, Brandon. 2018. "Relative Costs of Driving Electric and Gasoline Vehicles in the Individual US States," Sustainable Worldwide Transportation Paper SWT-2018-1

(January). Accessed January 12, 2021, at http://umich.edu/~umtriswt/PDF/SWT-2018-1.pdf

Sweeney, Erica. 2019. "Are Beyond Meat and Impossible Burgers better for you? Nutritionists weigh in." *Huffington Post,* July 10. Accessed September 15, 2019, at https://www.huffpost.com/entry/beyond-meat-impossible-burger-healthy_l_5d164ad1e4b07f6ca57cc3ed

Tallman, Nicole. 2013. "4 US cities committed to reducing their carbon footprint." *Safewise,* August 14.

T&D World. 2018. "Number of electric cars rises from 2 million to more than 3 million." Accessed December 9, 2020, at https://www.tdworld.com/grid-innovations/distribution/article/20970872/number-of-electric-cars-rises-from-2-million-to-more-than-3-million.

United Nations Habitat. 2016. *The World's Cities in 2016: Urbanization and Development—Emerging Futures.* United Nations Habitat, New York. Accessed July 3, 2020, at https://unhabitat.org/world-cities-report

Voltaire. 1759 (1998 ed.). *Candide.* Electronic Scholarly Publishing. Accessed June 13, 2020, at http://www.esp.org/books/voltaire/candide.pdf

Wapner, Paul. 2020. *Is Wilderness Over?* Polity Press, Medford, MA.

Case Study 9.1

WHAT DOES HURRICANE KATRINA REVEAL ABOUT ENVIRONMENTAL JUSTICE IN CLIMATE-RELATED DISASTERS?

GABRIELLA A. SIEGFRIED[1]

Rufus Burkhalter, 61, and Bobby Brown, 58, were water-pump operators when Hurricane Katrina ravaged New Orleans in 2005. Twenty-year veterans of their jobs, both men told the media that they would be the last to leave their posts during any disaster, because their job was to ensure the water-pumps worked until the absolute last moment. Katrina howled around them: "Half of our roof was off the station . . . we had to scuffle back to cover up the electrical switch gearbox [to avoid an explosion] . . . after we couldn't pump no more, I thought I was dreaming . . . I thought I saw dead bodies in the water," Burkhalter said. "That was no dream," Brown added (StoryCorps 2006).

John Taylor spent his childhood in New Orleans and referred to the city as his "homeport." After Katrina, he spent seven days sleeping in the back of his truck before receiving any government-issued aid (Federal Emergency Management Agency [FEMA] 2006, Rice 2018, StoryCorps 2006). Taylor recounted how his sister died in a nearby hospital due to power outages and stated that his city appeared to have sustained "chemical warfare" (FEMA 2006, Rice 2018, StoryCorps 2006).

Douglas P. deSilvey of Gulfport, Mississippi, was the only one in his family to survive the storm surge. As deSilvey stated, "The water came up so fast . . . the roof came down on all of us and I kept asking Jesus to not let me go . . . I got out and they didn't . . . I could not do a thing." deSilvey lost his mother-in-law, father-in=law, wife, and daughter to the rushing waters of Hurricane Katrina (StoryCorps 2006).

These oral histories were recorded by National Public Radio a year after the storm, and over 15 years later, the memories are still fresh among those who bore witness to the storm's ravages. In total, Katrina damaged approximately 90,000 square miles of housing—an area the size of the United Kingdom—in the Gulf Shores region in August 2005. The storm surge displaced approximately 1.5 million people, and of those affected, 1.36 million requested federal disaster assistance (Elliott & Pais 2006, 302; Graif 2016; Groen & Polivka 2008, 32). *USA Today* and the *New York Times* reported 160 billion dollars in economic damages, with over 1,800 causalities (Dewan & Schwartz 2017, FEMA 2006, Rice 2018). Hurricane Katrina was the most destructive and costly storm in the history of the United States, and Hurricane Harvey (August 2017), which had approximately 40 deaths and 125 billion dollars in damages, is second (Dewan & Schwartz 2017, Rice 2018).[2]

The Census Bureau's 2006 Current Population Survey documented that Katrina evacuees fled to some 45 states, with the vast majority relocating to Houston, Dallas, Atlanta, and Memphis (Groen & Polivka 2008, 43). Katrina survivors who did not evacuate and watched the damage firsthand listed the following reasons for staying behind: lack of shelter [elsewhere], lack of transportation, inability to leave their money/property/jobs, poor health, few interstate social networks, miscommunication by local authorities, low-risk perception due to previous storms, and lack of trust in the government (Eisenman et al. 2007). Before Katrina, there were "no precedents for this degree of sudden devastation and out-migration from a major urban region," and by far, the largest portion of those displaced due to the hurricane came from New Orleans (Elliott & Pais 2006, 302; Groen & Polivka 2008, 36). As reported by Figure 9.3, those who initially returned to the impacted Gulf Shores area were older, married, more educated, and predominantly white (Groen & Polivka 2008).

Climate Vulnerability and Environmental Justice

The story of disaster-related mass migration is a recurring one, specifically as the United Nations and other international agencies consider whether climate migrants should warrant refugee status (see Chapter 8). In the United States alone, 632 "major disaster declarations" were issued in the decade after Katrina (Center for American Progress 2015). "Climate vulnerability" has emerged as a term used to describe the human frailty of hard-hit groups. A broad definition of climate vulnerability. as published by the *Journal of Sustainability,* is "the propensity or predisposition of human societies to be negatively impacted by climate hazards" (Grecequet et al. 2017). While hurricanes are not directly caused by climate change, their severity is greatly exacerbated by climate change, as discussed in Chapter 2.[3]

One well-documented example of forced migration due to climate hazards was that of millions of Americans from Oklahoma, Texas, Arkansas, and Missouri due to the epic dust storms during the Great Depression (Paris 2018).[4] Unlike Katrina, the 1930s Dustbowl, was a crisis forged by drought in an already-depressed economy, with no modern technology to understand possible

Characteristic	Returned to residence	Returned to country
Age, years		
16 to 19	63.5	71.2
20 to 24	44.5	56.8
25 to 34	53.8	64.8
35 to 44	65.0	70.6
45 to 54	68.9	77.5
55 to 69	78.3	82.6
70 and older	76.0	80.0
Race		
White	73.2	81.9
Black	48.5	53.8
Asian	80.7	84.5
Other race	42.9	58.1
Hispanic ethnicity		
Non-Hispanic	65.2	72.5
Hispanic	57.1	72.2
Gender		
Female	63.7	71.1
Male	66.3	74.1
Education[1]		
Less than high school	59.3	67.9
High school	72.5	77.9
Some college	64.9	74.9
College	70.3	74.7
Marital status[2]		
Married	72.8	78.2
Was married[3]	64.3	73.4
Never married	50.7	61.1
Number of childeren[4]		
0	66.1	74.0
1	61.3	68.8
2	64.7	71.9
3 or more	62.2	66.9

[1] For persons aged 25 years and older.
[2] For persons aged 20 years and older.
[3] Widowed, divorced, or separated.
[4] Number of own children under age 18, for persons aged 20 years and older.

Source: Current population Survey.

Figure 9.3 Percent of evacuees who returned to their pre-Katrina residences or countries, by personal characteristics, October 2005-October 2006 (Groen & Polvika 2008 via the Bureau of Labor Statistics Monthly Labor Review https://www.bls.gov/opub/mlr/2008/03/art3full.pdf).

warning signs of the disaster. In recent years, US residents have withstood Hurricane Sandy, Hurricane Harvey, Hurricane Irma, and Hurricane Maria. These extreme weather events tested climate-vulnerable areas with over 300 billion dollars in damage between them, hundreds of thousands of people displaced, power outages for prolonged periods, and hundreds of casualties (Dewan & Schwartz 2017, FEMA 2006, Rice 2018).[5]

As mentioned in Chapter 9, extreme weather events are projected to increase internationally, while vital resources, like potable water, diminish in the world's most vulnerable areas. The majority of countries least vulnerable to climate change are in North America, Europe, and East Asia; countries with the highest vulnerability are mostly in sub-Saharan Africa, South America, Southeast Asia, and Melanesia (Grecequet et al. 2017). Findings like these fuel arguments by environmental justice advocates that exposures to pollution and climate hazards are unequally distributed by socioeconomic standing, gender, and race (Mohai et al. 2009). Indeed, environmental justice is defined by the US Environmental Protection Agency as "the fair treatment and meaningful involvement of people . . . [meaning] that no population, due to policy or economic disempowerment, is forced to bear a disproportionate share of the negative human health or environmental impacts of pollution or environmental consequences resulting from industrial, municipal, and commercial operations" (Mohai et al. 2009). The residents of New Orleans, although not the most vulnerable internationally, endured severe inequality prior to Katrina, and the damages the tempestuous storm brought did fall disproportionately along these preexisting discriminatory lines.

Racial Segregation in New Orleans Pre-Katrina

In 1718, the city of La Nouvelle-Orléans was founded by Jean Baptiste le Moyne, Sieur de Bienville, and after being passed between the French and the Spanish, in 1803 the United States acquired New Orleans via the Louisiana Purchase. The Louisiana Purchase required the development of what would become 15 present-day US states and increased the demand for labor in the Deep South. The abolishment of the international slave trade in 1808 caused the US slave trade to grow exponentially to satisfy labor demands and "New Orleans' slave trading enterprise trumped that of all other American cities for most of the antebellum era" (Campanella 2013).[6] New Orleans was "perfectly positioned as a transshipment point along the watery intercourse between the slave-supply regions of the Upper South and the labor-demanding plantations of the lower Mississippi Valley" (Campanella 2013). Slave auctions were a public affair in New Orleans, as world-renowned hotels such as the St. Charles and St. Louis hotels kept dozens of slave pens around their blocks and held famously large auctions to sell enslaved people (Spain 1979, 85).[7]

"When cotton was King, New Orleans was its Queen city" (Abrams 2015), and as the city underwent rapid economic growth, so did New Orleans' slave market. From 1804 to 1862, there were approximately 135,000 bills of sale for slaves in New Orleans, with approximately one million slaves passing through the port city before relocation in other slave states (Abrams 2015, Levendis 2007, 162).

In 1830, New Orleans had a population of over 45,000 people; of these, 43 percent were white, 24 percent were free blacks, and 33 percent were enslaved black people (Berry 2018, 143; Plyer & Gardere 2018). In the 1840s, free blacks in New Orleans had basic levels of education, were highly skilled artisans, and were primarily located in Faubourg Tremé (Berry 2018; Spain 1979).[8] The lives of free black people in New Orleans soon changed, however, when Louisiana passed legislation in 1857 criminalizing the emancipation of slaves (Sumpter 2018).[9] This was the beginning of local discriminatory legislation, and a primary example of segregation in New Orleans was the splitting of the city into three municipalities: one governed primarily by French Creoles, one by Americans, and one primarily by European immigrants. Free blacks lived on the east side of the city in areas like the Creole Faubourgs (Tremé and Marginy) and the French Quarter to avoid American practices of segregation on the west side of the city (Spain 1979; Sumpter 2018).[10] Enslaved black people were forced to form communities near their masters' homes. On their day off, hundreds of slaves gathered at Congo Square, near Tremé, to sing, dance, and relive the lives they left behind in Africa (Tulane School of Architecture n.d.). These gatherings attracted thousands of Creoles, whites, blacks, and immigrants, and they have been connected to the birth of jazz music and the celebratory Mardi Gras traditions of today.

New Orleans fell to the Union after the Civil War, and from 1860 to 1870, the black population doubled, from 25,000 to approximately 50,000, as newly freed black people from the upper South immigrated to New Orleans (Spain 1979). As their numbers grew, black residents fought for civil rights in the 1870s, until New Orleans police "shot, clubbed, and slaughtered thirty-four blacks and three whites" for supporting black suffrage at a convention (Berry 2018). By the 1890s, segregation was federally authorized by *Plessy v. Ferguson*'s "separate but equal" Supreme Court decision, fortifying the segregated municipalities established in the 1860s and paired with terror tactics of lynching and arson by white supremacists.[11] The Jim Crow laws of the 1890s established statewide codes to ensure distance between races, legacies of which created the racial vulnerabilities underscored by Katrina over a hundred years later (Spain 1979).[12]

In the early 20th century, New Orleans was a sprawling urban city with Creole, Spanish, French, Irish, German, Native American, Jewish, Catholic, black, and white residents living in its metropolitan area. During the 1920s, blacks and immigrants were required to abandon the municipality structure and allow redevelopment of their neighborhoods. Faubourg Tremé, the historically diverse Creole neighborhood, saw spurs of redevelopment initiated by city planners. The city eventually cleared out historical parts of Tremé and the rear of Congo Square to build a municipal auditorium. The redevelopment of this area pushed wealthy Creoles out of the neighborhood, and by 1941, the Lafitte public housing project stood where the cottages of freed black slaves once were. This pattern occurred across New Orleans and shifted the makeup of the inner city to predominantly black residents with high levels of poverty, while affluent residents were able to move outside of the city center (Tulane School of Architecture n.d.). Suburban enclaves formed as "millions of dollars in federal funds directed Louisianans away from the high ground near the Mississippi River and into the drained swamps near Lake Pontchartrain" (Horowitz 2020, 70).

During the Great Depression, the local government began decreasing the value of land along the Mississippi River, and neighborhoods like the current Ninth Ward, where Katrina hit hardest, became much cheaper to develop on (Horowitz 2020, Spain 1979). Urban renewal, like the construction of Interstate 10 and the Superdome, displaced blacks from the inner city of New Orleans. For example, Tremé became desperately impoverished via the redevelopment of the early 20th century, and by the 1960s, the city's urban renewal plan ran Interstate 10 through the heart of Tremé. This construction project displaced over a thousand predominantly black residents, and tens of historical buildings were demolished for its creation. "House-types and styles, like those considered worthy of National Register status. . . . were razed, along with 150 years of history" (Tulane School of Architecture n.d.).

Displaced black families moved along the Mississippi River to areas like the Ninth Ward because of the black housing projects and low rent there (Spain 1979). By the end of the 1960s, 40 percent of blacks in New Orleans concentrated in the Seventh and Ninth Wards on the drained and lesser valued land (Spain 1979). In 1970, New Orleans was 45 percent black, with 44 percent of black people earning incomes below the poverty line, compared to only 10 percent of whites (Spain 1979, 94). From 10 to 35 percent of black people were unemployed (varying by categories according to age, gender, etc.), compared to between three and seven percent of whites, and poverty in New Orleans was the fifth worst of all US cities (Spain 1979).

The Warning Signs of Hurricane Katrina

Levees and floodwalls have lined most of New Orleans' flood-vulnerable waterfront along Lake Pontchartrain, the Mississippi River, the various canals in Metro New Orleans, and the Gulf Outlet/Intercoastal Waterways. Hurricane Betsy of 1965 toppled levees and floodwalls, which caused immense property damage, loss of life, and flooded primarily the Lower Ninth Ward (foreshadowing Katrina). Congress passed the Flood Control Act of 1965, which required the Army Engineer Corps to provide a hurricane protection plan to raise and strengthen levees and floodwalls, as well as gates at the canal entrances to the lake. Unfortunately, even then, completion of this plan was set for decades hence—namely, for 2015, half a century after Betsy and a decade after Katrina (Sills, et al. 2008).

Four years prior to Katrina, FEMA concluded that a "catastrophic hurricane in New Orleans was 'among the three likeliest . . . disasters facing this country'" (Dyson 2007, 78).[13] *Scientific American* (2001) was quoted stating that "only massive reengineering of southeastern Louisiana can save [New Orleans]" and "New Orleans is a disaster waiting to happen" (Dyson 2007, 80). The warning signs were clear, but there was little efficacy in terms of hurricane prevention for New Orleans.

Hurricane Katrina Crashes Through New Orleans

Mayor Ray Nagin (2002–2010) declared a State of Emergency on the Saturday before Katrina arrived. By Sunday morning, there were warnings that a storm surge would cause the levee system to topple, and the National Weather Service stated that this storm would cause "human suffering [deemed] incredible by modern standards" (Horowitz 2020, 117). Although 25 percent of New Orleans residents lived below the poverty line and likely had no means to evacuate, the warnings prompted Nagin to issue the first ever citywide mandatory evacuation notice in New Orleans (Dyson 2007, Horowitz 2020).

One million residents fled the New Orleans Metro Area, but 130,000 residents remained in the city, of which 10,000 residents took refuge in the Superdome (Horowitz 2020, 117). With up to 50 levee breaches throughout the city, "the Mississippi River–Gulf Outlet[,] and the Intracoastal Canal funneled Hurricane Katrina's storm surge towards the eastern side of the city" (Horowitz 2020, 118). Since most of New Orleans was a drained swamp, the breaches paired with the bowl-like shape of the city filled New Orleans to the same water level as Lake Pontchartrain.

New Orleans' death toll was 682, and black people accounted for 459 of those casualties (Horowitz 2020, 119).[14] With 372,000 water-damaged houses, the Superdome and Convention Center were the two main gathering sites for those left in New Orleans when the levees breached (Horowitz 2020, 118). Angela Perkins stayed on the third floor of a public housing development with 13 of her relatives to brave the storm. Perkins trekked through the flooded streets of New Orleans to get to the Superdome, where she was turned away and walked two miles to the Convention Center. She reported: "There was no electricity. The plumbing did not work, and the bathrooms quickly became unbearable. There was no food. There was no water. There were no police officers or relief workers in sight" (Horowitz 2020, 120). The survivors endured days after the August 29th storm surge with no sign of government aid. On September 1st, buses arrived at the Superdome to evacuate New Orleans residents, and two days later buses arrived at the Convention Center alongside 7,200 active-duty troops from Iraq sent to aid the national crisis (Dyson 2007, Horowitz 2020). Troops roamed the streets with authorization to "shoot and kill hoodlums" via the Governor of Louisiana (Committee on Homeland Security and Governmental Affairs [CHSGA] 2006, Dyson 2007).[15]

The governor's message may have spread locally, as at least nine New Orleans residents were shot within seven days of the storm surge (Horowitz 2020, 123). Henry Glover, a black man who was a father of four, was shot and killed by police. His body was then lit on fire and thrown over a levee. Danny Brumfield, a 45-year-old black man, took refuge at the Convention Center with his family and flagged down an officer for help. But that same officer shot him in the back as his family watched (Horowitz 2020). Perkins (from the earlier story) stated that the police "made us [black residents] look like animals, like we deserved whatever was happening to us" (Horowitz 2020).

The US media outlets added fuel to the fire by publishing stories with disgusting imagery of the scene: "black hurricane victims . . . have begun eating corpses to survive," and "people were 'in an almost animalistic state . . . watching hooligans killing people, raping people" (Horowitz 2020, 122). Meanwhile, an on-site National Guardsman stated that these stories were not true and that 99 percent of those taking refuge at the Superdome were well behaved: "'some of these guys look[ed] like thugs . . . but they were working their asses off, grabbing litter and running people' to the medical triage area of the Superdome" (Horowitz 2020). Due to the overall hostility against Katrina survivors, however, mayors of predominantly white nearby parishes, such as St. Bernard and Jefferson, passed resolutions to restrict New Orleans evacuees from entering and allowed the blocking of pedestrians from bridges by police, some of whom fired warning shots to inflict fear (Horowitz 2020).

The Lens of Race and Class During Hurricane Katrina

Analysts consider poverty concentration and class distinction to be determinants in how effectively people are able to respond in emergencies. For example, annual income correlates with how effectively people respond to disasters/emergencies (Elliott & Pais 2006). When Hurricane Katrina made landfall, New Orleans had a poverty rate of 23 percent (10 percent higher than the national average of 13 percent) (Dyson 2007, 5).[16] New Orleans' median household income was 31,369 dollars, while the national median was 44,684 dollars. In wards of high poverty concentration (where many of New Orleans' black inner-city residents had been concentrated), such as the Lower Ninth, incomes were barely above 20,000 dollars, which is 50 percent less than the national average and a third lower than the New Orleans median income.

Poverty plays a large role in whether people can effectively flee natural disasters. Nearly 10 percent of households in New Orleans did not "own or have access to a vehicle" (Dyson 2007, 5), and 21 percent of those who did not evacuate stated that they did not have transportation to leave or could not afford to evacuate. The Lower Ninth Ward was completely submerged when Katrina hit, and it maintained the largest concentration of poverty in New Orleans, with 36 percent of residents below the poverty line prior to Katrina (14 percent more than the national average) (Dyson 2007, 12).[17] A Gallup survey showed that New Orleans residents with incomes at or above 40,000 to 50,000 dollars were twice as likely to evacuate than those residents with incomes of 10,000 to 20,000 dollars (Elliott & Pais 2006, FEMA 2006).

Under the Bush administration, the Secretary of the Department of Homeland Security stated that "local and state officials called for mandatory evacuation. Some people chose not to obey that order. That was a mistake on their part" (Horowitz 2020). Likewise, FEMA Director Michael Brown stated that "unfortunately, [the anticipated high death toll was] going to be attributable a lot to people who did not heed the advanced warnings" (Horowitz 2020).[18] In regards to race and class, the federal government's rhetoric toward the disaster align with claims that some members of the government blamed survivors for the losses and casualties. Most of these survivors were impoverished, elderly, or disabled, and were unable to evacuate without government aid (CHSGA 2006; Horowitz 2020). FEMA spent some 2 billion dollars per day during the days after the disaster, but had little to show for it. By September 12th, FEMA Director Michael Brown resigned due to his failure to adequately respond to the disaster.

Even with Brown's resignation, and with the eventual evacuation of all remaining New Orleans residents, there was little resolution for the survivors. A month passed before residents could return to the city, and when New Orleans was opened on October 5th, the Lower Ninth Ward was still flooded. The predominantly black community, with two government housing locations, had to wait another two months before returning home (Horowitz 2020).

Post-Katrina: Drain, Shrink, and Other Renewal Strategies

The "Bring New Orleans Back Commission" (BNOBC) was headed by 15 men and 2 women, of whom 7 were CEOs and 3 were bank presidents (Horowitz 2020, 141). The BNOBC and the Urban Land Institute (ULI) established three investment zones based on degree of flood damage. Zone C comprised mostly riverfront areas that had not flooded and were the primary focus of reconstruction. The ULI stated that "whether for initial economic recovery, historic preservation, temporary housing, or other short-term benefits, activity in [Zone C] should move forward . . . with all haste" (Horowitz 2020, 142). Zone A included the areas with the most damage, like the Lower Ninth Ward, and the ULI said that these areas should explore "open-space allocations for wetlands, recreational parks, or open water retention systems" (Horowitz 2020, 142). This plan sought to "shrink the city," meaning that they would "privilege neighborhoods that had not flooded," proposing that the lowest-lying, most affected, and subsequently, predominantly black neighborhoods should not be a priority for reconstruction efforts (Dyson 2007, Horowitz 2020).

The claim of the BNOBC was that "the sprawling metropolitan region [of New Orleans] was not sustainable" (Horowitz 2020, 141). Their method of shrinking the city was to utilize Katrina's flood line to define new limits of the city and introduce what Naomi Klein claimed to be "disaster capitalism" (Dyson 2007, 131). No-bid contracts were provided for corporations like Kellogg, Brown, & Root Services, Inc., and Shaw Group. Over 15 contracts were awarded for over 100 million dollars, five of which were for over 500 million dollars, for services like debris removal, purchasing mobile homes, and providing transportation. The Department of Homeland Security's inspector general at the time stated that "when you do something like this, you increase the vulnerability for fraud, plain waste, abuse and mismanagement" (Dyson 2007, 133).[19] The proposals that accompanied these contracts included waiving environmental regulations that prevent reconstruction, permitting an increase in offshore oil drilling, and making the entire affected New Orleans metropolitan area a flat-tax, free-enterprise zone (Dyson 2007, 205).[20]

In 2005, New Orleans had 25 percent of its residents living below the poverty line while simultaneously living in an extremely climate vulnerable region.[21] These reconstruction proposals might not have helped the average Katrina survivor, and the position of the BNOBC seems to acknowledge this. The co-chair of the BNOBC admitted "poor folks don't have the resources to go back to our city. . . . so we won't get all these folks back. That's just a fact" (Horowitz 2020, 142). This rhetoric supported plans to shrink the city, whereas many black political leaders insisted on a more equitable reconstruction of New Orleans. Representatives of Zone A led a community-wide movement to insist on rebuilding the entire city and stressed that "it [was] the intent of this Council to ensure that all neighborhoods be included in the timely and simultaneous rebuilding of New Orleans neighborhoods" (Horowitz 2020, 143).

The BNOBC nonetheless advised the city not to issue any permits to the heavily flooded areas of Zone A. For these areas, the BNOBC issued a map of new parks and urban wetlands, which would cover large sections of New Orleans East and the Lower Ninth. In comparison, affluent and predominately white areas like Lakeview, which saw the same amount of damage as the Lower Ninth, were not designated as greenspaces on the BNOBC reconstruction maps. Ironically, Bring New Orleans Back seemed to further diminish historically black neighborhoods, many of which had been the ones that made New Orleans distinctive. Progressives at the time called this a racist plot, while conservatives argued that this plan provided a solution for the unsafe and unsustainable version of pre-Katrina New Orleans (Horowitz 2020, 144–145),

Neighborhood-level organizing and activism spared many neighborhoods from such plans (Horowitz 2020, 144). New Orleans residents drove from Texas, Georgia, and Tennessee to advocate for Old New Orleans and put up signs in front of their flooded homes stating "I'm coming home!" and "I will rebuild!" (Horowitz 2020, 145). These efforts pushed Mayor Nagin to change the BNOBC plan to one that would focus on hurricane protection, levee planning, and equitable citywide redevelopment.

After the BNOBC's failure, discrimination persisted in other reconstruction proposals (Horowitz 2020, 147). For example, the Road Home Program, which is still in effect today, made seven percent of black homeowners eligible for grants, while 45 percent of white homeowners were eligible (Horowitz 2020, 153). The Road Home Program also permitted the demolition of government housing projects and redeveloped these properties as mixed-income housing. The B. W. Cooper housing project was the coming-out spot for the Black Eagle Mardi Gras Indians and their parade float. When it was set for demolition, the Black Eagles prayed and danced on Mardi Gras morning to honor the home that served their tribe since the 1950s (Horowitz 2020, 162).[22] The attachment to these projects is overtly apparent, and their demolition caused nearly 3,077 families to relocate or be homeless. By 2011, 1,512 of those families had returned to New Orleans, but only 200 of the families were able to move into the newly built apartments on the old government housing sites (Horowitz 2020, 162).

Four years after the storm, the historical black middle-class subdivision Pontchartrain Park was still in a state of disrepair, while predominantly white neighborhoods such as the Garden District were nearly 100 percent rebuilt (Gafford 2010, 386).[23] Pontchartrain Park had fewer than 50 percent of its residents return, whereas the upper-class and predominantly white Garden District had a higher population than reported pre-Katrina (Gafford 2010, 386). A decade after the storm, the Lower Ninth Ward remained 90 percent black, while household income dropped to 23,000 dollars in inflation-adjusted terms. Lakeview, a 90-percent-white community with flooding equivalent to that of the Ninth Ward, saw an inflation-adjusted increase of median income from 49,500 to 77,000 dollars (Horowitz 2020, 139). The two geographically similar neighborhoods also saw discrepancies of returned evacuees: Lakeview's population only decreased from 19,500 to 17,000, while the Lower Ninth decreased from 19,500 to 6,300 (Horowitz 2020, 139–140).[24]

A decade after the storm, 25 percent of Gulf Shores residents had relocated more than 450 miles away, and 10 percent had relocated at least 830 miles away (Center for American Progress 2015). Most of New Orleans' residents who did not return ended up in Houston, Texas, around 350 miles away from the communities they called their "homeport" (Dyson 2007, Groen & Polivka 2008). Texas native and First Lady Laura Bush said, "Almost everyone I've talked to says, 'We're moving to Houston,' What I'm hearing which is sort of scary is that they all want to stay in Texas . . . And so many of the people in the arena here, you know, were underprivileged anyway, so this—this [she chuckles slightly] is working very well for them" (Dyson 2007, 131). This is a representative snapshot of the rhetoric by affluent politicians and administrators when referring to Katrina survivors, who are principally black, low-income people (Dyson 2007, Elliott & Pais 2006, Horowitz 2020).

New Orleans was drained, and the city shrunk as people dispersed after Hurricane Katrina and narrow reconstruction decisions limited who was able to return. The Lower Ninth Ward and Tremé were the heart of New Orleans' black community for generations, and the post-Katrina reconstruction process threatened another version of rebuilding that would damage the entire network of black families that had been in New Orleans for centuries.

Conclusion

The "Big Easy," the "birthplace of jazz," "America's Soul Kitchen," and home to America's favorite carnival, Mardi Gras, New Orleans is arguably the most unique city in the United States, largely because a majority of residents are descendants of the Creoles and slaves who built the city (Horowitz 2020, 184).[25] These cliché characteristics of New Orleans drew 18.5 million tourists in 2018 and have been cultivated by the native black community in the port city (Tourism New Orleans 2019). While the immeasurable amount of culture is the first layer of New Orleans, the weight of slavery, Jim Crow, and natural disasters are the second, third, and fourth. There is no doubt the traditions of blues and the originating moments of jazz are interwoven with the "[poor] black identities in New Orleans" (Dyson 2007, 157). It is understood, some believe New Orleans' proud and internationally unique culture should be considered as part of the stakes in shrinking the city. Others argue that the struggle of poor black New Orleans residents seems pertinent only when linked to cultural tourism. Some of these residents refuse to use cultural preservation as means to save their neighborhoods and instead raise points of discrimination and inequality to argue against shrinking the city.

A survey in 2015 reported that white residents were twice as likely as black residents to say their quality of life had improved since the storm surge, and black residents were three times more likely to state their quality of life had worsened since Katrina's devastation (Horowitz 2020, 196). In 2020, there were four hurricanes in the Gulf of Mexico, which was only the sixth year in recorded history with that many storms in one hurricane season.[26] Adjacent to the Gulf of Mexico, there were seven named tropical storms from September 1 through 14, 2020, a record-breaking number for the Atlantic Basin.[27] These intense storm seasons are felt worldwide with increasing ferocity, as climate change dramatically increases the intensity and frequency of extreme weather events like Katrina (see Chapter 2). They are called hurricanes in the United States and typhoons in South Asia, where in May 2020 the Bay of Bengal endured Cyclone Amphan, the strongest storm ever recorded in the region of West Bengal, India (Berlinger et al. 2020).[28]

As of 2019, the US Army Corps of Engineers had finally completed one of the largest public works projects in modern US history by rebuilding the New Orleans hurricane protection system post-Katrina. This, after having spent untold billions pre-Katrina in hurricane protection systems over several prior decades. Regrettably, after 14 billion dollars spent in this rebuilding, the levees are already sinking due to sea-level rise and erosion, so additional funding will be required to repair and maintain the system (Frank 2019). There is little talk of neighborhood-specific planning or if allocation of funds for hurricane protection services will account for the most vulnerable people affected by storm surges. New Orleans, like many climate-vulnerable areas, faces the threat of natural disasters annually. Katrina tragically showed administrators the importance of having up-to-date emergency response systems that allocate services fairly in order to adequately provide for climate-vulnerable regions. But did they design these systems without regard to race and class, or are climate change and the wrath of Katrina laying bare a pattern of environmental injustice?

EXERCISE

SCENARIO FOR CLASSES

The US Army Corps of Engineers has spent 14 billion dollars on the current hurricane protection system in New Orleans. This included rebuilding levees, raising floodwalls, and installing new water pumps that can pump as much water as an Olympic-sized pool in five seconds. These updates are only the first step in maintaining safety for New Orleans residents, as a 2019 report from the Army Corps of Engineers states that "absent future levee lifts to offset consolidation, settlement, subsidence, and sea level rise, risk to life and property in the Greater New Orleans area will progressively increase" (Frank 2019). The Army Corps of Engineers predicts that by 2023 the integrity of the hurri-

Continued

cane protection system will be compromised, and therefore updates are required.

These updates to New Orleans' disaster preparation are estimated to cost hundreds of millions of dollars. In this hypothetical scenario for role playing, assume the federal government has announced it cannot provide, so the funds must be raised locally and from taxpayers. A local public hearing has been created to discuss and debate which updates are vital, what type of taxation should be imposed, and if there should be an expansion of the hurricane protection system at large. The class is divided in half into two groups, the New BNOBC and the Lower Ninth Coalition, with the instructor serving as moderator of the debate (in the capacity of mayor or city manager, who has convened this meeting of the two sides).

The New BNOBC consists of the bank presidents, corporate CEOs, and their boosters who invested predominantly in white, suburban New Orleans post-Katrina. This group will argue for a special election to vote in favor of a one-time flat tax for all of metropolitan New Orleans to acquire the funding for the hurricane system updates. This group will also argue to wait until 2023 to begin the repairs in order to analyze the damages and decide where to concentrate repairs as a continuation of their city "shrinkage" policy. This group will advocate to see which areas receive the most damage between now and 2023 to set new city limits that shrink the city and turn those extremely high-risk areas into urban wetlands or greenspaces. This group will also seek to concentrate new public education resources (whether for charter schools or public schools) in focus areas.

The Lower Ninth Coalition, with leadership from the city's historically black and Katrina-damaged neighborhoods, is made up of local business owners and New Orleans' community leaders. This group will argue for a special election to vote in favor of a citywide progressive tax to fund the updates to the hurricane system.[29] With mostly black leadership from the Lower Ninth Ward, Tremé, and other middle- and lower-income parts of town, this group represents a broader racial coalition and will advocate that this special election needs to take place immediately in order to start risk-assessment and repairs not only for the hurricane protection system but for the most vulnerable New Orleans neighborhoods. This group will advocate for allocation of these tax dollars and resources for physical repairs to the hurricane protection system as well as the development of infrastructure in communities that are the most exposed to storm surges. This group will also argue against shrinking the city in any way and allege that earlier post-Katrina policies were discriminatory, that a discrimination-proof structure needs to be put in place, and that any reconstruction funds should be raised progressively and spent equitably among the city's race- and class-stratified neighborhoods. This group is considering whether to file a federal lawsuit to try to start an investigation of possible discrimination in the urban planning of New Orleans going back generations, but which they say has been particularly pronounced since Katrina.

Each student should prepare a memo of talking points, and the students should be further divided into groups of four or six (two or three per preassigned group). Students should argue among themselves for 20 to 30 minutes, and then the "mayor" should convene a broader discussion/debrief, which might conclude by addressing some of the broader Discussion Questions given here.

THE NEW BNOBC PREPARATION QUESTIONS

1. Why is a flat tax a fair solution to acquire funding?
2. How does waiting till 2023 to do repairs save money, time, or other resources?
3. How was the BNOBCs plan to shrink the city successful?
4. What does shrinking the city entail, and how would it benefit New Orleans' residents?
5. Why is your council group best-suited to make decisions for New Orleans as a whole?
6. What are your best arguments?

LOWER NINTH COALITION PREPARATION QUESTIONS

1. Why is a progressive tax a fair solution to acquire funding?
2. Why should repairs begin immediately, and how is your reasoning connected to climate vulnerability?
3. How did Hurricane Katrina paint a picture of environmental injustice?
4. Which subset of New Orleans residents are most affected by shrinkage policies? Are they the same groups that are most affected by storm surges?
5. Why is your council group best-suited to make decisions for New Orleans as a whole?
6. What are your best arguments?

DISCUSSION QUESTIONS

1. During and after Katrina, were the policies and emergency response tactics of the local and federal governments purposefully discriminatory?

Continued

2. Which evidence, if any, determines this case to be one of environmental injustice? Does any evidence disconfirm the "environmental injustice" thesis?
3. Data on class disparities are often better than the data on racial disparities. How much overlap do you expect there to be between race and class in analyzing climate and/or environmental justice with regard to Katrina? How about with regard to other hurricanes and climate-related weather events, like Hurricane Maria, mentioned in Chapter 8 in relation to its impacts on Dominica?
4. Did the post-Katrina reconstruction diminish, maintain, or exacerbate inequalities between New Orleans residents?
5. Can you think of any natural disasters, in the United States or abroad, that arguably led to climate and/or environmental injustice?

CASE STUDY 9.1 NOTES

1. Gabriella Siegfried is an MPA student at American University's School of Public Affairs and has worked in sustainability and as a racial justice organizer in her hometown of Orlando, Florida.
2. Hurricane Katrina was the third-deadliest hurricane in US history to date.
3. Hurricanes are also known as typhoons, cyclones, or monsoons.
4. Graphically depicted in John Steinbeck's *The Grapes of Wrath*.
5. Areas in the United States that could be considered the most vulnerable to Katrina-scale disasters are New York City, Miami, and New Jersey (Climate Central 2017).
6. The antebellum period references the period of time before the Civil War.
7. See Figure 9.1.
8. Faubourg Tremé is thought by many to be the oldest African American neighborhood in the United States. It served as an enclave for Creoles and free people of color during its conception.
9. Legislation also passed that required freed slaves to leave the state if they arrived in New Orleans after 1825.
10. Americans would go on to develop the Garden District due to these municipalities, which provided public amenities solely for white people and were heavily segregated.
11. In New Orleans, Homer Plessy, who was one-eighth black, notified the train conductor of his lineage in order to be arrested so that his case could be used by black Creole activists to challenge segregation laws.
12. Jim Crow laws were named after a Black minstrel character and were a collection of regulations that legalized racial segregation at the federal, state, and local levels.
13. A terrorist attack on New York City and an earthquake in San Francisco were the other two likeliest disasters.
14. New Orleans' death toll was nearly a third of the state's total death toll.
15. Democratic Governor Kathleen Blanco (2004–2008).
16. New Orleans also had a population that was 67.9 percent black.
17. The Lower Ninth Ward was 90 percent black prior to Katrina.
18. Prior to the full public's awareness of Brown's ineffective operations, President Bush offered an oft-quoted praise to the beleaguered FEMA director: "Brownie, you're doing a heck of a job" (C-SPAN 2005).
19. Inspector general Richard L. Skinner.

20. Also known as a special economic zone, where economic regulations, business and trade laws, and tax rules are different than other parts of the country/region in order to attract investment.
21. The nationwide poverty rate was approximately 13 percent at the time of the storm.
22. Black Eagle Mardi Gras Indians was established in 1951. Mardi Gras Indian Nation is some 40 self-proclaimed "tribes" of black carnival participants whose tribes and culture date as far back as the late 1880s, when New Orleans' Native Americans confronted freed slaves/ Buffalo Soldiers during the first year of Mardi Gras. At the time of this writing, information was not available on whether the group may consider changing its name in deference to Native Americans.
23. Pontchartrain Park was known as the first black suburb in New Orleans and one of the first black suburban subdivisions of the segregated South.
24. Recall Figure 9.3, which showed that mostly white, affluent, and educated folks were able to return to their homes post-Katrina.
25. In 2000, 8 out of 10 people in Louisiana were born in the state, with 9 out of 10 black people in New Orleans being born in the city. For comparison, Atlanta's black residents have a 6 out of 10 nativity rate (Horowitz 2020).
26. As of October 4, 2020.
27. As of October 4, 2020.
28. This region has typhoon season every year, similarly to the Atlantic Basin's hurricane season. Cyclone Amphan required the evacuation of 2.4 million people, with 12,000 cyclone shelters housing West Bengal citizens.
29. Such taxes are levied at a higher rate against more affluent taxpayers, who are deemed more able to pay, as opposed to value added taxes (VATs), which are "flat" taxes like the sales tax consumers pay for gasoline or purchases at retail stores, which are the same percentage of the total purchase for everyone and hence not progressive. Critics sometimes label these taxes as "regressive," meaning they benefit the wealthy disproportionately as the poor have to pay greater shares of their income in these situations.

CASE STUDY 9.1 REFERENCES

Abrams, Eve. 2015. "Remembering New Orleans' overlooked ties to slavery." *Weekend Edition*, July 18. Accessed September 11, 2020, at https://www.npr.org/2015/2007/2018/423803204/remembering-new-orleans-overlooked-ties-to-slavery.

Berlinger, Joshua, Helen Regan, and Zamira Rahim. 2020. "May 20 Cyclone Amphan news." *CNN*, May 20. Accessed October 3, 2020, at https://www.cnn.com/asia/live-news/live-updates-cyclone-amphan-intl-hnk/index.html.

Berry, Jason. 2018. "The burial master." Accessed September 12, 2020, at http://www.jstor.org.proxyau.wrlc.org/stable/10.5149/9781469647166_berry.12

Blackwell, Jasmine. 2020. "Record-breaking Atlantic hurricane season draws to an end." *NOAA*, November 24. Accessed January 9, 2020, at https://www.noaa.gov/media-release/record-breaking-atlantic-hurricane-season-draws-to-end

C-SPAN. 2005. "Katrina 10 years later: 'Brownie, you're doing a heck of a job.'" Accessed December 9, 2020, at https://www.c-span.org/video/?c4548480/katrina-10-years-later-brownie-heck-job

Campanella, Richard. 2013. "Cityscapes of the New Orleans slave trade: Part II." *Tulane School of Architecture*. Accessed October 1, 2020, at https://richcampanella.com/wp-content/uploads/2020/02/article_Campanella_Preservation-in-Print_April2013_Cityscapes-of-Slave-Trade-Part-II.pdf.

Center for American Progress. 2015. "When you can't go home: The Gulf Coast 10 years after Katrina." Accessed September 5, 2020, at https://www.americanprogress.org/issues/green/reports/2015/2008/2018/119511/when-you-cant-go-home/

Climate Central. 2017. "These US cities are most vulnerable to major coastal flooding and sea level rise." Accessed September 6, 2020, at https://sealevel.climatecentral.org/research/reports/these-us-cities-are-most-vulnerable-to-major-coastal-flooding-and-sea-lev/

Committee on Homeland Security and Governmental Affairs (CHSGA). 2006. "Hurricane Katrina: A nation still unprepared." Accessed September 20, 2020, at https://www.govinfo.gov/content/pkg/CRPT-109srpt322/pdf/CRPT-109srpt322.pdf

Dewan, Shaila, and John Schwartz. 2017. "How does Harvey compare with Hurricane Katrina? Here's what we know." *New York Times*, August 28. Accessed September 10, 2020, at https://www.nytimes.com/2017/08/28/us/hurricane-katrina-harvey.html

Dyson, Michael Eric. 2007. *Come hell or high water: Hurricane Katrina and the color of disaster*. Basic Books, New York.

Eisenman, David P., Kristina M. Cordasco, Steve Asch, Joya F. Golden, and Deborah Glik. 2007. "Disaster planning and risk communication with vulnerable communities: Lessons from Hurricane Katrina." *American Journal of Public Health* 97(suppl. 1):S109–S115. DOI: 10.2105/ajph.2005.084335

Elliott, James R., & Jeremy Pais. 2006. "Race, class, and Hurricane Katrina: Social differences in human responses to disaster." *Social Science Research* 35(2):295–321. Accessed September 4, 2020, at https://doi.org/10.1016/j.ssresearch.2006.02.003.

Federal Emergency Management Agency (FEMA). 2006. "Summary report on building performance." Accessed September 10, 2020, at https://www.fema.gov/sites/default/files/2020-08/fema_548_summary_report_building_performance_hurricane_katrina_2005.pdf

Frank, Thomas. 2019. "After a $14-billion upgrade, New Orleans' levees are sinking." *Scientific American*, April 11. Accessed October 3, 2020, at https://www.scientificamerican.com/article/after-a-14-billion-upgrade-new-orleans-levees-are-sinking/

Gafford, Farrah D. 2010. "Rebuilding the park: The impact of Hurricane Katrina on a black middle-class neighborhood." *Journal of Black Studies* 41(2):385–404.

Graif, Corina. 2016. "(Un)natural disaster: Vulnerability, long-distance displacement, and the extended geography of neighborhood distress and attainment after Katrina." *Population and Environment* 37(3):288–318.

Grecequet, Martina, Jack DeWaard, Jessica J. Hellmann, and Guy J. Abel. 2017. "Climate vulnerability and human migration in global perspective." *Sustainability* 9(5):720. DOI: 10.3390/su9050720

Groen, Jeffrey A., & Anne A. Polivka. 2008. "Hurricane Katrina evacuees: Who they are, where they are, and how they are faring." *Monthly Labor Review*. Accessed September 7, 2020, at https://www.bls.gov/opub/mlr/2008/2003/art2003full.pdf

Horowitz, Andy. 2020. *Katrina: A history, 1915–2015*. Harvard University Press, Cambridge, MA.

Levendis, John. 2007. "Qualities and effective-quantities of slaves in New Orleans." Accessed September 20, 2020, at http://www.ser.tcu.edu/2007/SER2007%20Levendis%20161-178.pdf

Mohai, Paul, David Pellow, & Roberts, J. Timmons. 2009. "Environmental justice." *Annual Review of Environment and Resources* 34(1):405–430. DOI: 10.1146/annurev-environ-082508-094348

Paris, Francesca. 2018. "What migrants displaced by the Dust Bowl and climate events can teach us." *Here and Now*, October 20. Accessed September 6, 2020, at https://www.npr.org/2018/2010/2020/659074873/what-migrants-displaced-by-the-dust-bowl-and-climate-events-can-teach-us

Plyer, Allison, & Lamer Gardere. 2018. "The New Orleans Prosperity Index: Tricentennial edition." *The Data Center*, April 11. Accessed October 2, 2020, at https://www.datacenterresearch.org/reports_analysis/prosperity-index/

Rice, Doyle. 2018. "2017's three monster hurricanes—Harvey, Irma and Maria—among five costliest ever." *USA Today*, January 30. Accessed September 7, 2020, at https://www.usatoday.com/story/weather/2018/2001/2030/2017-s-three-monster-hurricanes-harvey-irma-and-maria-among-five-costliest-ever/1078930001/.

Sills, George L., Noah D. Vroman, Ronald E. Wahl, & Neil T. Schwanz. 2008. "Overview of New Orleans levee failures: Lessons learned and their impact on national levee design and assessment." *Journal of Geotechnical and Geoenvironmental Engineering* 134(5):556–565. Accessed September 14, 2020, at https://www.tulane.edu/~sanelson/New_Orleans_and_Hurricanes/overviewofleveefailures.pdf

Spain, Daphne. 1979. "Race relations and residential segregation in New Orleans: Two centuries of paradox." *The Annals of the American Academy of Political and Social Science* 441:82–96.

StoryCorps. 2006. "Survivors' stories: Living through Katrina." *NPR Morning Edition*, August 25. Accessed September 4, 2020, at https://www.npr.org/templates/story/story.php?storyId=5704652

Sumpter, Amy R. 2018. "Segregation of the free people of color and the construction of race in antebellum New Orleans." 19–37. *Southeastern Geographer*, vol. 48 no. 1, 2008, p. 19-37. *Project MUSE*, doi:10.1353/sgo.0.0010. Accessed October 3, 2020 at https://muse.jhu.edu/article/240387

Tourism New Orleans. 2019. "New Orleans tourism visitation and visitor spending break records in 2018." *NewOrleans.com*, November 22. Accessed October 4, 2020, at https://www.neworleans.com/articles/post/new-orleans-tourism-visitation-and-visitor-spending-break-records-in-2018/

Tulane School of Architecture. n.d. "Faubourg Tremé." Accessed October 3, 2020, at https://architecture.tulane.edu/preservation-project/place/359.

APPENDIX

Solutions

Instructor Guide to Exercise 1.1: Incentivizing Divergent Outcomes (to Contrast Cooperation and Discord)

The likely outcome is that the team incentivized to think collectively and toward the future will cooperate, whereas the individualistic team incentivized to the present will not be able to cooperate. The three ways we know of to immediately help one team cooperate are (1) to have its chips (money) count 10-fold if they reach an agreement, (2) to distribute the chips equally among that team (rather than giving all of them to one or two players), and (3) to make it about individual chip tallies rather than about "solving the problem" of making coffee. The cooperative team's points (money) should also only have value if they are able to cooperate. This is the "absolute gains" team: the one which will be able to cooperate.

The "zero-sum gains" team will have contrary incentives. The chips and coffee parts will be given to just a few members of that group (making them strong at the expense of others), whoever has the most points in this group wins (regardless of whether they make coffee), and their chips (money) are worth their face value—no more and no less—whether or not they achieve cooperation.

The debrief should discuss the three critical issues: (1) the role of inequality, (2) the incentives of individuals versus groups, and (3) the incentives to think about a future outcome (cooperation) rather than just the present value of their labor and resources. This exercise can be referenced later in the course, especially in discussions about the Paris Agreement.

Solutions to Exercise 2.2

The goal of the exercise is to model how land storage of carbon can change depending on the degree of future temperature change and the amount of atmospheric carbon in the future.

Equation 2.3 is used to solve for land carbon storage. As a reminder, this equation is

$$C_{eq\,t} = II * (C_a/(C_a + C_{0.5})) * (C_{a(0)}/C_a)^{\mu} * (f(T)/r_0)$$

The μ term in the equation (μ = system sensitivity to temperature change) has to be solved for four different temperature changes that might occur if atmospheric CO_2 is doubled. We are going to use 2.0, 3.0, 3.5, and 4.0°C as our temperature changes, but students can use different temperature change values. The μ term itself is found by using equation 2.2:

$$\mu = (\Delta T_{2xCO2}/10) * (\log Q_{10}/\log 2)$$

where Q_{10} is given as 2.5, and the four estimates of are ΔT_{2xCO2} 2.0, 3.0, 3.5, and 4.0 (again, these can differ student to student).

Then, for ΔT_{2xCO2} = 2.0:

$$\mu = (2.0/10) * (\log 2.5/\log 2)$$
$$\mu = (0.2) * (0.398/0.301)$$
$$\mu = 0.264$$

For ΔT_{2xCO2} = 3.0:

$$\mu = (3.0/10) * (\log 2.5/\log 2)$$
$$\mu = (0.3) * (0.398/0.301)$$
$$\mu = 0.367$$

For ΔT_{2xCO2} = 3.5:

$$\mu = (3.5/10) * (\log 2.5/\log 2)$$
$$\mu = (0.35) * (0.398/0.301)$$
$$\mu = 0.462$$

And lastly, for ΔT_{2xCO2} = 4.0:

$$\mu = (4.0/10) * (\log 2.5/\log 2)$$
$$\mu = (0.4) * (0.398/0.301)$$
$$\mu = 0.595$$

Now that we have established some μs for ΔTs, we are ready for the terrestrial carbon storage model shown in equation 2.3:

$$C_{eq\,t} = II * (C_a/(C_a + C_{0.5})) * (C_{a(0)}/C_a)^{\mu} * (f(T)/r_0)$$

Solving for $C_{eq\,t}$ (terrestrial carbon storage) requires all ppm CO_2 values be converted to Gt C by multiplying ppm by 2.13 Gt C:

$$C_{0.5} = 500\text{ ppm} * 2.13\text{ Gt C/ppm} = 1{,}065\text{ Gt C}$$

$$C_{a(0)} = 280\text{ ppm} * 2.13\text{ Gt C/ppm} = 596\text{ Gt C}$$

C_a is carbon content in the atmosphere at any given time "a".

Next, increase atmospheric carbon concentrations in 20- or 30-ppm increments between 190 to 1,100 ppm. So for 190 ppm:

$$C_a = 190\text{ ppm} * 2.13\text{ Gt C/ppm} = 404\text{ Gt C}$$

For 1,100 ppm:

$$C_a = 1{,}100\text{ ppm} * 2.13\text{ Gt C/ppm} = 2{,}343\text{ Gt C}$$

In addition:

$$\text{II} = 120\text{ GtC}$$

$$f(T) = 1$$

$$r_0 = 0.034$$

Note that all variables are in Gt C or are unitless. Therefore, you can do the math without worrying about units, and the answer will be in Gt C.

Now we are set to solve equation 2.3. We will use C_a of 190 ppm (converted to 404 Gt C as an example, but students will solve equation 2.3 for concentrations up to 1,100 ppm in 20- or 30-ppm increments as mentioned earlier). In this example solution, the μ will be that calculated when $\Delta T = 3.5°C$ (μ = 0.462). To complete the exercise, all four μs will have to be utilized, and those other solutions follow the same format as shown here:

$$C_{eq\,t} = \text{II} * (C_a/(C_a + C_{0.5})) * (C_{a(0)}/C_a)^{\mu} * (f(T)/r_0)$$

$$C_{eq\,t} = 120 * (404/(404 + 1{,}065)) * (596/404)^{0.462} * (f(T)/r_0)$$

$$C_{eq\,t} = 120 * (404/1{,}469) * (1.48)^{0.462} * (1/0.034)$$

$$C_{eq\,t} = 120 * 0.275 * 1.11 * 29.4$$

$$C_{eq\,t} = 1{,}164\text{ (with units of Gt C)}$$

If you use ΔTs of 2.0, 3.0, 3.5, and 4.5 and solve the model for concentrations of atmospheric CO_2 from 190 to 1,100 ppm, you should be able to generate curves that resemble those shown in Figure 2.7.

Among the important take home lessons from this exercise are:

1. Terrestrial carbon storage is expected to increase for the near term; however, there is an inflection point where it will start to decline. This means the soils start releasing CO_2 (respiration of the microbes releases more carbon than is taken up through leaf litter and other organic matter).
2. Where the inflection point occurs depends on how much warming we can expect from CO_2 doubling since pre-industrial levels (280 ppm). Where the inflection point occurs also depends on the rate that CO_2 increases in the atmosphere.
3. Emissions of CO_2 from burning fossil fuels are completely controlled by humans and their policies. If the soils start emitting CO_2, however, there will be little humans could do to stop it.

Solutions to Exercise 5.2

Find royalties for the three different production rates (in cubic feet [cf]) and times given. Royalty rate is 12.4 percent.

FIRST SIX MONTHS: WELLHEAD PRICE AND ROYALTY

Find the flow rate over time period (1) and the price per unit volume (2). Then, find the wellhead price by multiplying price by volume over time period (3). Lastly, find the royalties that would be paid if 12.4 percent is the rate (4). For simplicities sake, assume 30 day months.

$$1.1 \text{ million cf/day} * 6 \text{ months} * 30 \text{ days/month} = 198 \text{ million cf} \tag{1}$$

$$\$3.35/\text{thousand cf} * 1000 \text{ thousand cf}/1 \text{ million cf} = \$3{,}350/\text{million cf} \tag{2}$$

$$\$3{,}350/\text{million cf} * 198 \text{ million cf} = \$663{,}300 \left(\text{the wellhead price}\right) \tag{3}$$

$$\$663{,}300 * 12.4\% = \$82{,}249 \text{ paid to Mrs. Davis the first six months} \tag{4}$$

SECOND SIX MONTHS: WELLHEAD PRICE AND ROYALTY

Find the flow rate over time period (1) and the price per unit volume (2). Then, find the wellhead price by multiplying price by volume over time period (3). Lastly, find the royalties that would be paid if 12.4 percent is the rate (4).

$$0.5 \text{ million cf/day} * 6 \text{ months} * 30 \text{ days/month} = 90 \text{ million cf} \tag{1}$$

$$\$3.35/\text{thousand cf} * 1000 \text{ thousand cf}/1 \text{ million cf} = \$3{,}350/\text{million cf} \tag{2}$$

$$\$3350/\text{million cf} * 90 \text{ million cf} = \$301{,}500 (\text{the wellhead price}) \tag{3}$$

$$\$301{,}500 * 12.4\% = \$37{,}386 \text{ paid to Mrs. Davis the second six months} \tag{4}$$

NEXT FIVE YEARS: WELLHEAD PRICE AND ROYALTY

Find the flow rate over time period (1) and price per unit volume (2). Then, find the wellhead price by multiplying price by volume over time period (3). Lastly, find the royalties that would be paid if 12.4 percent is the rate (4).

$$0.2 \text{ million cf/day} * 5 \text{ years} * 365 \text{ days/year} = 365 \text{ million cf} \quad (1)$$

$$\$3.35/\text{thousand cf} * 1000 \text{ thousand cf}/1 \text{ million cf} = \$3{,}350/\text{million cf} \quad (2)$$

$$\$3350/\text{million cf} * 365 \text{ million cf} = \$1{,}222{,}750 \text{ (the wellhead price)} \quad (3)$$

$$\$1{,}222{,}750 * 12.4\% = \$151{,}621 \text{ paid to Mrs. Davis over the next 5 years} \quad (4)$$

Total value to Mrs. Davis $271,256, so the offer of $200,000 should be rejected.

If the wellhead price rose to $5.70 per thousand cf, then equation 2 becomes

$$\$5.70/\text{thousand cf} * 1{,}000 \text{ thousand cf/million cf} = \$5{,}700/\text{million cf}$$

In this case, adjust the price in equation 3, and royalties change to a total of $461,540.

Solutions to Exercise 6.2

Two equations are used to solve the problems. First:

$$P = 1{,}000 * \mu * Q * g * H \quad (6.2)$$

where P is power in watts, "μ" (pronounced "mu") is loss from friction or drag (as a decimal, not aq percentage), Q is discharge per second (converted from $l^3 s^{-1}$ to $m^3 s^{-1}$' there are 1,000 l^3 in a m^3), g is acceleration of gravity (9.81 m s^{-2}), and H is hydraulic head in meters. The 1,000 is just a scaling multiple. (If you are using $m^3 s^{-1}$ for discharge and want output in watts, you need to multiply by 1,000. If you want kilowatts, do not multiply by 1,000.) Second:

$$Q = A * \sqrt{2 * g * H} \quad (6.7)$$

where A is the area of the nozzle discharging the water into a Pelton turbine.

The term Q from equation 6.7 is plugged into equation 6.2 to find power. Interestingly, for Pelton turbines, additional jets can be added to a single wheel. To find the new power output, you just multiply the power generated for one jet by the number of jets. Some Pelton arrangements can have eight jets or more! To find the amount of power, simply multiply the Q from equation 6.7 by the number of jets (provided jet area is the same).

The quantitative objective of this exercise is to show how variable hydroelectric power can be depending on the height of the dam, the type of turbine, the number of turbines, efficiency, and seasonal precipitation (hence, flow rate). Another objective is to show how adding additional turbines to existing power plants can add to capacity.

Using the appropriate equations outlined earlier, calculate the power capacity of the four hydropower stations.

GALLOWAY, SCOTLAND, HYDROELECTRIC POWER STATION

For the Galloway hydroelectric power station, hydraulic head (H) = 25 m, Francis turbine efficiency is 88 percent, flow rate (Q) = 600 l^3/min, and g = 9.81 m s^{-2}. Convert to meters per second, and your answer will be in kilowatts.

To solve, use equation 6.2:

$$P\,(kW) = \mu * g * H * Q$$

where Q= 0.6 m^3/min (divide by 60 = 0.01 m^3/sec), H = 25 m, g = 9.8, and μ = 0.88.

Answer: P = 2.15 kilowatts.

NIAGARA FALLS HYDROELECTRIC POWER STATION

For the Niagara Falls hydroelectric power station, hydraulic head (H) = 100 m, Francis turbine efficiency is 83 percent, flow rate (Q) = 360,000,000 l^3/min, and gravity is 9.81 m s^{-2}. Convert to meters per second, and your answer will be in kilowatts.

To solve, first convert cubic liters/min to cubic meters/sec, which gives 360,000 m^3/min or 6,000 m^3/sec. Then:

$$P(kW) = 0.83 * 6{,}000 * 9.8 * 100$$

$$P(kW) = 4{,}880{,}400$$

or

$$P(W) = 1{,}000 * 0.83 * 6{,}000 * 9.8 * 100$$

$$P(W) = 4{,}880{,}400{,}000$$

PELTON HYDROELECTIC POWER STATION

For the Pelton hydroelectric power station, hydraulic head (H) = 181 m (from Three Gorges), and efficiency is 90 percent. What is the KW output from one jet, A = 1.3m? And what if you had three jets on the same Pelton wheel?

$$Q = A * \sqrt{2 * g * H}$$
$$Q = A * \sqrt{2 * 9.8 * 181}$$
$$Q = A * \sqrt{3{,}547}$$
$$Q = 1.3 * 59.56$$
$$Q = 77.43$$

and

$$P = \mu * Q * g * H$$
$$P = 0.9 * 77.4 * 9.8 * 181$$
$$P(\text{kW}) = 123{,}562.9\,\text{kW}$$

Then, multiply P by three, for 370,688 kW.

ITAIPU POWER STATION

Itaipu Power Station is shared between Paraguay and Brazil. It supplies 95 percent of Paraguay's power and up to 80 percent of Brazil's power! The flow of water into it is seasonal, however, so output can be variable. What are the low- and high-power possibilities for Itaipu? Again, flow depends on season, so Q = 9,000–36,000 m^3/sec, and H = 118.4 m. Francis turbine efficiency is 90 percent.

$$P = \mu * Q * g * H$$
$$P = 0.9 * 9{,}000 * 9.8 * 118.4 = 9{,}398{,}592\,\text{kW at low flow}$$
$$P = 0.9 * 36{,}000 * 9.8 * 118.4 = 37{,}594{,}368\,\text{kW at high flow}$$

Index[1]

[1] The authors thank Karl VonZabern for constructing this index, although they take responsibility for any oversights or errors.